GLENCOE MATHEMATICS

MathScape

Seeing and Thinking Mathematically

Course 1

McGraw-Hill Glencoe

New York, New York   Columbus, Ohio   Chicago, Illinois   Peoria, Illinois   Woodland Hills, California

Glencoe

The McGraw-Hill Companies

Send all inquiries to:
Glencoe/McGraw-Hill
8787 Orion Place
Columbus, OH 43240

ISBN: 0-07-860466-4

9 079/058 09

TABLE OF **CONTENTS**

The Language of Numbers 48

Inventing and Comparing Number Systems

From Wholes to Parts 92
Operating with Factors, Multiples, and Fractions

Designing Spaces 162
Visualizing, Planning, and Building

Beside the Point 206
Operating with Decimals, Percents, and Integers

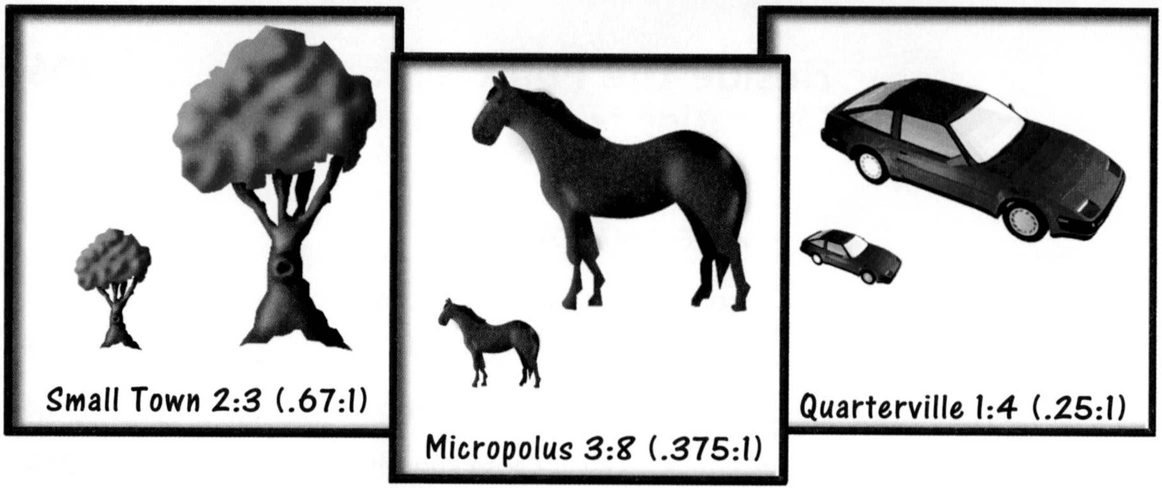

Gulliver's Worlds 276
Measuring and Scaling

Patterns in Numbers and Shapes . .320
Using Algebraic Thinking

PHASE**ONE**
Measures of Central Tendency

Super Data Company collects data, shows data on graphs, and analyzes data. As an apprentice statistician, you will conduct a survey, create a graph of the data you collect, and analyze it. These are important skills for statisticians to know.

How can we use data to answer questions about the world around us?

WHAT DOES THE DATA SAY?

PHASE**TWO**

Representing and Analyzing Data

You will explore bar graphs. You will create single bar graphs and identify errors in several bar graphs. You will also investigate how the scale on a bar graph can affect how the data is interpreted. At the end of the phase, you will conduct a survey of two different age groups. Then you will make a double bar graph to represent the data and compare the opinions of the two groups.

PHASE**THREE**

Progress over Time

Start off with a Memory Game in which you find out if your memory improves over time and with practice. You will create a broken-line graph to show your progress. Then you will analyze the graph to see if you improved, got worse, or stayed the same. The phase ends with a project in which you measure your progress at a skill of your choice.

PHASE**FOUR**

Probability and Sampling

You will investigate the chances of choosing a green cube out of a bag of green and yellow cubes. Then you will explore how changing the number of cubes in the bag can change the probability of picking a green cube. At the end of the phase, you will apply what you've learned so far by helping one of Super Data Company's clients solve the Jelly Bean Bag Mix-Up.

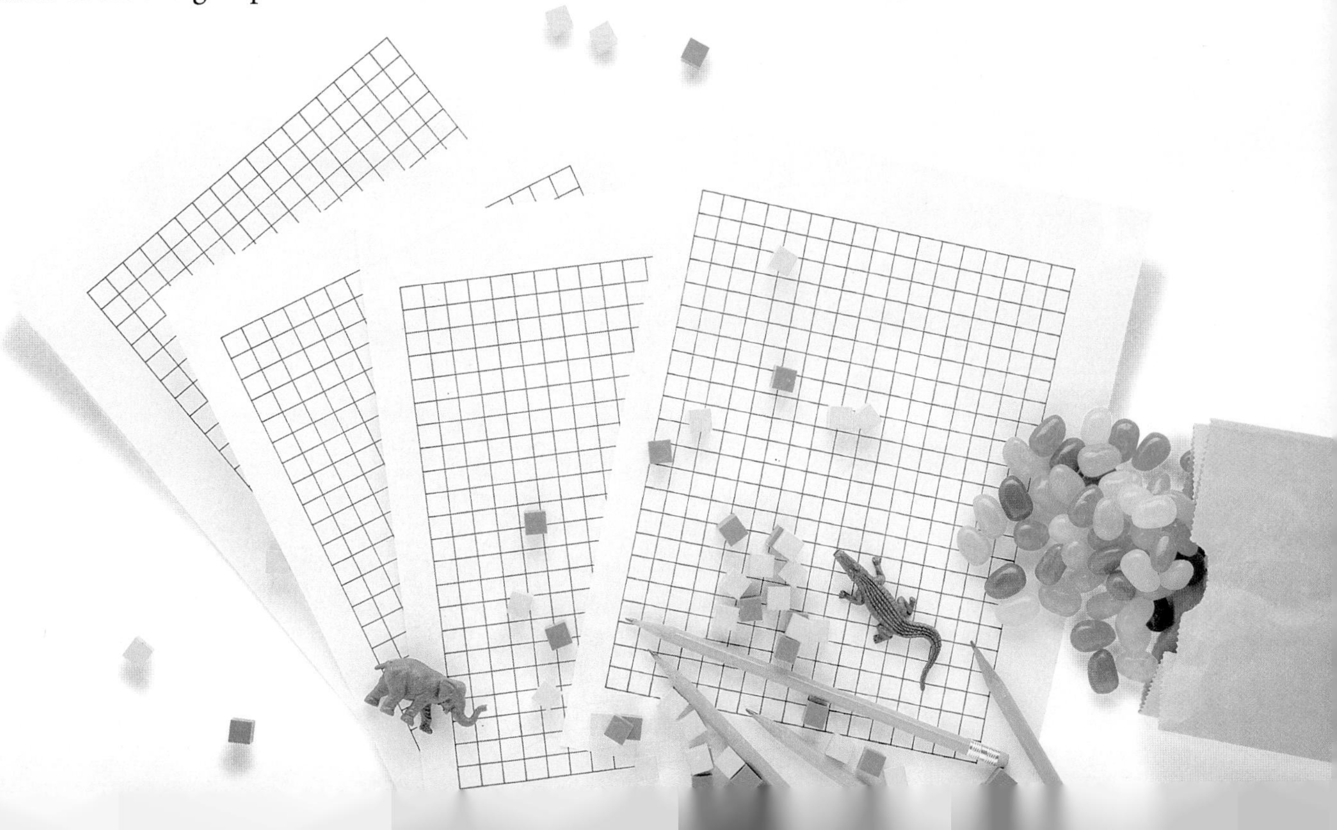

PHASE ONE

To: Apprentice Statisticians
From: President, Super Data Company

Welcome to Super Data Company! In your new job as an apprentice statistician, you will conduct and analyze surveys for our many customers.

There are different kinds of questions that you can ask in a survey. There is also a lot of information that you can get from the results of surveys. In your first assignment, you will conduct a survey and collect data about your classmates.

A statistician collects and organizes data. Surveys, questionnaires, polls, and graphs are tools that statisticians use to gather and analyze the information.

In Phase One, you will begin your new job as an apprentice statistician by collecting data about your class. You will learn ways to organize the data you collect. Then you will analyze the data and present your findings to the class.

Measures of Central Tendency

WHAT'S THE MATH?

Investigations in this section focus on:

COLLECTING DATA

- Conducting surveys to collect data
- Collecting numerical data

GRAPHING

- Making and interpreting frequency graphs

ANALYZING DATA

- Finding the mean, median, mode, and range of a data set
- Using mean, median, mode, and range to analyze data

MathScape Online
mathscape1.com/self_check_quiz

1 Class Survey

ANALYZING DATA TO FIND MODES AND RANGES

How well do you know your class? Taking a survey is one way to get information about a group of people. You and your classmates will answer some survey questions. Then you will graph the class data and analyze it. You may be surprised by what you find out about your class.

Find the Mode and Range

How can you find the mode and range for a set of data?

The data your class tallied from the Class Survey Questions is a list of numbers. The number that shows up most often in a set of data like this is called the *mode.* The *range* is the difference between the greatest number and the least number in a set of data. For the data in the class frequency graph you see here, the mode is 10. The range is 9.

Look at the frequency graph your class created for Question A. Find the mode and the range for your class data.

How Many Glasses of Soda We Drink

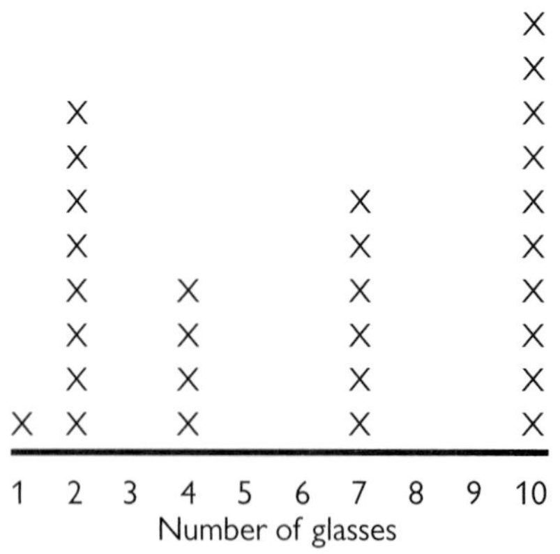

Analyze the Class Data

Your teacher will give your group the class's responses for one of the survey questions you answered at the beginning of the lesson. Follow these steps to find out everything you can about your class.

1 Create a frequency graph of the data.

- **a.** Include everyone's answer on your graph.
- **b.** Don't forget to label the graph and give it a title.

2 Analyze the data from your graph.

- **a.** Find the mode.
- **b.** Find the range.

How can you use mode and range to analyze data?

Write About the Class Data

Write a summary that clearly states what you learned about your class from the data. Be sure to include answers to the following questions:

- What does the data tell you about the class? Make a list of statements about the data. For example, "Only one student in the class has 7 pets."
- What information did you find out about the class from the mode and range?

hot words
mode
range
frequency graph

Homework
page 36

2 Name Exchange

EXPLORING MEANS AND MEDIANS

One of the questions often asked about a set of data is, "What is typical?" You have learned to find the mode of a list of numbers. Two other measures of what's typical are the *mean* and the *median.* Here you will use mean, median, and mode to analyze data on the names in your class.

How can you find the mean length of a name in the class?

Find the Mean

One way to find the mean length of a set of first names is to do the Name Exchange. Follow these steps to find the mean, or average, length of the first names of members in your group.

1. Write each letter of your first name on a different sheet of paper.
2. Members of your group should exchange just enough letters so that either:

   a. each member has the same number of letters, or

   b. some group members have just *one* letter more than other members.

   You may find that some members of the group do not need to change letters at all.
3. Record the mean, or average, length of the first names in your group.

> There are 5 girls in the group and these are their names.
> Sherry 6 Lorena 6 Natasha 7 Daniella 8 Ann 3
> Natasha and Daniella gave letters to Ann, so everyone would have 6 letters.
> Sherry 6 Lorena 6 Natash-a 6 Daniel-la 6 Ann+ala
> The mean length for the group is 6.

How does the mean for your group compare to the mean for the class?

Find the Median

How can you find the median for a data set?

When the numbers in a set of data are arranged in order from least to greatest, the number in the middle is the median. If there is an even number of numbers in a set of data, the median is the mean of the two middle numbers. Use the frequency graph your class made for the lengths of names to answer these questions.

- What is the median length of a first name for your class?
- What does the median tell you?

How does the mean compare to the median for the class?

Write About the Class Data

You have learned about the mean, median, mode, and range. Think about what you have learned to answer the following questions about your class:

- What do each of the measures (mean, median, mode, and range) tell you about the lengths of the first names in the class?
- Which of the measures (mean, median, or mode) do you think gives the best sense of what is typical for the class? Why?
- What are some situations where it would be helpful to know the mean, median, mode, or range for the class?

Ms. Bryan's Class

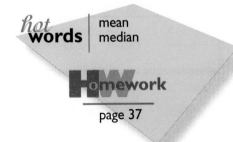

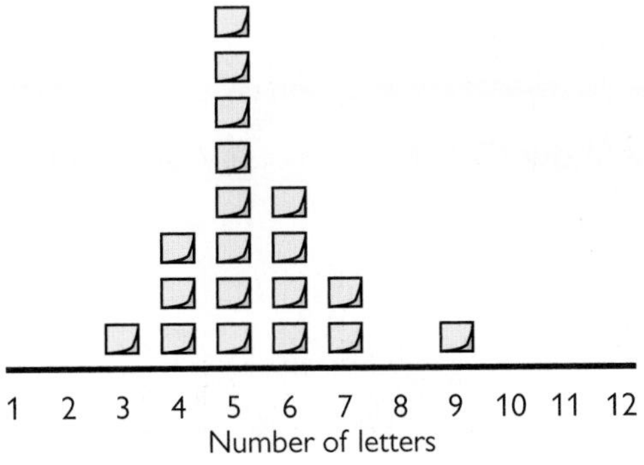

hot words
mean
median

Homework
page 37

3 TV Shows

INVESTIGATING RATINGS AND DISTRIBUTIONS

Rating scales are often used to find out about people's opinions. After your class rates some television shows, you will look at some data on how another group of students rated other television shows. Then you will apply everything you have learned so far to conduct and analyze a survey of your own.

Analyze Mystery Graphs

What information can you get by analyzing the distribution of data in a graph?

The graphs below show how some middle school students rated four TV shows. Use the information in the graphs to answer the following questions:

- Overall, how do students feel about each show?
- Do the students agree on their feelings about each show? Explain your answer.
- Which TV shows that *you* watch might give the same results if your classmates rated the shows?

Mystery Graphs

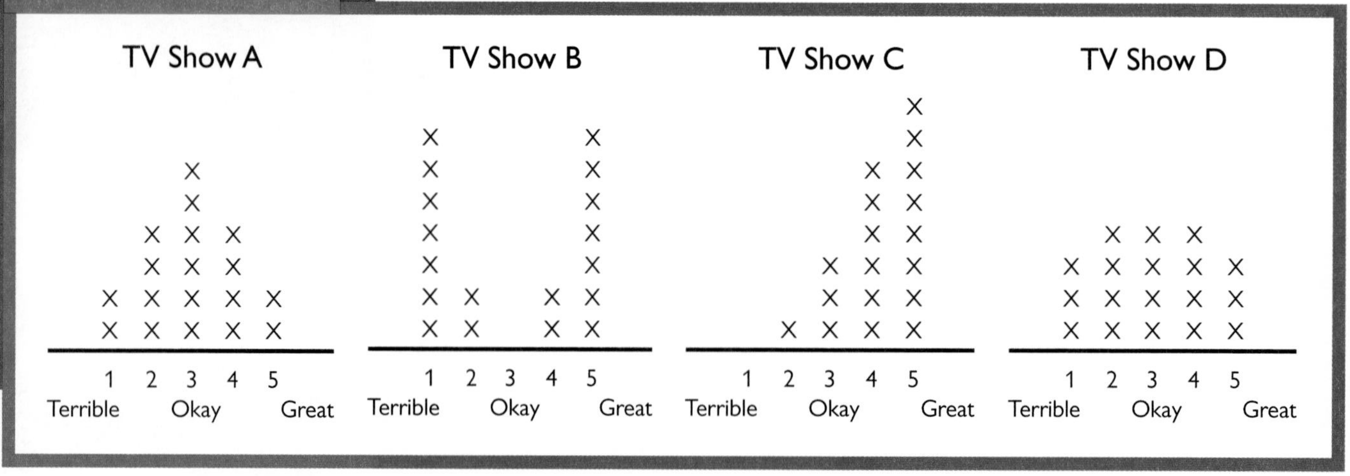

Collect and Analyze Data, Part 1

How can you conduct and analyze a survey?

Now it's your turn! You will apply what you have learned about collecting, representing, and analyzing data to find out about a topic of your choice. Follow these steps.

1 Make a data collection plan

Choose a topic	On what topic would you like to collect data?
Choose a population to survey	Whom do you want to survey? For example, do you want to ask 6th graders or 1st graders? How will you find at least 10 people from your population to survey?
Write survey questions	Write four different survey questions that can be answered with numbers. At least one of the questions should use a rating scale. Make sure that the questions are easy to understand.
Identify an audience	Who might be interested in the information you will collect? Why might they be interested?

2 Collect and represent data

Collect data	Collect data for just one of the survey questions. Ask at least 10 people from the population you chose. Record your data.
Graph data	Create an accurate frequency graph of your data.

3 Analyze the data

Write a report that answers these questions:	What are the mean, median, mode, and range? How would you describe the distribution, or shape, of the data? What did you find out? Make a list of statements that are clearly supported by the data.

hot words: frequency graph, distribution

Homework
page 38

PHASE TWO

To: Apprentice Statisticians
From: President, Super Data Company

I hope you are enjoying your work with Super Data Company!

Super Data Company is working on a new exhibit for the local zoo. The exhibit will make it easy for visitors to find out and compare interesting facts about animals. A bar graph is a useful way to represent data. In your next assignment, you will make bar graphs about zoo animals.

In this phase, you will investigate and create bar graphs. You will also learn how the scales on a bar graph can affect how the information is interpreted.

Graphs are used to represent information about many things, such as advertisements, test results, and political polls. Where have you seen bar graphs? What type of information can be shown on a bar graph?

Representing and Analyzing Data

WHAT'S THE MATH?

Investigations in this section focus on:

COLLECTING DATA

- Conducting surveys to collect data
- Collecting numerical data

GRAPHING

- Making and interpreting single and double bar graphs

ANALYZING DATA

- Comparing data to make recommendations

MathScape Online
mathscape1.com/self_check_quiz

4 Animal Comparisons

INVESTIGATING BAR GRAPHS AND SCALES

Graphs are used in many different ways, like showing average rainfall or describing test results. Here you will explore how the scale you choose changes the way a graph looks as well as how it affects the way people interpret the data.

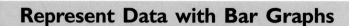

Represent Data with Bar Graphs

How can you choose scales to accurately represent different data sets in bar graphs?

1 Choose two sets of data about zoo animals to work with.

2 Make a bar graph for each set of data you chose. Follow the guidelines below when making your graphs:

a. All the data must be accurately represented.

b. Each bar graph must fit on an $8\frac{1}{2}$" by 11" sheet of paper. Make graphs large enough to fill up at least half the paper.

c. Each bar graph must be labeled and easy to understand.

3 After you finish making the bar graphs, describe how you chose the scales.

What is important to remember about showing data on a bar graph?

Animal	Weight (pounds)
Sea cow	1,300
Saltwater crocodile	1,100
Horse	950
Moose	800
Polar bear	715
Gorilla	450
Chimpanzee	150

Animal	Weight (ounces)
Giant bat	1.90
Weasel	2.38
Shrew	3.00
Mole	3.25
Hamster	4.20
Gerbil	4.41

Animal	Typical Number of Offspring (born at one time)
Ostrich	15
Mouse	30
Python	29
Pig	30
Crocodile	60
Turtle	104

Animal	Speed (miles per hour)
Centipede	1.12
House spider	1.17
Shrew	2.5
House rat	6.0
Pig	11.0
Squirrel	12.0

Investigate Scales

The scale you choose can change the way a graph looks. To show how this works, make three different graphs for the data in the table.

Number of Visitors per Day at the Zoo

Zoo	Visitors per Day
Animal Arc Zoo	1,240
Wild Animal Park	889
Zooatarium	1,573

1 Make one graph using a scale that gives the most accurate and fair picture of the data. Label this Graph A.

2 Make one graph using a scale that makes the differences in the numbers of visitors look smaller. Label this Graph B.

3 Make one graph using a scale that makes the differences in the numbers of visitors look greater. Label this Graph C.

How does changing the scale of a graph affect how people interpret the data?

Write About Scales

Look at the three graphs you made for Number of Visitors per Day at the Zoo.

- Explain how you changed the scale in Graphs B and C.
- Describe some situations where someone might want the differences in data to stand out.
- Describe some situations where someone might want the differences in data to be less noticeable.
- List tips that you would give for choosing accurate scales when making a graph.
- Explain how you check a bar graph to make sure it accurately represents data.

hot words | bar graph

Homework page 39

5 Double Data

CREATING AND INTERPRETING DOUBLE BAR GRAPHS

A double bar graph makes it easy to compare two sets of data. After analyzing a double bar graph, you will make recommendations based on the data shown. Then you will be ready to create and analyze your own double bar graph.

How can you use a double bar graph to compare two sets of data?

Analyze the Double Bar Graph

The student council at Brown Middle School surveyed forty-six students and forty-six adults to find out about their favorite lunches. The double bar graph shows the results.

- What is the most popular lunch for students? for adults?
- What is the least popular lunch for students? for adults?
- Why is there no bar for adults where hamburgers are shown?

Make Recommendations

Use the data in the double bar graph to make recommendations to the student council about what to serve for a parent-student luncheon. Be sure to use fractions to describe the data.

Favorite Lunches of Middle School Students and Adults

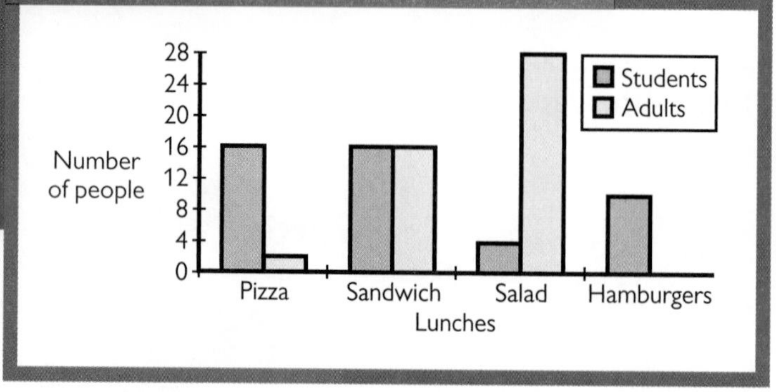

- What would you recommend that the student council serve at the parent-student luncheon?
- What other things should the student council think about when choosing food for the luncheon?

Make Double Bar Graphs

How can you create a double bar graph to compare two sets of data?

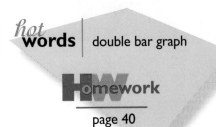

The table shown gives data on the number of hours people in different age groups sleep on a typical night.

Hours of Sleep	6-Year-Olds	12-Year-Olds	14-Year-Olds	Adults
5	0	0	3	9
6	0	2	7	15
7	0	16	24	20
8	0	19	36	20
9	3	25	5	9
10	4	12	5	3
11	31	4	0	4
12	37	2	0	0
13	5	0	0	0

1. Choose two columns from the table.
2. Make a double bar graph to compare the data. Be sure your graph is accurate and easy to read.
3. When you finish, write a summary of what you found out.

Write About Double Bar Graphs

Students in Ms. Taylor's class came up with this list of topics.

Answer these questions for each topic:

- Could you make a double bar graph to represent the data?
- If it is possible to make a double bar graph, how would you label the axes of the graph?

A. Number of students and teachers at our school this year and last year.

B. Heights of students in the same grade.

C. Time students spend doing homework and time they spend playing sports.

D. Number of hours students watch television and number of televisions in their homes.

E. Number of miles students travel to school.

hot words | double bar graph

Homework

page 40

6 Across the Ages

MAKING COMPARISONS AND RECOMMENDATIONS

Do you think middle school students and adults feel the same about videos? In this lesson, you will analyze the results of a survey about videos. Then you will conduct your own survey to compare the opinions of two age groups.

How can you use double bar graphs to compare the opinions of two age groups?

Compare Opinion Data from Two Age Groups

The parent-student luncheon was a huge success. The student council has decided to hold a parent-student video evening. They surveyed about 100 middle school students and about 100 adults to find out their opinions of four videos. The double bar graphs on the handout Video Rating Scale show how students and adults rated each video. Use the graphs to answer these questions:

- How do middle school students feel about each video? Adults?
- Which video do students and adults disagree about the most? Explain your thinking.
- Which video do students and adults agree about the most? Explain your thinking.

Make Recommendations

Make recommendations to the Student Council at Brown Middle School. Be sure the data supports your recommendations.

- Which video should the council choose for a students-only video evening? Why?
- Which video should the council choose for an adults-only video evening? Why?
- Which video should the council choose to show at a parent-student video evening? Why?

Collect and Analyze Data, Part 2

How can you compare data from two different groups?

In Lesson 3, you conducted your own survey. This time you will conduct the survey again, but you will collect data from a different age group. Then you will compare the results of the two surveys. You will need your survey from Lesson 3 to complete this activity.

1 Choose a new age group.

- **a.** What age group do you want to survey?
- **b.** How will you find people from the new age group to survey? In order to compare the two groups fairly, you will need to use the same number of people as you did for the first survey.

2 Make a prediction about the results.

- **a.** How do you think people in the new group are likely to respond to the survey?
- **b.** How similar or different do you think the responses from the two groups will be?

3 Collect and represent the data.

- **a.** Collect data from the new group. Ask the same question you asked in Lesson 3. Record your data.
- **b.** Create a single bar graph to represent the data from the new group.
- **c.** Create a double bar graph to compare the data from the two groups.
- **d.** Explain how you chose the scales for the two graphs.

4 Analyze the data.

Write a report that includes the following information:

- **a.** What are the mean, median, mode, and range for the new age group? What do these measures tell you?
- **b.** How do the responses for the two groups compare? Make a list of comparison statements that are clearly supported by the data.
- **c.** How do the results compare with your predictions?

hot words | survey
double bar graph

Homework
page 41

PHASE THREE

To: Apprentice Statisticians
From: President, Super Data Company

You have been doing a great job conducting surveys and making graphs. Next, you are going to analyze progress over time.

When people learn new skills, like typing or playing a musical instrument, they need to practice a lot. It's hard work, so they want to know whether the practicing is paying off. Statistics can help measure progress. In your next assignment, you will collect data and analyze your performance at several skills.

Have you ever tried to learn a new skill, such as typing, juggling, or making free throws? How can you tell if you are improving? How can you tell if you are getting worse?

In Phase Three, you will measure your progress at some skills. Then you will use statistics to help you see whether you have been improving, staying the same, going up and down, or getting worse.

Progress over Time

WHAT'S THE MATH?

Investigations in this section focus on:

DATA COLLECTION

- Collecting numerical data

GRAPHING

- Making and interpreting broken-line graphs
- Using broken-line graphs to make predictions

DATA ANALYSIS

- Finding mean, median, mode, and range in a data set
- Using mean, median, mode, and range to analyze data

MathScape Online
mathscape1.com/self_check_quiz

7 Are You Improving?

USING STATISTICS TO MEASURE PROGRESS

Learning new skills takes a lot of practice. Sometimes it's easy to tell when you are improving, but sometimes it's not. Statistics can help you measure your progress. To see how this works, you are going to practice a skill and analyze how you do.

Graph and Analyze the Data

How can a broken-line graph help you see whether you improved from one game to the next?

After you play the Memory Game with your class, make a broken-line graph to represent your data.

1. For each game, or trial, plot a point to show how many objects you remembered correctly. Connect the points with a broken line.
2. When your graph is complete, analyze the data. Find the mean, median, mode, and range.
3. Write a summary of your findings. What information do the mean, median, mode, and range tell you about your progress? Overall, do you think you improved? Use data to support your conclusions. How do you predict you would do on the 6th game? the 10th one? Why?

The Memory Game

How to play:

1. You will be given 10 seconds to look at pictures of 9 objects.
2. After the 10 seconds are up, write down the names of the objects you remember.
3. When you look at the pictures again, record the number you remembered correctly.

Sample Broken-Line Graph for the Memory Game

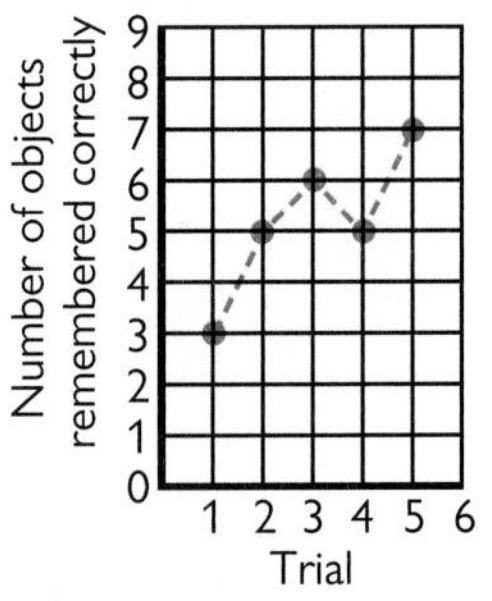

Compare Data

How can you compare two sets of data?

Two students played 5 trials of the Memory Game. In each trial, they looked at pictures of 9 objects for 10 seconds. The table shows their results. Use the table to compare Tomiko's and Bianca's progress.

1 Make broken-line graphs to show how each student did when playing the Memory Game.

2 Find the mean, median, mode, and range for each student.

3 Analyze the data by answering the following questions:

a. Which student do you think has improved the most? Use the data to support your conclusions.

b. How many objects do you think each student will remember correctly on the 6th game? the 10th game? Why?

Tomiko's and Bianca's Results

Number of Objects Remembered Correctly

Trial Number	Tomiko	Bianca
1	3	4
2	3	6
3	4	5
4	6	9
5	8	7

Design an Improvement Project

Now it's time to apply what you have learned. Think of a skill you would like to improve, such as juggling, running, balancing, or typing. The Improvement Project handout will help you get started.

1 Write a plan for a project in which you will practice the skill (see Step 1 on the handout).

2 Over the next 5 days, practice the skill and record your progress (see Step 2 on the handout).

hot words
broken-line graph
predict

Homework
page 42

8 How Close Can You Get?

GRAPHING AND ANALYZING ERRORS

In the last lesson, you kept track of your progress in the Memory Game. Now you will play an estimation game in which you try to get closer to the target with each turn. You will keep track of your progress and graph your errors to see if you improved.

Graph and Analyze Progress

How can you represent your errors so you can easily see the progress you have made?

After you play the Open Book Game, make a broken-line graph to represent your data. Be sure to label both axes.

1. For each trial, plot a point to represent the error. Connect the points with a broken line.
2. When your graph is complete, analyze the data. Find the mean, median, mode, and range.
3. Write a summary of your progress. What information do the mean, median, mode, and range tell you about your progress? Overall, do you think you improved? Make sure to use data to support your conclusions. How do you predict you would do on the 6th game? the 10th one? Why? How would you describe your progress from game to game?

The Open Book Game

How to play:

1. Your partner will tell you a page to turn to in the book. Try to open the book to that page without looking at the page numbers.
2. Record the page number you tried to get (Target) and the page you opened the book to (Estimate).
3. Figure out and record how close you were to the target page (Error) by finding the difference between the Target and the Estimate. Subtract the lesser number from the greater number.
4. For each trial, your partner will tell you a different page number. After 5 trials, switch roles.

Investigate Mixed-up Data

Kim loves to run and wants to get faster. Every day for 20 days she ran around her block and timed how long it took. She kept track of her progress on a broken-line graph and wrote about it in her journal. Unfortunately, her journal fell apart, and all the entries are out of order. Can you figure out which journal entry goes with which days?

How can you use what you have learned about graphing to sort out some mixed-up information?

Kim's Journal Entries

A Practicing is paying off. I'm making steady progress.

B I'm disappointed because I'm not making progress. At least I'm not getting worse.

C Wow. I've made my biggest improvement yet.

D I've been doing worse. I hope it's because I have a bad cold.

E I don't know what's going on. The time it takes me to run has been going up and down from one day to the next.

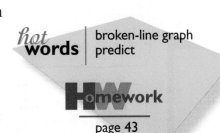

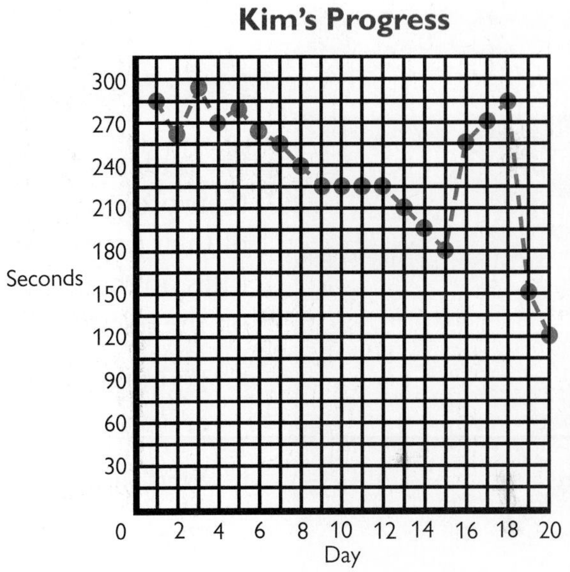

Read over the entries and examine the graph to help you answer these questions:

- Which days do you think each journal entry describes? Why? Tip: Each entry describes Kim's progress over 2 or more days.
- How did you figure out which entry went with which days on the graph?
- How would you describe Kim's overall progress for 20 days?
- How do you think Kim would do on days 21, 22, 23, and 24? Why? If you were Kim, what would you write in your journal about your progress on those days?

hot words broken-line graph, predict

Homework page 43

9 Stories and Graphs

INTERPRETING MULTIPLE REPRESENTATIONS OF DATA

Can you look at a graph and figure out what story it tells? In this lesson, you will interpret unlabeled graphs to figure out which ones match different people's descriptions of learning skills. Then you will compare and analyze graphs of progress and predict future performance.

Match Descriptions to Graphs

How can you figure out what stories a graph might represent?

Six students worked on improving their skills. They measured their progress by timing themselves. Then they wrote descriptions of their progress. They also graphed their data, but forgot to put titles on their graphs. Read the descriptions and study the graphs on the handout of Graphs of Students' Progress.

- Figure out which graph goes with which student. Explain your reasoning.
- Write a title for each graph.
- The extra graph belongs to Caitlin. Choose a skill for Caitlin. Then write a description of her progress.

Descriptions of Students' Progress

I've been practicing running fast around my block. I keep getting faster and faster. Tian

I've been practicing balancing on one foot with my eyes closed. I did better for a few days, then I did worse, then I did better again. Natasha

I'm working on rollerblading fast around my block. I'm not improving at all. I had my best time on my first day and the worst time on my last day. Miguel

Oops. I forgot to write a description. Caitlin

I'm practicing balancing on one foot with my eyes closed. I've been making steady progress. Keishia

I'm practicing juggling as long as I can without dropping a ball. I did great the first day, then I got worse. Then I stayed the same. At the end I was doing better. Terrence

Analyze Data from the Improvement Project

After practicing a skill for 5 days, how can you tell if you improved?

Now it's time to look at the data you've been collecting for the last 5 days. You will need the Improvement Project handout. Enter your data for each day in the table on your handout. Then analyze the data.

- Complete the table by entering the range, mean, mode, and median for each day's data.
- Look at your completed table. Use data you choose from the table to make two graphs. For example, you may want to graph the mean, or the highest score, or the total score for each day.

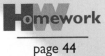

Write About the Improvement Project

Write a report to share your project with the class. Include answers to the following questions in your report:

- How did you decide which data to graph? Does one graph show more improvement than the other?
- Describe your progress from day to day.
- Overall, do you think you improved? Use the data to support your conclusions.
- How do you think you will do on the 6th day? Why?
- What mathematics did you use in the project?

hot words: predict, broken-line graph

Homework
page 44

PHASE FOUR

To: Apprentice Statisticians
From: President, Super Data Company

I hope you have enjoyed your work at Super Data Company so far!

People often talk about the likelihood of things happening (there's an 80% chance of rain; the Lions are favored to win). Being able to predict the probability that a particular event will occur can be very useful. In your next assignment, you will find the probability of picking a green cube out of a bag of colored cubes. You will need to use your data-collection and analysis skills in the investigation.

Have you ever decided to wear a raincoat because the weather report said it was likely to rain? Have you ever bought a raffle ticket because you thought your chances of winning were good? If you have, then you were basing your decisions on the probability that a specific event would occur.

Probability is the mathematics of chance. In this phase, you will investigate probability by playing some games of chance.

Probability and Sampling

WHAT'S THE MATH?

Investigations in this section focus on:

DATA COLLECTION

- Collecting and recording data
- Sampling a population

DATA ANALYSIS

- Using the results of sampling to make a hypothesis
- Using bar graphs to make an informed prediction

DETERMINING PROBABILITY

- Describing probabilities
- Calculating theoretical and experimental probabilities

MathScape Online
mathscape1.com/self_check_quiz

10 What Are the Chances?

DETERMINING PROBABILITY

Probability is the mathematics of chance. Raffles involve chance. The tickets are mixed together, and one ticket is picked to be the winner. Here you will play a game that is similar to a raffle. Then you will figure out the probability of winning.

How can you use data to predict the probability of picking a green cube?

Analyze Data from a Game of Chance

The Lucky Green Game is a game of chance. The chances of winning could be very high or very low. You will conduct an investigation to find out just how good the chances of winning really are.

1. Play the game with your group. Be sure each player takes 5 turns. Record each player's results in a table.
2. After each player has taken 5 turns, answer the following questions:
   a. How many greens do you think are in the bag? Use your group's results to make a hypothesis. Be sure to explain your thinking.
   b. Which of these words would you use to describe the probability of picking a green cube?

Never • Very Unlikely • Unlikely • Likely • Very Likely • Always

How can you analyze the class results?

The Lucky Green Game

Your group will be given a bag with 5 cubes in it. Do not look in the bag! Each player should do the following steps 5 times.

1. Pick one cube from the bag without looking.
2. If you get a green, you win. If you get a yellow, you lose. Record your results.
3. Put the cube back and shake the bag.

Analyze Data for a Different Bag of Cubes

What conclusions can you draw from another class's data?

Ms. Ruiz's class did an experiment with a bag that had 100 cubes in two different colors. Each group of students took out 1 cube at a time and recorded the color. Then they put the cube back in the bag. Each group did this 10 times. The groups put their data together in a class table shown on the handout, A Different Bag of Cubes. Help Ms. Ruiz's class analyze the data by answering these questions:

1. What are the mode, mean, median, and range for the numbers of cubes of each color that were drawn?
2. Based on the whole class's data, what is the experimental probability of picking each color?
3. Here is a list of bags that the class might have used in the experiment. Which bag or bags do you think the class used? Explain your thinking.

   a. 50 red, 50 blue
   b. 20 red, 80 blue
   c. 80 red, 20 blue
   d. 24 red, 76 blue
   e. 70 red, 30 blue
   f. 18 red, 82 blue

Types of Probabilities

Experimental probabilities describe how likely it is that something will occur. Experimental probabilities are based on data collected by conducting experiments, playing games, and researching statistics in books, newspapers, and magazines.

The experimental probability of getting a cube of a particular color can be found by using this formula:

$$\frac{\text{Number of times a cube of a particular color was picked}}{\text{Total number of times a cube was picked}}$$

Theoretical probabilities are found by analyzing a situation, such as looking at the contents of the bag.

The theoretical probability of getting a cube of a particular color can be found by using this formula:

$$\frac{\text{Number of cubes of that color in the bag}}{\text{Total number of cubes in the bag}}$$

hot words
experimental probability
theoretical probability

Homework
page 45

11 Changing the Chances

EXPERIMENTING WITH PROBABILITY

Does having more cubes in the bag improve your chances of winning? In this lesson, you will change the number of green and yellow cubes. Then you will play the Lucky Green Game to see if the probability of winning has gotten better or worse.

How does changing the number of cubes in the bag change the probability of winning?

Compare Two Bags of Cubes

In Lesson 10, you found the probability of winning the Lucky Green Game. Now, you'll conduct an experiment to see how changing the number of cubes in the bag changes the chances of winning. Follow these steps to find out:

1. Change the number of cubes in the bag you used in Lesson 10 (Bag A), so that it contains 6 green cubes and 4 yellow cubes. Call this new bag, Bag B.
2. Make a hypothesis about which bag (Bag A or Bag B) gives you a better chance of picking a green cube. Explain your reasoning.
3. Collect data by playing the Lucky Green Game (see page 30). Make sure each player takes 5 turns! Record your results on the handout Changing the Cubes Recording Sheet.

Summarize the Data

After your group finishes the experiment, write a summary of your data that includes the following information:

- What were the range, mode, and mean number of greens?
- What was the experimental probability of picking a green cube?
- Did your results support your hypothesis about which bag (A or B) gives you a better chance of picking a green cube? Why or why not?

Rank the Bags

Mr. Chin's class wants to investigate the chances of winning with different bags of cubes. The table below shows the numbers of cubes in the bags Mr. Chin's class plans to use.

How can you compare bags with different numbers of cubes?

More Bags of Cubes

Bag	Green Cubes	Yellow Cubes	Total Number of Cubes
B	6	4	10
C	7	13	20
D	14	6	20
E	13	27	40
F	10	30	40

1 Choose one of the bags (except Bag B). If you picked a cube from that bag 100 times, how many times do you think you would get a green cube? Why?

2 For each bag, find the theoretical probability of picking a green cube. Explain how you figured it out.

3 Rank the bags from the best chance of getting a green cube to the worst chance of getting a green. (Best = 1, Worst = 5) Be sure to explain your answer.

4 After you finish ranking the bags, make a new bag of cubes that will give you a better chance of getting a green than the second-best bag, but not as good as the best bag. How many green and yellow cubes are in the new bag? Explain.

Make Generalizations

Use the results of your data to answer these questions:

- A class did an experiment with one of the bags shown in the table. In 100 turns, they got 32 yellow cubes. Which bag or bags do you think it is most likely that they used? Why?
- What generalizations would you make about how to determine which bag of cubes gives you a better chance of picking a green cube?

hot words
chance
probability

Homework
page 46

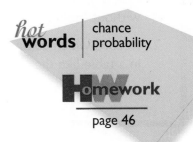

12 Which Bag Is Which?

APPLYING PROBABILITY AND STATISTICS

In the last two lessons, you used a method called *sampling* when you made predictions. Here you will use sampling again to predict what's in the bag, but this time you will need to share your findings with the rest of the class in order to be sure.

How can you use what you have learned about sampling to make predictions?

Investigate the Jelly Bean Bag Mix-Up

The graphs on the handout Jelly Bean Bag Combinations show how many jelly beans are in each bag. Each group in your class will get one of the bags to sample. Can you tell which graph matches your bag?

1 Collect data by sampling your bag.

2 Compare your data to the bar graphs on Jelly Bean Bag Combinations. Which bag do you think you have? Write down why you think your group has that bag. If you are not sure, explain why.

Sampling the Jelly Bean Bags

How to sample:

Each student should do the following steps 6 times (that is, take 6 samples):

1. Pick one cube from the bag without looking.
2. Record which color you got in a table like the one shown.
3. Put the cube back and shake the bag before taking the next sample.

Student	Cherry (Red)	Blueberry (Blue)	Lemon (Yellow)	Lime (Green)
Marie Elena	I	I I I	I	I
Ricardo	I I	I I I	I	
Myra	I	I I	I I	I
Ursula	I	I I	I	I I

Analyze and Compare Bags

After the class has solved the Jelly Bean Bag Mix-up, write about the investigation by answering these questions.

1 Write about your group's bag.

- **a.** Which bag did your group have? What strategies did your group use to try to figure this out?
- **b.** Use your group's data to figure out the experimental probability of picking a jelly bean of each color from the bag.
- **c.** What is the theoretical probability of picking a jelly bean of each color from the bag?

2 Compare the five bags.

- **a.** Rank the five bags from best to worst theoretical probability of picking red. (Best = 1, Worst = 5)
- **b.** Rank the five bags from best to worst theoretical probability of picking green. (Best = 1, Worst = 5)
- **c.** Explain how you figured out how to rank the bags.
- **d.** Fiona took many samples from one of the bags. She got 62 reds, 41 blues, 8 yellows, and 9 greens. Which bag or bags do you think she had? Why?

hot words | sampling with replacement
probability

Homework
page 47

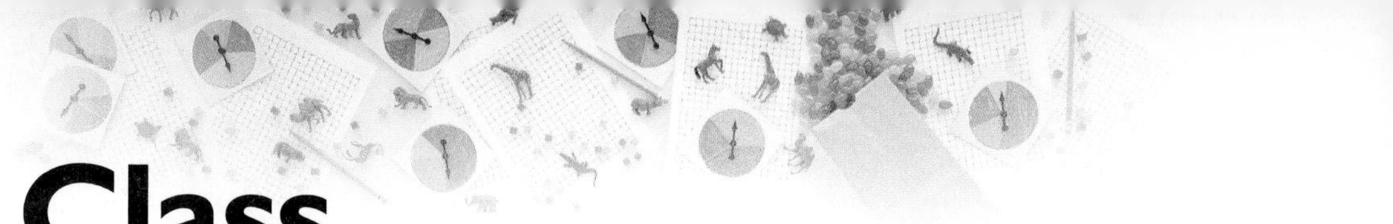

Homework 1: Class Survey

Applying Skills

In items 1–5, find the range and mode (if any) for each set of data. Be sure to express the range as a difference, not as an interval.

1. 14, 37, 23, 19, 14, 23, 14

2. 127, 127, 117, 127, 140, 133, 140

3. 93, 40, 127, 168, 127, 215, 127

4. 12, 6, 23, 45, 89, 31, 223, 65

5. 1, 7, 44, 90, 6, 89, 212, 100, 78

6. Mr. Sabot's class took a survey in which students were asked how many glasses of water they drink each day. Here are the results:

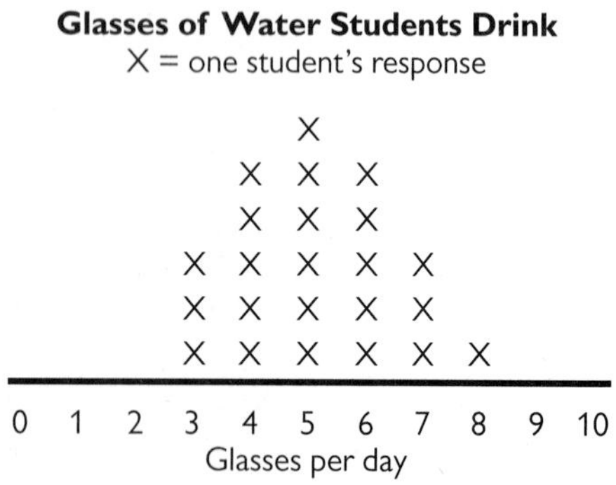

What are the range and mode of the data?

7. Ms. Feiji's class took a survey to find out how many times students had flown in an airplane. Below is the data. Make a frequency graph for the survey and find the mode and range.

- Nine students had never flown.
- Ten students had flown once.
- Six students had flown twice.
- One student had flown five times.

Extending Concepts

8. Ms. Olvidado's class took this survey, but they forgot to label the graphs. Decide which survey question or questions you think each graph most likely represents. Explain your reasoning.

Question 1: How many hours do you sleep on a typical night?

Question 2: How many times do you eat cereal for breakfast in a typical week?

Question 3: In a typical week, how many hours do you watch TV?

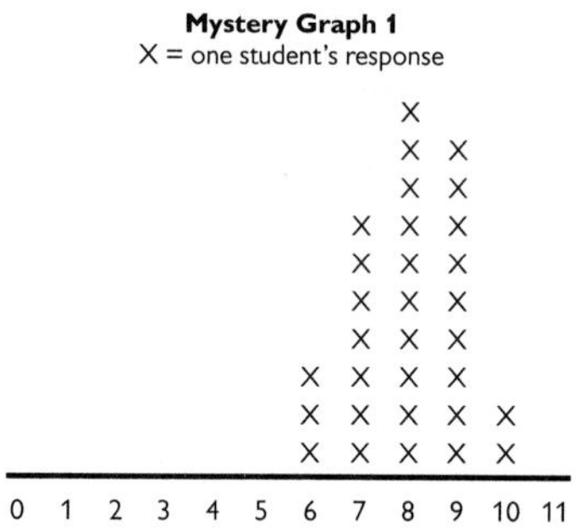

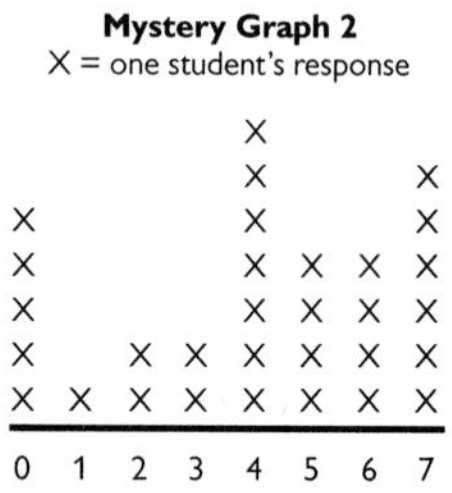

Writing

9. Answer the letter to Dr. Math.

Dear Dr. Math:

We tried to survey 100 sixth graders to find their preferences for the fall field trip, but somehow we got 105 responses. Not only that, some kids complained that we forgot to ask them. What went wrong? Please give us advice on how to conduct surveys.

Minnie A. Rohrs

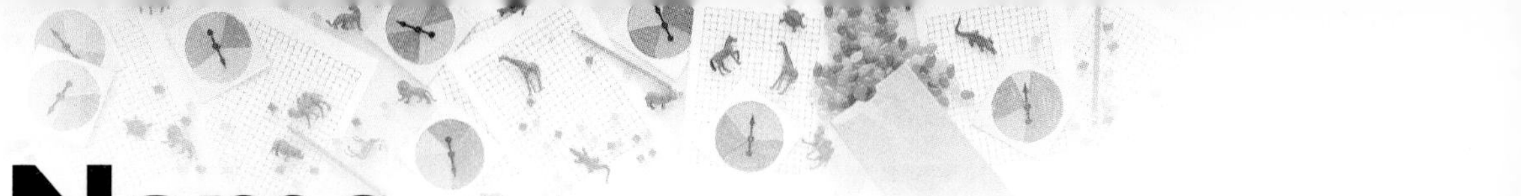

Homework 2

Name Exchange

Applying Skills

Find the mean and median of each data set.

1. 10, 36, 60, 30, 50, 20, 40

2. 5, 8, 30, 7, 20, 6, 10

3. 1, 10, 3, 20, 4, 30, 5, 2

4. 18, 22, 21, 10, 60, 20, 15

5. 29, 27, 21, 31, 25, 23

6. 3, 51, 45, 9, 15, 39, 33, 21, 27

7. 1, 4, 7, 10, 19, 16, 13

8. 10, 48, 20, 22, 57, 50

A study group has the following students in it:

Girls: Alena, Calli, Cassidy, Celina, Kompiang, Mnodima, and Tiana

Boys: Dante, Harmony, J. T., Killian, Lorn, Leo, Micah, and Pascal

9. Find the mean and median number of letters in the girls' names.

10. Find the mean and median number of letters in the boys' names.

11. Find the mean and median number of letters in *all* the students' names.

12. Find the mean and median numbers of pretzels in a bag of Knotty Pretzels, based on this graph of the results of counting the number of pretzels in 10 bags.

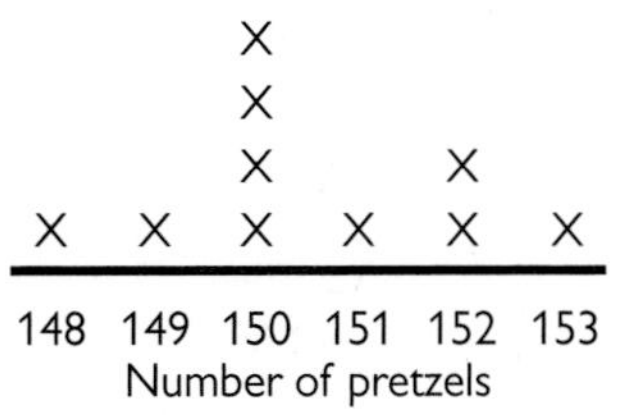

Extending Concepts

Professor Raton, a biologist, measured the weights of capybaras (the world's largest rodent) from four regions in Brazil.

Weights of Capybaras

Region	Weights (kg)
A	6, 21, 12, 36, 15, 12, 27, 12
B	18, 36, 36, 27, 21, 48, 36, 33, 21
C	12, 18, 12, 21, 18, 12, 21, 12
D	30, 36, 30, 39, 36, 39, 36

13. Find the mean and median weight of the capybaras in each region.

14. Find the mode and range of each data set. For each set explain what the range tells us that the mode doesn't.

Writing

15. Answer the letter to Dr. Math.

Dear Dr. Math:
When we figured out the mean, median, mode, and range for our survey, some answers were fractions or decimals, even though we started with whole numbers. Why is this? If the numbers in the data set are whole numbers, are any of those four answers *sure* to be whole numbers?
Frank Shun and Tessie Mahl

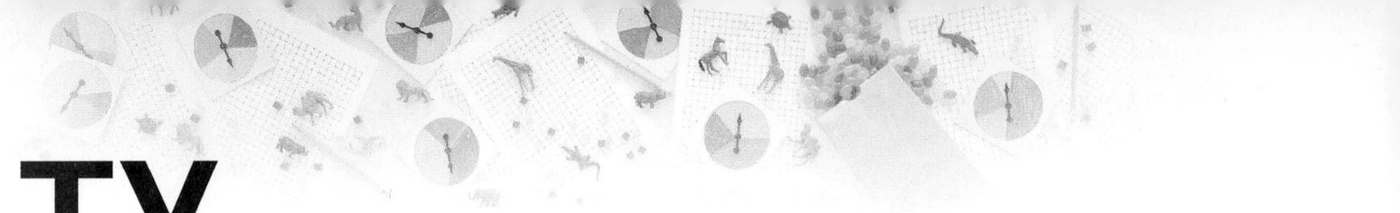

Homework 3: TV Shows

Applying Skills

Jeff conducted a survey rating TV shows on a scale of 1 ("bo-o-o-oring") to 5 ("Excellent, dude!"). Here are the results:

Show 1

Rating	Number of Students
1	0
2	3
3	7
4	3
5	7

Show 2

Rating	Number of Students
1	2
2	2
3	4
4	7
5	5

1. Draw a frequency graph of the results of each survey.
2. Find the mode(s) for the ratings, if any, of each survey.
3. Find the median rating for each survey.

Lara's class took a survey asking students to rate four different activities on a scale of 1 ("Yuck!") to 5 ("Wowee!").

Activity A
X = one student's response

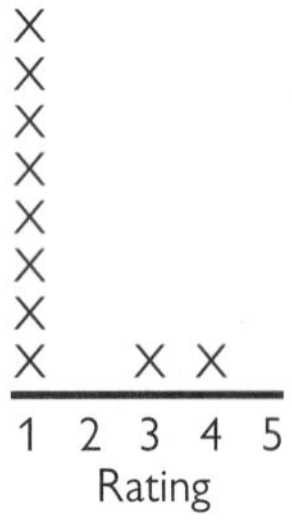

Activity B
X = one student's response

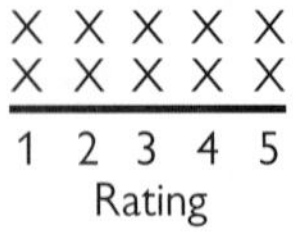

Activity C
X = one student's response

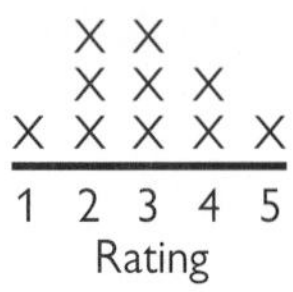

Activity D
X = one student's response

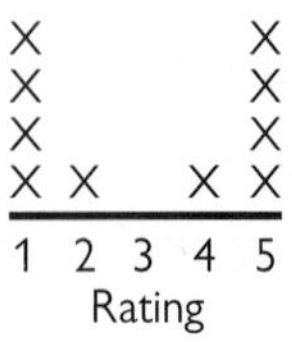

4. Find the median rating for each survey.
5. Find the mean rating for each survey.

Extending Concepts

Statisticians describe graphs of data sets by using four different types of distributions.

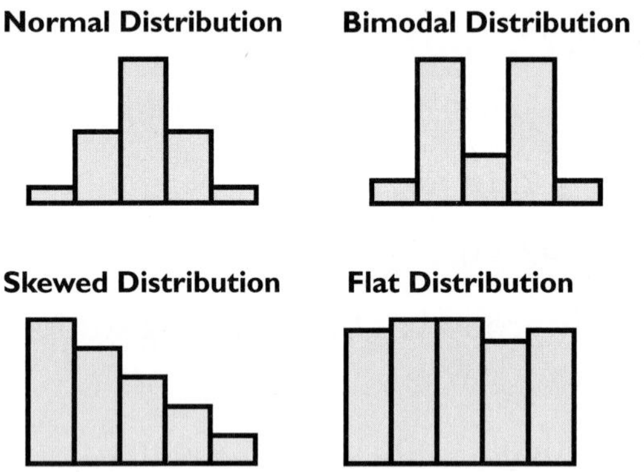

6. Describe the shape of each graph showing the data from Lara's class.
7. Here are four activities: going to the dentist, listening to rap music, taking piano lessons, and rollerblading. Tell which activity you think goes with each graph for Lara's class, and why.

Homework 4

Animal Comparisons

Applying Skills

Here is some information about dinosaurs. "MYA" means "Millions of Years Ago."

Dinosaur	Length (ft)	Height (ft)	Lived (MYA)
Afrovenator	27	7	130
Leaellynasaura	2.5	1	106
Tyrannosaurus	40	18	67
Velociraptor	6	2	75

1. Make a bar graph showing the length of each dinosaur.

2. Make a bar graph showing the height of each dinosaur.

3. Make a bar graph showing how many millions of years ago the dinosaurs lived.

Here are one student's answers to items **1–3:**

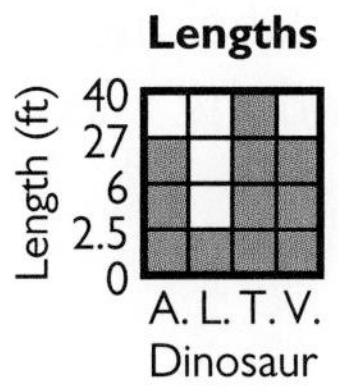

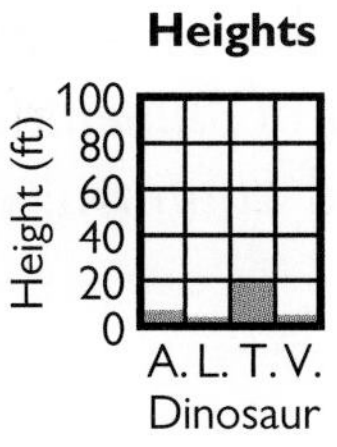

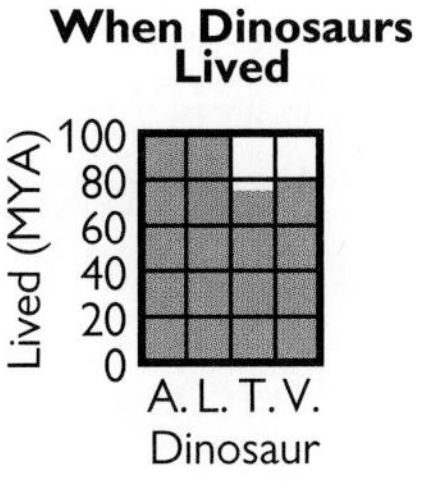

4. What is wrong with the bar graph of the dinosaurs' lengths?

5. What is wrong with the bar graph of the dinosaurs' heights?

6. What is wrong with the graph showing how long ago the dinosaurs lived?

Extending Concepts

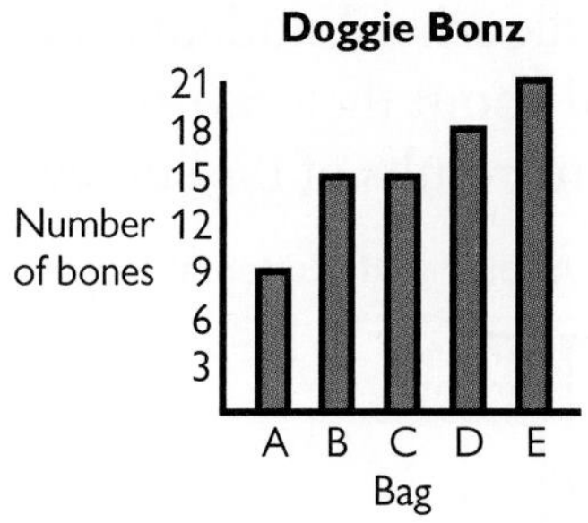

7. Use the graph that shows the number of dog bones in five different bags of Doggie Bonz to make a table of the data.

8. Find the range, mean, and median number of bones in a bag of Doggie Bonz.

Making Connections

Some scientists think that the size of the largest animals on land has been getting smaller over many millions of years. Here are the weights of the largest *known* animals at different periods in history.

Animal	MYA	Estimated Weight (tons)
Titanosaur	80	75
Indricothere	40	30
Mammoth	3	10
Elephant	0	6

9. Draw a bar graph of this information.

10. Does the graph seem to support the conclusion that the size of the largest animals has been getting smaller? What are some reasons why this conclusion might *not* actually be true?

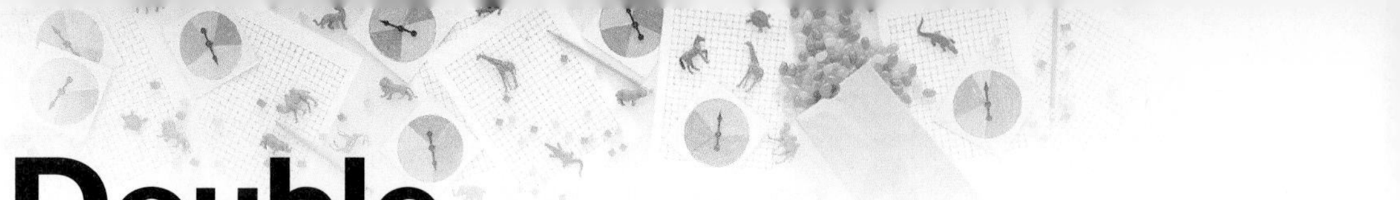

Homework 5: Double Data

Applying Skills

Forty middle school students and 40 adults were asked about their favorite activities. Here are the results of the survey.

Adults' and Students' Favorite Activities

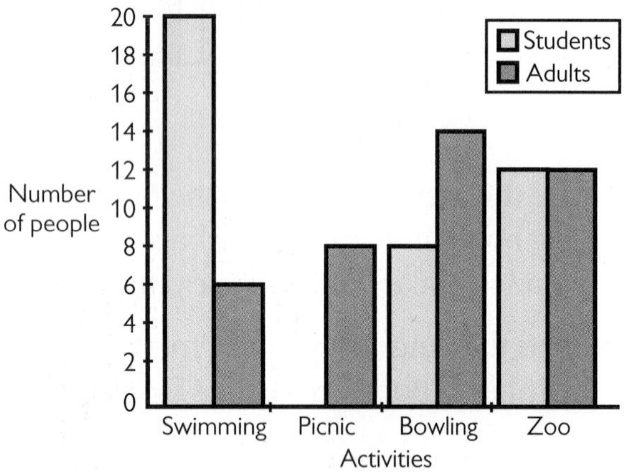

1. Show the survey results in a table (like the one for items 5–7 below).

2. What is the most popular activity for students?

3. What is the most popular activity for adults?

4. Why is there no student bar for Picnic?

Eighty 6-year-olds and eighty 12-year-olds were asked how many hours a day they usually watch TV. Here are the results.

Hours	6-Year-Olds	12-Year-Olds
0	16	4
1	23	9
2	26	15
3	15	38
4	0	14

5. Make a double bar graph of the data.

6. Calculate the mean, median, and mode for the number of hours each group watches TV.

7. Which group on average watches more TV?

Extending Concepts

Signorina Cucina's cooking class rated pies made with 1 cup of sugar, 2 cups of sugar, or 3 cups of sugar. Here are the results.

	Yucky	OK	Yummy
1 cup	2	4	14
2 cups	7	9	4
3 cups	11	6	3

8. Make a triple bar graph of the results. Label the y-axis *Number of Students* and the x-axis *Number of Cups.*

9. Now make another triple bar graph with the survey results. This time, label the y-axis *Number of Students* and the x-axis *Yucky, OK,* and *Yummy.*

Writing

10. Tell whether you could make a double bar graph for each set of data. If you *could* make one, tell what the labels on the axes would be. If not, explain why not.

- **a.** Heights of students at the beginning of the year and at the end of the year.
- **b.** Ages of people who came to see the school show.

Homework 6: Across the Ages

Applying Skills

Here are the ratings given to two different bands by 100 students and 100 adults.

Band A

Rating	Number of Students	Number of Adults
Terrible	3	1
Bad	5	3
OK	40	27
Good	43	60
Great	9	9

Band B

Rating	Number of Students	Number of Adults
Terrible	2	26
Bad	10	22
OK	14	20
Good	21	18
Great	53	14

1. Make a double bar graph for the ratings of Band A. Use different colors for students and adults.

2. Make a double bar graph for the ratings of Band B. Use different colors for students and adults.

3. Make a double bar graph for the ratings by students. Use different colors for Band A and Band B.

4. Which band would be best for a party for students?

5. Which band would be best for a party for adults?

6. Which band would be best for a party for students and adults?

Extending Concepts

7. Here are the results of a survey a student did on the number of glasses of milk 35 sixth graders and 35 adults drink in a typical week. Describe what's wrong with the graph and make a correct one.

Glasses per Week	Number of 6th Graders	Number of Adults
0	2	7
1	1	0
2	0	1
3	2	7
4	1	0
5	4	7
6	5	4
7	10	6
8	4	2
9	6	1

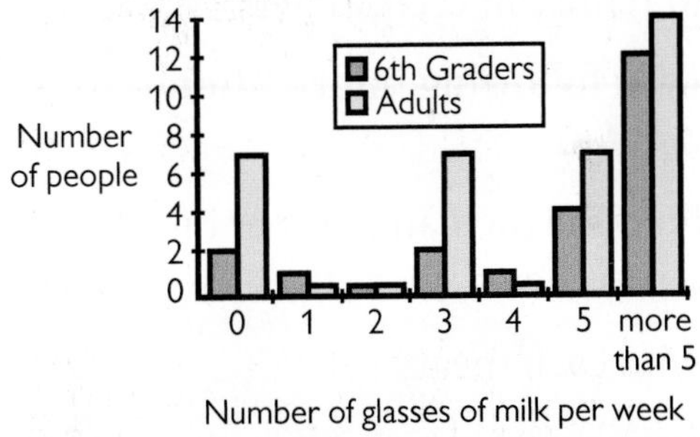

Writing

8. Pat wanted to compare how many seventh graders and kindergartners have pets. She found that of 25 seventh graders, 15 had pets. She didn't know any kindergartners, so she questioned 5 younger brothers and sisters of the seventh graders. Two of them had pets. What do you think about Pat's survey?

Homework 7

Are You Improving?

Applying Skills

Our Data for the Memory Game

Trial	Number of Objects Remembered	
	Ramir	Anna
1	3	6
2	4	5
3	6	5
4	7	8
5	8	7

The table shows how Ramir and Anna did when they played the Memory Game.

1. Draw a broken-line graph to show each student's progress. Use a different color to represent each student.

Find the following information for Ramir and for Anna.

2. Find the median number of objects remembered correctly.

3. Find each mean.

The graph shows Caltor's progress while playing the Memory Game.

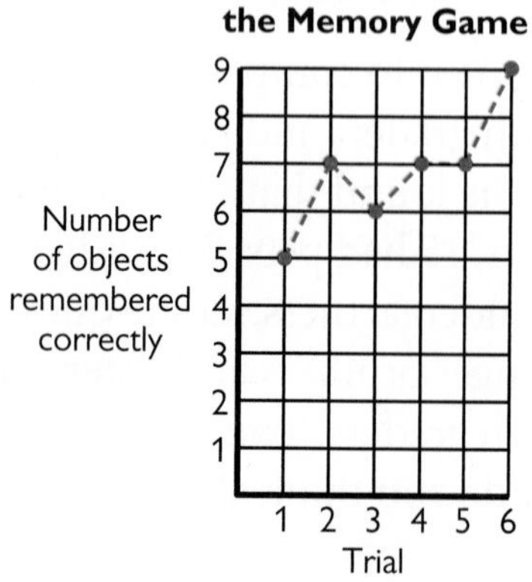

4. How many objects did Caltor remember correctly on Trial 3?

5. How many more objects did Caltor remember correctly on the 6th trial than on the 1st trial?

6. What is the mode?

Extending Concepts

Katia is trying to learn Spanish. Her teacher gave her worksheets with pictures of 20 objects. She has to write the Spanish word for each object. Then she checks to see how many words she got correct. The graph shows her progress.

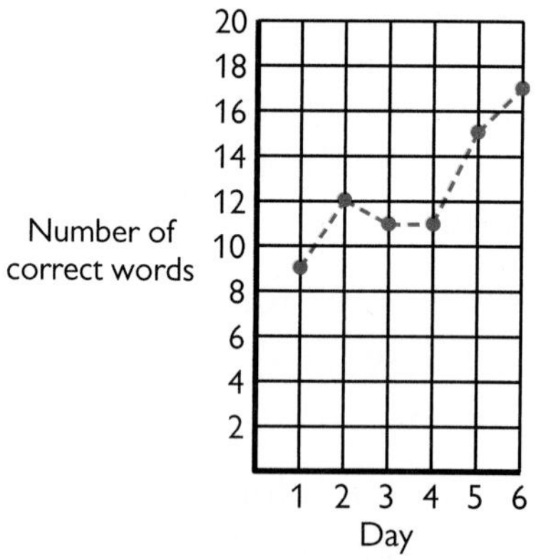

7. Create a table of the data.

8. Find the mean, median, mode, and range for Katia's data.

Writing

9. Give examples of five different types of data for which you might use a broken-line graph. Tell how you would label the *x*-axis and the *y*-axis for each graph.

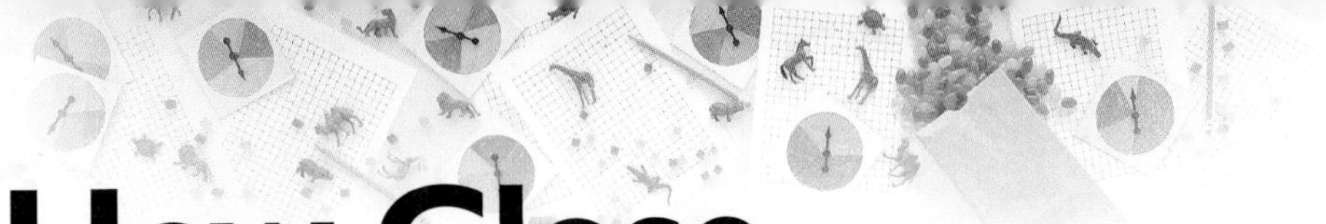

How Close Can You Get?

Homework 8

Applying Skills

The table below shows the results for a student who played the Open Book Game five times.

Dolita's Data for the Open Book Game

Trial	Target	Estimate	Error
1	432	334	
2	112	54	
3	354	407	
4	247	214	
5	458	439	

1. Complete the table by finding the Error for each trial.
2. Make a broken-line graph of Dolita's errors.
3. Find the mean, median, mode (if any), and the range of Dolita's errors.

Every day for 12 days, Tomas runs down the block and times how long it takes.

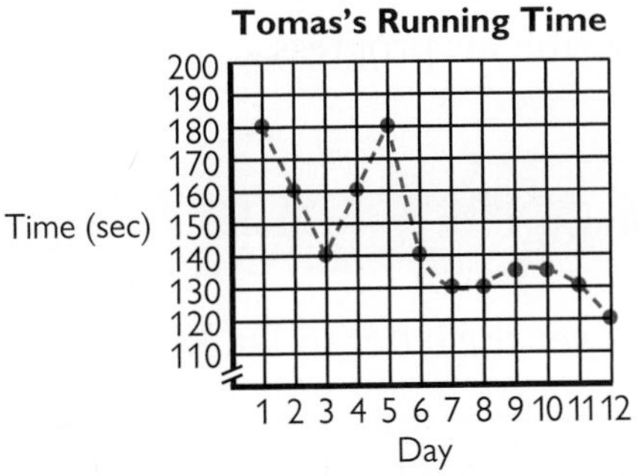

4. Use the graph to figure out the mode, median, and range for Tomas's running times.
5. Use a calculator to figure out Tomas's mean time for running down the block.
6. Make a prediction for how fast you think Tomas would run on Day 15. Explain how you made your prediction.
7. Look at the data for the first 5 days only. On which days did Tomas get faster?
8. On which day(s) did Tomas run the fastest?

Extending Concepts

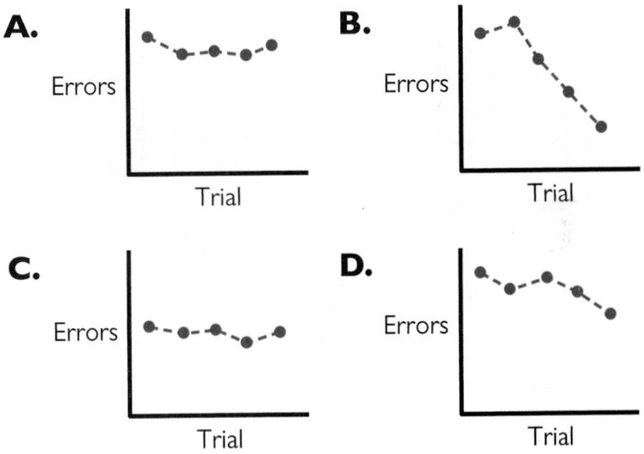

9. The graphs shown represent four students' progress in the Open Book Game. For each graph, write a description of the student's progress.
10. Which graph shows the least improvement? Explain.
11. Which graph shows the most improvement? Explain.

Writing

12. Dottie noticed that when her scores in the Memory Game improved, her graph kept going up. But when she played the Open Book Game, her graph went down, although she was sure she was improving. Explain to Dottie how to read a broken-line graph.

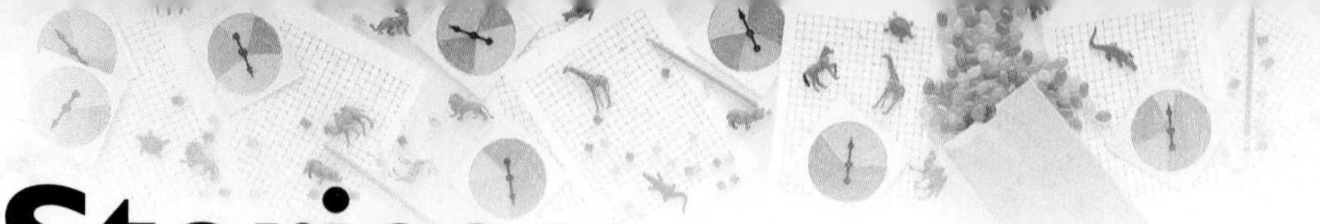

Homework 9: Stories and Graphs

Applying Skills

A.
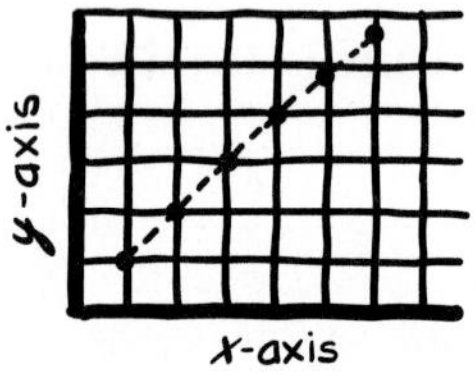

B.
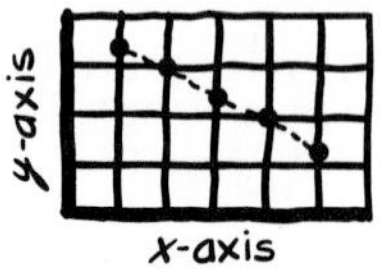

C.
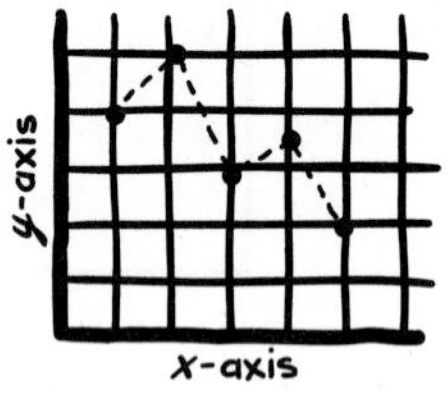

D.
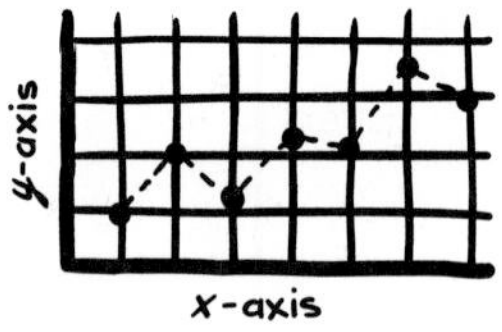

Match each description with the appropriate graph. Then tell how you would label the *x*-axis (across) and the *y*-axis (up and down).

1. "I've made steady improvement in the Memory Game."

2. "I've been running faster every day."

3. "My errors for the Open Book Game have been going up and down. Overall, I've gotten better."

4. "My swimming speed has been going up and down. Overall, I don't seem to be improving!"

5. "I've been timing how long I can stand on my head. I've had good days and bad days, but mostly I've increased my time."

Extending Concepts

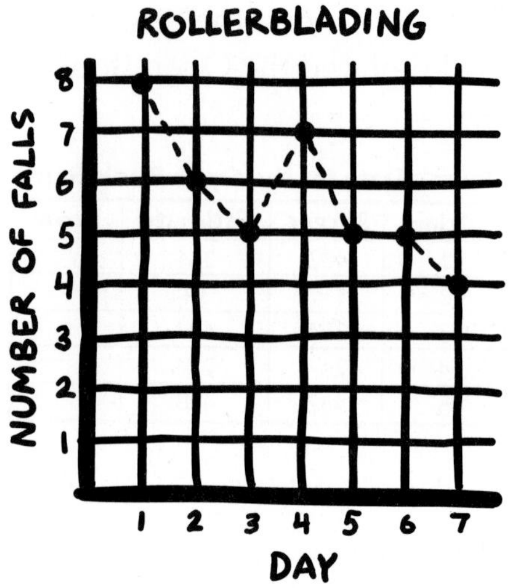

6. Make a table for the data represented on the graph "Rollerblading."

7. Calculate the mean, median, mode (if any), and range for number of falls.

8. Did this student improve at rollerblading? Write a sentence to describe his progress.

Writing

9. Answer the letter to Dr. Math.

Dear Dr. Math,
I'm confused. I don't know how you can look at a broken-line graph that has no numbers and figure out whether it shows that a student is or isn't improving.
Reada Graph

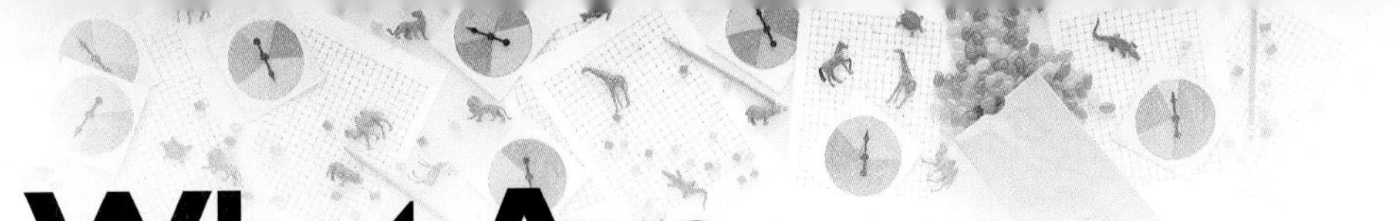

Homework 10

What Are the Chances?

Applying Skills

Each student picked a cube from a bag twenty times. After each turn, the cube was returned to the bag. Results for each student were recorded in the table.

Data From Our Experiment

Students	Number of Greens	Number of Yellows
Anna	15	5
Bina	18	2
Carole	12	8
Dan	16	4
Elijah	14	6

Use the data on green and yellow cubes to find the following values for each color.

1. mode **2.** range

3. mean **4.** median

Use fractions to describe each student's experimental probability of getting a green cube.

Example: Anna: $\frac{15}{20}$

5. Bina **6.** Carole **7.** Dan

8. Elijah

Use fractions to describe each student's experimental probability of getting a yellow cube.

9. Anna **10.** Bina **11.** Carole

12. Dan **13.** Elijah

14. Combine the data for the whole group. What is the whole group's experimental probability for picking a **green cube**?

Extending Skills

Here is a list of bags that the students might have used to collect the data shown in the table. For each bag, decide whether it is likely, unlikely, or impossible that students used that bag. Explain your thinking.

15. 28 green, 12 yellow

16. 10 green, 30 yellow

17. 8 green, 2 yellow

18. 16 green, 4 blue

19. 15 green, 15 yellow

20. If students used a bag with 100 cubes in it, how many green and yellow cubes do you think it contained? Explain your thinking.

Making Connections

21. Frequently on TV the weather reporter gives the chance of rain as a percentage. You might hear, "There's a 70% chance of rain for tomorrow afternoon, and the chances increase to 90% by tomorrow night." What does this mean? Why do you think this kind of language is used to talk about weather? In what other situations do people talk about the chances of something happening?

Homework 11

Changing the Chances

Applying Skills

Bag	Blue	Red	Total Number of Cubes	Theoretical Probability of Picking Blue	Theoretical Probability of Picking Red
A	9	1	10	$\frac{9}{10}$	$\frac{1}{10}$
B	7	13			
C	16	4			
D	15	15			
E	22	8			
F	30	10			

1. Copy the table and fill in the missing information. Use the first row as an example.

2. Which bag gives you the highest probability of getting a blue cube?

3. Which bag gives you the highest probability of getting a red cube?

4. Which bag gives you the same chance of picking a blue or a red cube?

5. Yasmine has a bag with 60 cubes that gives the same probability of picking a blue cube as Bag C. How many blue cubes are in her bag?

6. How many red cubes are in Yasmine's bag?

7. Rank the bags from the best chance of getting a blue cube to the worst chance of getting a blue cube.

Extending Concepts

Students did experiments with some of the bags shown in the table. The results of these experiments are given below. For each of the results, find the indicated experimental probability. Which bag or bags do you think it is most likely that the students used? Why?

8. In 100 turns, we got 20 reds.

9. We got 44 blues and 46 reds.

10. In 100 turns, we got 75 blues.

11. In 5 turns, we got 0 reds.

Writing

12. Suppose Sandy's bag has 2 purple cubes out of a total of 3 cubes and Tom's bag has 8 purple cubes out of 20 cubes. Explain how to figure out which bag gives you the best chance of picking a purple cube if you pick without looking.

Homework 12

Which Bag Is Which?

Applying Skills

This bar graph shows the number of cubes of different colors that are in a bag of 20 cubes.

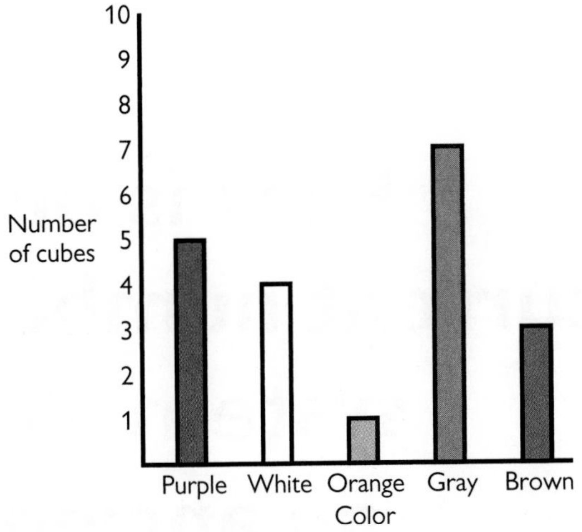

Use the graph to figure out the theoretical probability of picking each color. Be sure to write the probability as a fraction.

1. a purple cube

2. a white cube

3. an orange cube

4. a grey cube

5. a brown cube

Students each took 10 samples from the bag and recorded their data in the table shown.

Data From Our Experiment

Student	Purple Cubes	White Cubes	Orange Cubes	Gray Cubes	Brown Cubes
Miaha	2	2	0	4	2
Alec	3	1	1	5	0
Dwayne	3	2	1	3	1
SooKim	2	3	0	3	2

6. Combine the data for all the students to figure out the group's **experimental probability** of picking each color. Write the probability as a fraction.

7. For each color, find the mean number of times it was picked.

Extending Skills

8. A box of Yummy Chewy Candy has 30 pieces of candy. The pieces of candy are blue, green, red, and pink. The theoretical probability of picking a blue piece is $\frac{1}{3}$, a green piece is $\frac{1}{6}$, and a red piece is $\frac{1}{5}$. How many pieces of pink candy are in the box? Explain.

Writing

9. Answer the letter to Dr. Math.

> Dear Dr. Math,
> I was looking at the results of the Jelly Bean Supreme Investigation and I'm confused. The theoretical probability of picking a blueberry from Bag A is $\frac{7}{12}$. My group picked 24 times from Bag A and got 16 blueberries. Is that more or less blueberries than you would expect? Why didn't our results match the theoretical probability exactly?
> Beanie

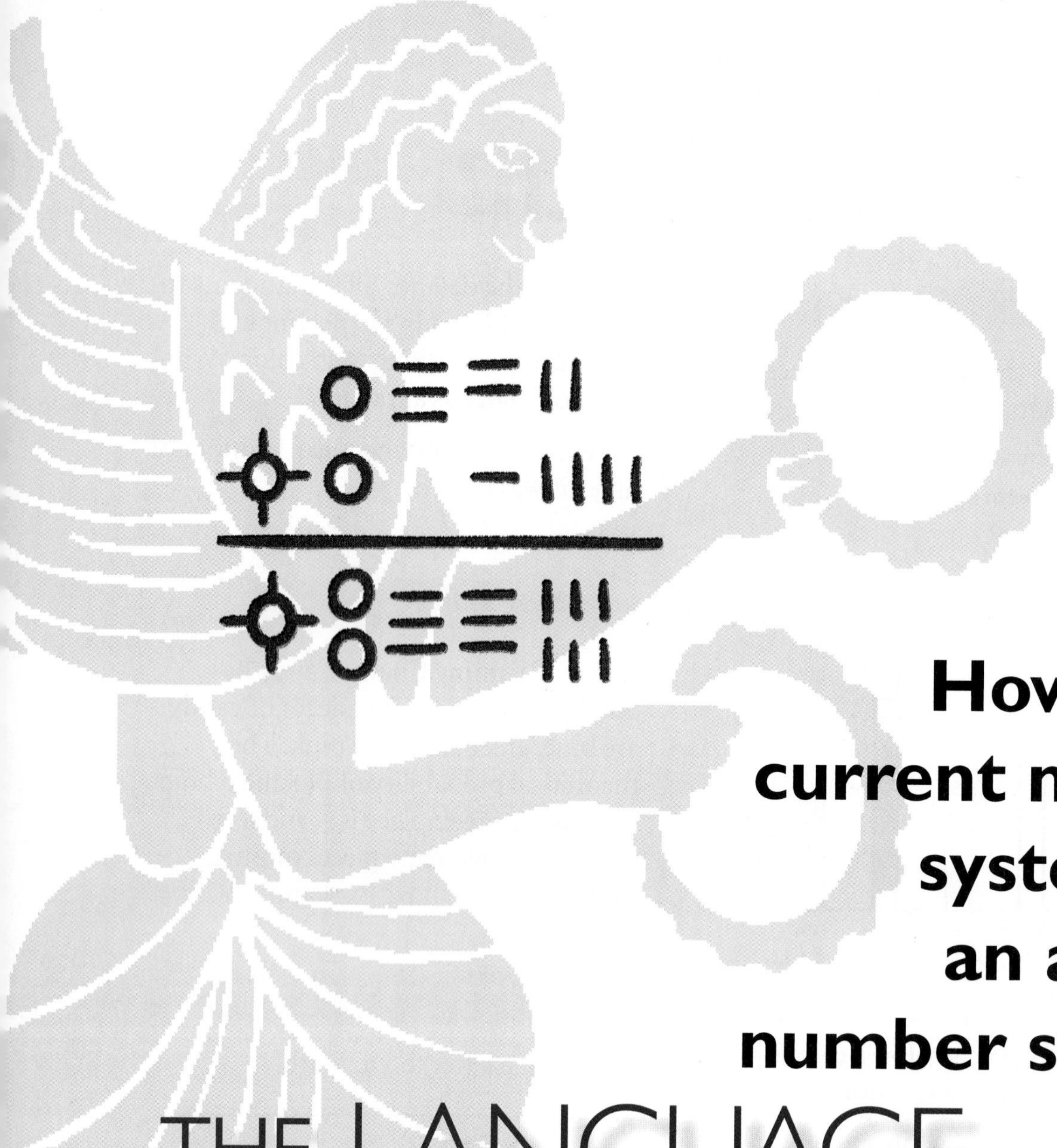

How is our current number system like an ancient number system?

THE LANGUAGE OF NUMBERS

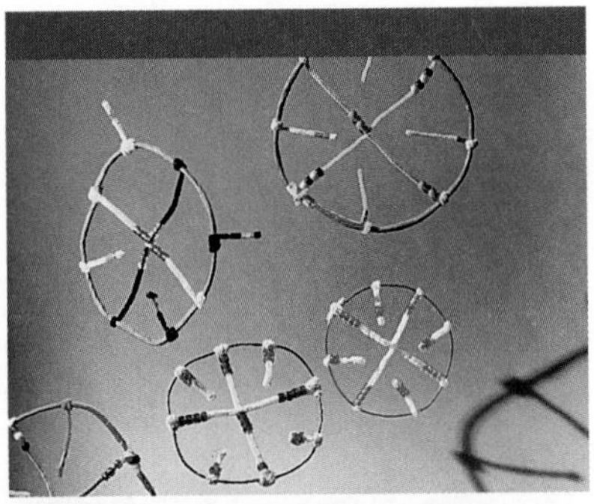

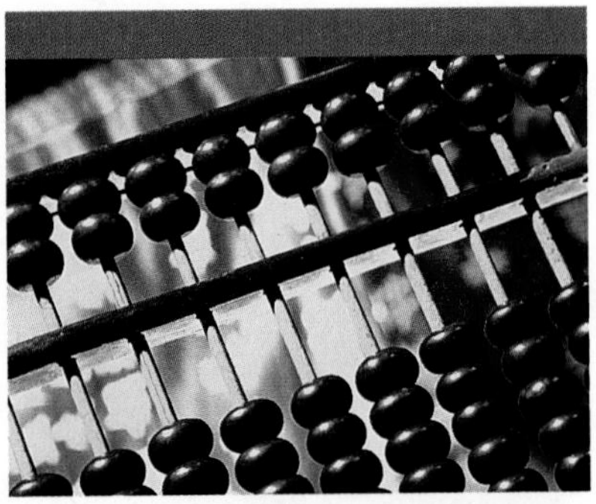

PHASE**ONE**

Mystery Device™

Our everyday number system is one of humanity's greatest inventions. With just a set of ten simple digits, we can represent any amount from 1 to a googol (1 followed by 100 zeros) and beyond. But what if you had to create a new system? In Phase One, you will investigate the properties of a number system. To do this, you will be using a Mystery Device to invent a new system.

PHASE**TWO**

Chinese Abacus

The Chinese abacus is an ancient device that is still used today. You will use the abacus to solve problems such as: What 3-digit number can I make with exactly three beads? You will compare place value in our system to place value on the abacus. This will help you to better understand our number system.

PHASE**THREE**

Number Power

In this phase, you will test your number power in games. This will help you see why our number system is so amazing. You will explore systems in which place values use powers of numbers other than 10. You will travel back in time to decode an ancient number system. Finally, you will apply what you have learned to create the ideal number system.

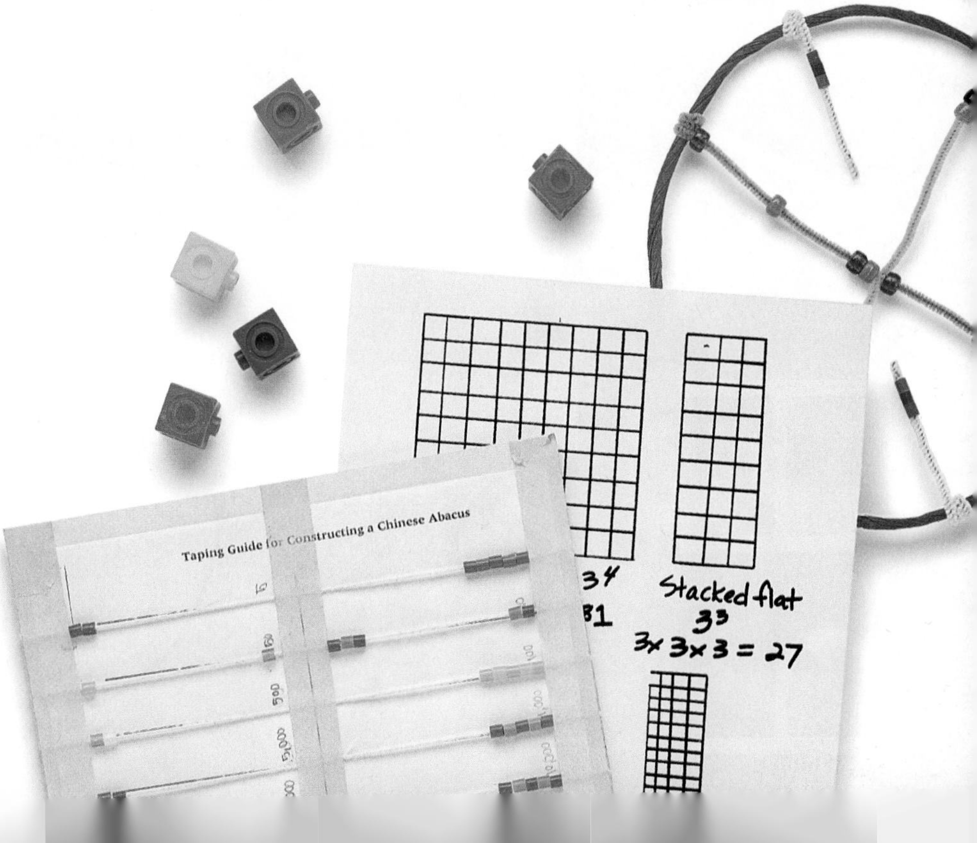

PHASE ONE

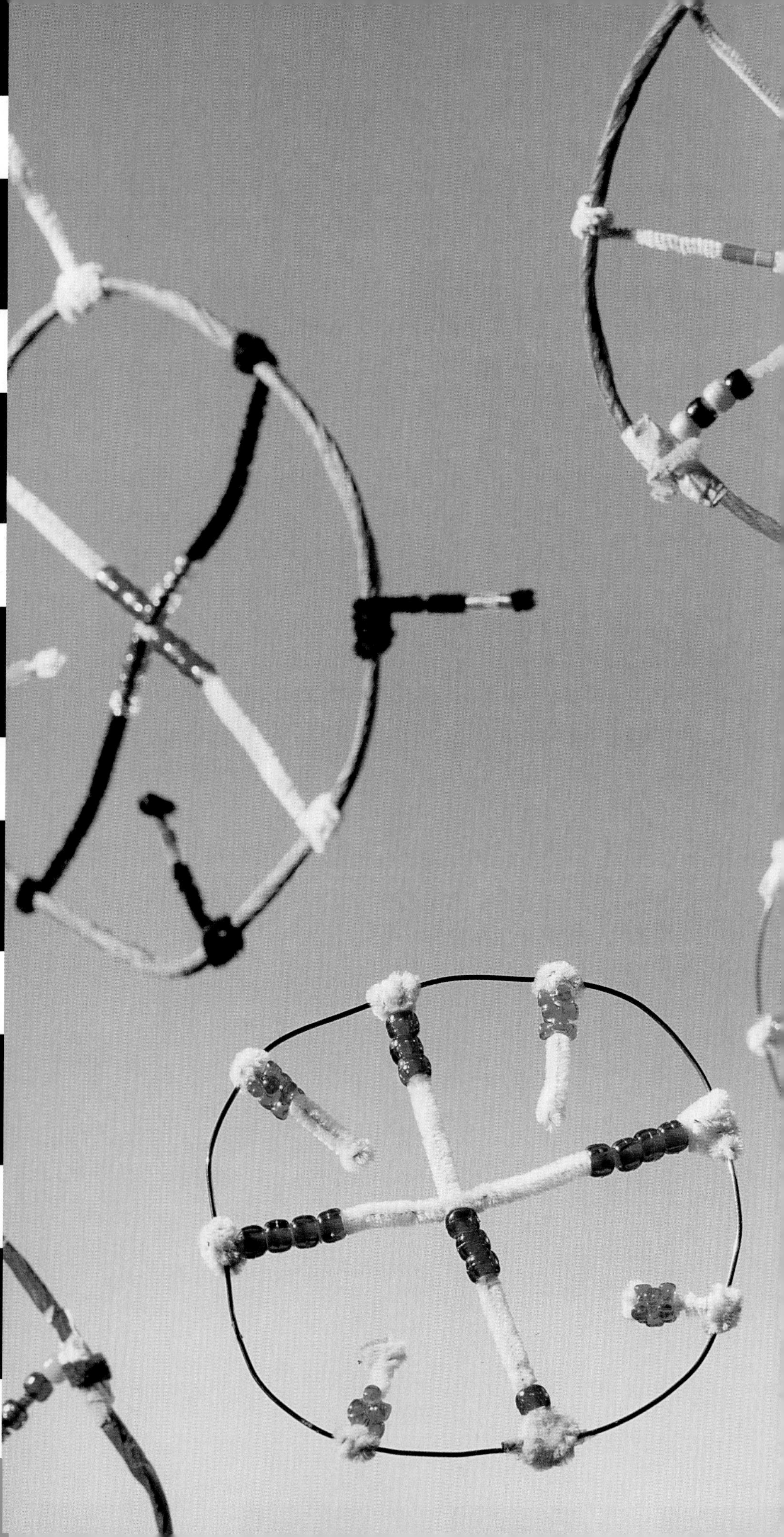

Imagine a mysterious number-making device has been discovered. The device does not work with our everyday number system. Only you can unlock its secrets.

What do computer programmers and experts in cracking codes have in common? For people in these careers, understanding number systems is an important skill. Can you think of other careers in which number systems are important?

Mystery Device

WHAT'S THE MATH?

Investigations in this section focus on:

PROPERTIES of NUMBER SYSTEMS

- Identifying different properties of a number system
- Analyzing a new number system, and comparing it to our everyday number system
- Making connections between number words and a number system's rules
- Describing a number system as having symbols, rules, and properties

NUMBER COMPOSITION

- Using expanded notation to show how numbers are made in different systems
- Writing arithmetic expressions for number words
- Recognizing that the same number can be written in different ways
- Finding arithmetic patterns in number words

MathScape Online
mathscape1.com/self_check_quiz

1 Inventing a Mystery Device System

REPRESENTING NUMBERS IN DIFFERENT WAYS

Some pipe cleaners and beads are all you need to make your own Mystery Device. You will use it to invent your own system for making numbers. Can you make rules so that others will be able to use your system?

What would you need to invent a number system?

Create a Mystery Device

Use the Mystery Device Assembly page to make your own Mystery Device. Your Mystery Device will look like this when it is complete. Make sure that the short "arms" can be turned outward as well as inward.

Make Numbers Using the Mystery Device

How can we use a Mystery Device to represent numbers?

Find a way to make all the numbers between 0 and 120 on your Mystery Device. See if you can create one set of rules for making all of the numbers on the Mystery Device. Use the following questions to test your new system.

- How does my system use the beads to make a number? Does the size of a bead or the position of a bead or an arm make a difference?
- Does my system work for large numbers as well as small numbers? Do I have to change my rules to make any number?
- How could I explain to another person how to use my system? Would it make a difference which part of the device was at the top?

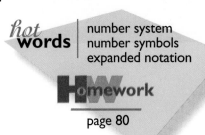

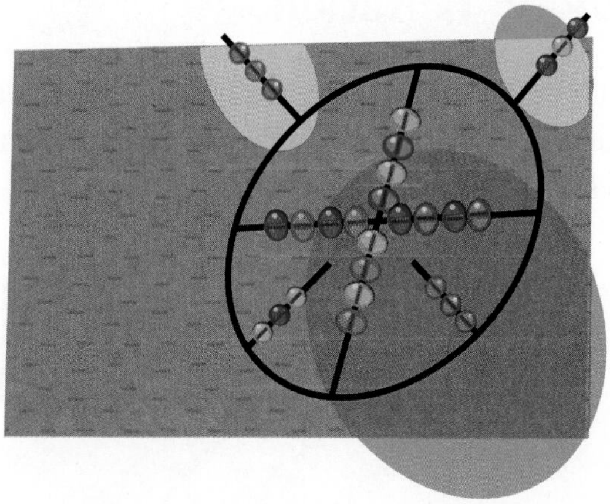

Describe the Invented Number System

Describe the rules for your Mystery Device number system. Use drawings, charts, words, or numbers. Explain your number system so that someone else could understand it and use it to make numbers.

1. Explain in words and drawings how you used your system to make each of the following numbers: 7, 24, 35, 50, 87, and 117.
2. Explain in words and pictures how you made the largest number it is possible to make in your system.
3. Explain how you can use expanded notation to show how you composed a number in your system.

How is your system different from the other systems in the classroom?

hot words
number system
number symbols
expanded notation

Homework
page 80

2 Comparing Mystery Device Systems

USING EXPANDED NOTATION

What are the "building blocks" of a number system? To find out, you will make different numbers on the Mystery Device. You will invent your own way to record them. See how the building blocks of your Mystery Device system compare to those in our number system.

How can you use expanded notation to show how you made a number in your system?

Explore Expanded Notation

Use your Mystery Device to make these numbers. Come up with a system of expanded notation to show how you made each number on the Mystery Device.

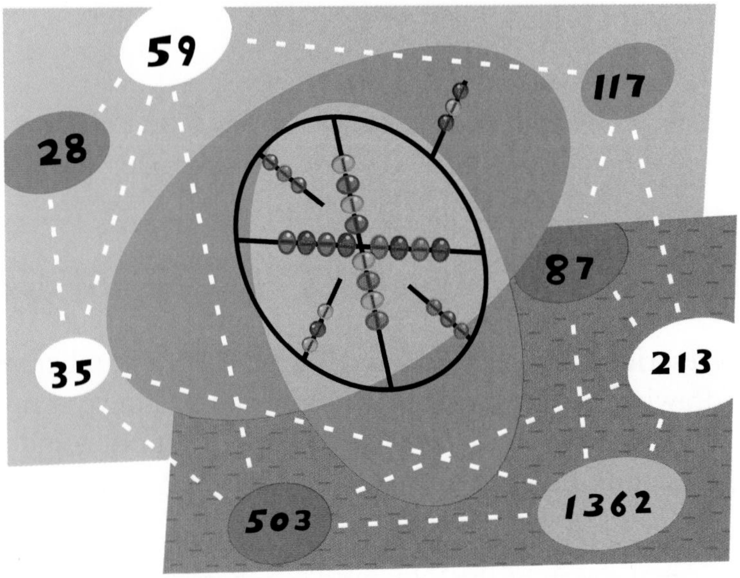

Why is it important for the class to agree on one method of expanded notation?

Investigate the Building Blocks of Number Systems

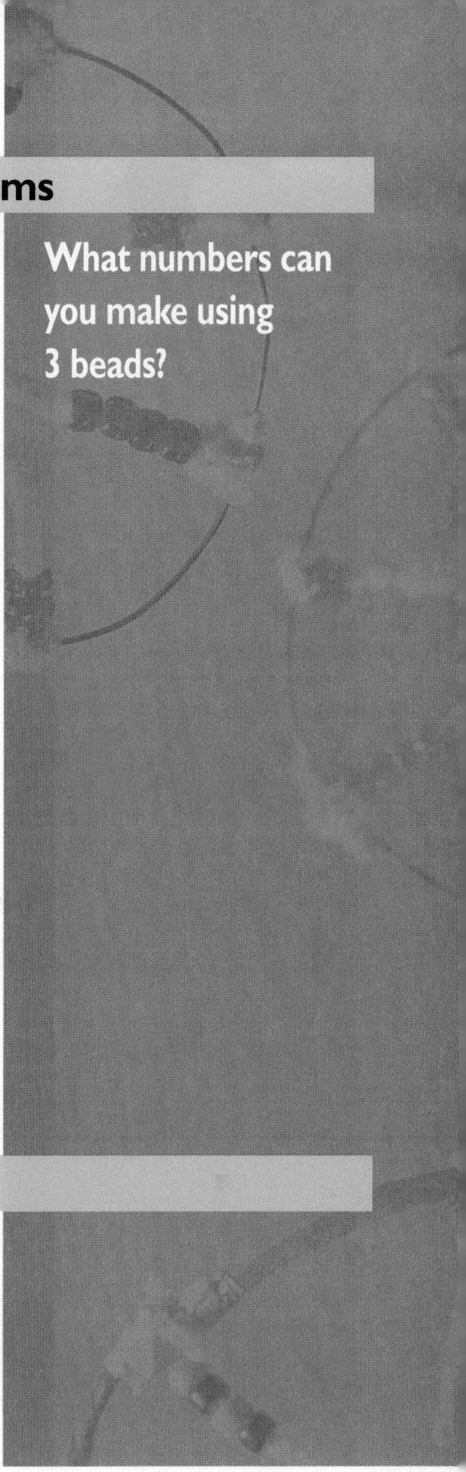

What numbers can you make using 3 beads?

Figure out the different numbers you can make on your Mystery Device using 3 beads. The beads you use can change from number to number, but you must use only 3 for each number. Keep a record of your work using expanded notation.

- What do you think is the least number you can make with only 3 beads? the largest?
- If you could use any number of beads, could you make a number in more than one way?
- Do you think there are numbers that can be made in only one way?

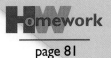

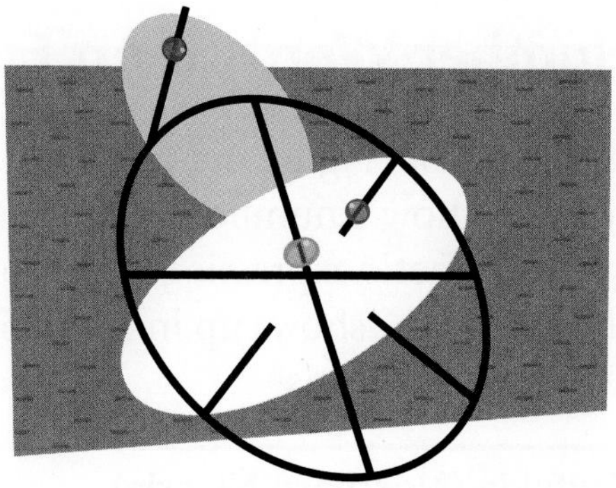

$20 + 4 + 1 = 25$

Compare Number Systems

Answer the five questions below to compare your Mystery Device system to our number system. Then make up at least three of your own questions for comparing number systems.

1. Can you make a 3-bead number that can be written with exactly 3 digits in our number system?
2. Can you make a 3-bead number with more than 3 digits?
3. Can you make a 3-bead number with fewer than 3 digits?
4. What are the building blocks of the Mystery Device system?
5. What are the building blocks of our number system?

hot words | arithmetic expression, expanded notation

Homework page 81

3 Number Words in Many Languages

FINDING ARITHMETIC PATTERNS

Patterns in the number words of other languages can help you see how numbers can be made. Here you will search for patterns in number words from different languages. This will help you understand the arithmetic behind some English number words.

What can you learn about number systems from looking at number words in many languages?

Find Patterns in Number Words from Fulfulde

Look at the Fulfulde words for 1–100. Figure out how each of the Fulfulde number words describes how a number is made. Beside each number word, write an arithmetic expression that shows the building blocks for that number. The *e* shows up in many of the number words. What do you think *e* means?

Number Words in Fulfulde (Northern Nigeria)

1	go'o	15	sappo e joyi
2	ɗiɗi	16	sappo e joyi e go'o
3	tati	17	sappo e joyi e ɗiɗi
4	nayi	18	sappo e joyi e tati
5	joyi	19	sappo e joyi e nayi
6	joyi e go'o	20	noogas
7	joyi e ɗiɗi	30	chappan e tati
8	joyi e tati	40	chappan e nayi
9	joyi e nayi	50	chappan e joyi
10	sappo	60	chappan e joyi e go'o
11	sappo e go'o	70	chappan e joyi e ɗiɗi
12	sappo e ɗiɗi	80	chappan e joyi e tati
13	sappo e tati	90	chappan e joyi e nayi
14	sappo e nayi	100	teemerre

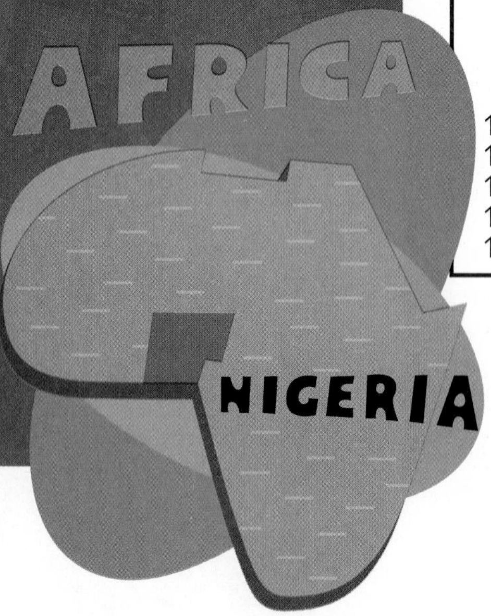

How are Fulfulde number words similar to English number words?

Decode Number Words from Another Language

Work as a group to complete each of the following steps. Decode the number words using either a Hawaiian, Mayan, or Gaelic number words chart.

1. Write an arithmetic expression for each word on the number chart.
2. Predict what the number words for 120, 170, 200, and 500 would be in the language that you selected.
3. Write an arithmetic expression for each new number word and explain how you created the new number word.

What do the different languages have in common in the way they make number words?

How can you use arithmetic expressions to compare number words in different languages?

Create a Mystery Device Language

Invent a Mystery Device language that follows the rules of at least one number system you have decoded. Use the chart below as an example. Make up number words for 1–10 in your own Mystery Device language.

1. Using your new Mystery Device language, try to create words for the numbers 25, 43, 79, and 112. The words should describe how these numbers would be made on your Mystery Device.
2. Write an arithmetic expression to show how you made each number.

1	en
2	sessi
3	soma
4	vinta
5	tilo
6	chak
7	bela
8	jor
9	drona
10	winta

hot words: multiple, pattern

Homework page 82

4 Examining Alisha's System

ANALYZING A NEW NUMBER SYSTEM

How well does the number system invented by Alisha work? You will use what you have learned to analyze Alisha's number system and language. See if Alisha's system works well enough to become the Official Mystery Device System!

Analyze Alisha's Mystery Device System

How does Alisha's system work?

Use the chart below to figure out Alisha's system. As you answer each question, make a drawing and write the arithmetic expression next to it. Only show the beads you use in each drawing.

1. How would you make 25 in Alisha's system, using the least number of beads?
2. Choose two other numbers between 30 and 100 that are not on the chart. Make them, using the least number of beads.
3. What is the greatest number you can make?

Alisha's Mystery Device System

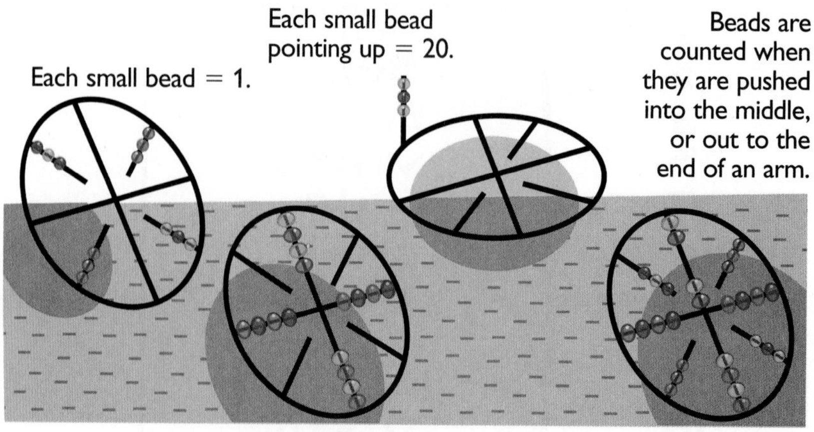
Each small bead = 1.
Each small bead pointing up = 20.
Beads are counted when they are pushed into the middle, or out to the end of an arm.
Each large bead = 4.
This shows 11.

Make Number Words in Alisha's System

Alisha also made up number words to go with her Mystery Device system. They are shown in the table. Answer the questions below to figure out how Alisha's system works.

1 Tell what number each number word represents and write the arithmetic expression.

a. soma, sim-vinta, en

b. set-soma, vintasim

c. sim-soma, set

d. vinta-soma, set-vinta

e. vintaen-soma, set-vinta, sim

2 Write the word in Alisha's system for 39, 95, and 122.

1	en	11	set-vinta, sim	30	soma, set-vinta, set
2	set	12	sim-vinta	40	set-soma
3	sim	13	sim-vinta, en	50	set-soma, set-vinta, set
4	vinta	14	sim-vinta, set	60	sim-soma
5	vintaen	15	sim-vinta, sim	70	sim-soma, set-vinta, set
6	vintaset	16	vinta-vinta	80	vinta-soma
7	vintasim	17	vinta-vinta, en	90	vinta-soma, set-vinta, set
8	set-vinta	18	vinta-vinta, set	100	vintaen-soma
9	set-vinta, en	19	vinta-vinta, sim		
10	set-vinta, set	20	soma		

How is Alisha's system like our number system?

Evaluate Number Systems

Use these questions to evaluate your Mystery Device system and Alisha's system. Decide which one should become the Official Mystery Device System. Explain your reasons.

- What are two things that an Official Mystery Device System would need to make it a good number system?
- Which of the two things you just described does Alisha's system have? Which of them does your system have? Give examples to show what you mean.
- What is one way you would improve your system to make it the Official Mystery Device System?

hot words rule, arithmetic expression

Homework
page 83

PHASE TWO

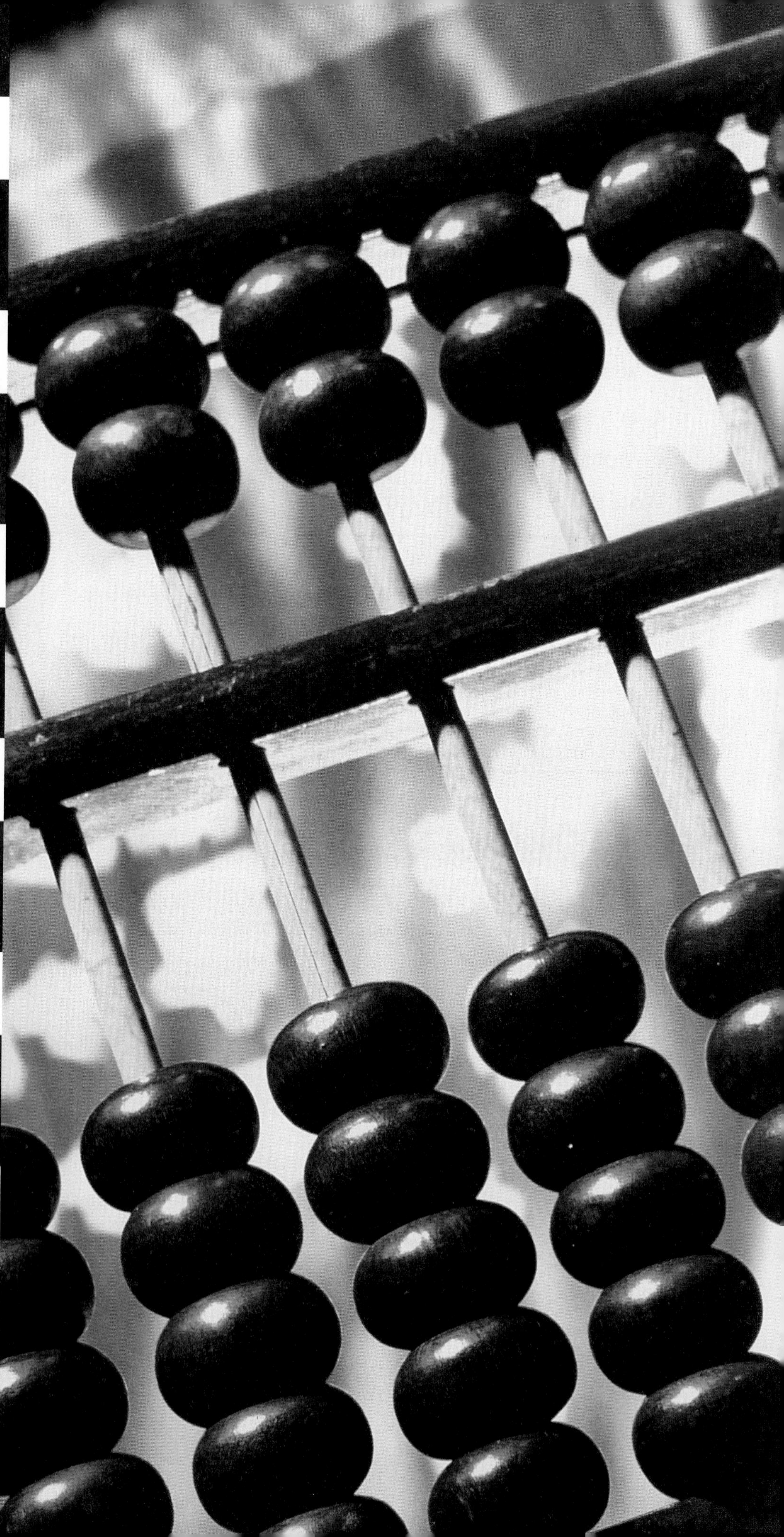

This counting instrument is called *choreb* in Armenian. In Japanese, it is a *soroban,* and the Turks know it as the *coulba.* The Chinese call it a *suan pan* or *sangi.* Most of us know it by the Latin name *abacus.*

Different forms of the abacus have developed in different cultures around the world. Many are still widely used today. You may be familiar with the Chinese, Japanese, Russian, or other abaci. The abacus helps us to see how place value works in a number system.

Chinese Abacus

WHAT'S THE MATH?

Investigations in this section focus on:

PROPERTIES of NUMBER SYSTEMS

- Representing and constructing numbers in a different number system
- Investigating and contrasting properties of number systems
- Understanding the use and function of place value in number systems

NUMBER COMPOSITION

- Understanding the connection between trading and place value in number systems
- Recognizing patterns in representing large and small numbers in a place-value system
- Understanding the role of zero as a place holder in our own place-value number system

MathScape Online
mathscape1.com/self_check_quiz

5 Exploring the Chinese Abacus

INVESTIGATING PLACE VALUE

As on the Mystery Device, you move beads to show numbers on the Chinese abacus. But you will find that in other ways the abacus is more like our system than the Mystery Device. Can you find the ways that the abacus system is like our system?

Make Numbers on a Chinese Abacus

How does an abacus make numbers?

The columns on this abacus are labeled so that you can see the values. See if you can follow the Chinese abacus rules to make these numbers: 258; 5,370; and 20,857.

Chinese Abacus Rules

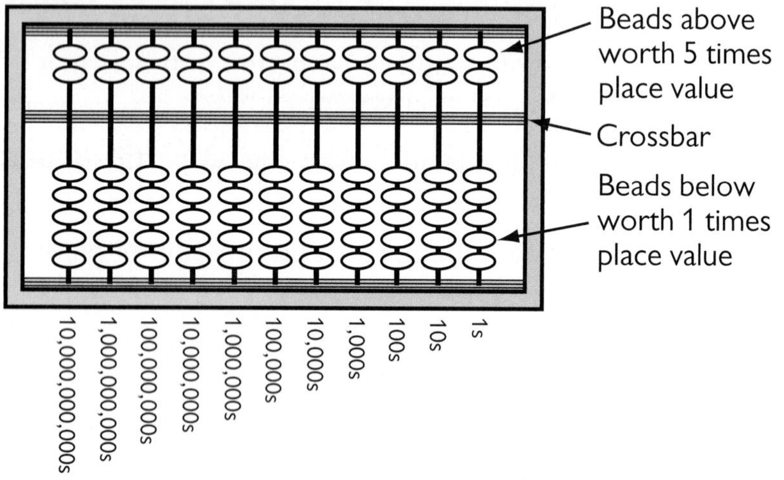

- Each column on the Chinese abacus has a different value.
- A crossbar separates the abacus into top and bottom sections.
- Each bead above the crossbar is worth 5 times the value of the column if pushed toward the crossbar.
- Each bead below the crossbar is worth 1 times the value of the column if pushed toward the crossbar.
- A column shows 0 when all the beads in the column are pushed away from the crossbar.

Here are a few numbers. Only part of the abacus is shown.

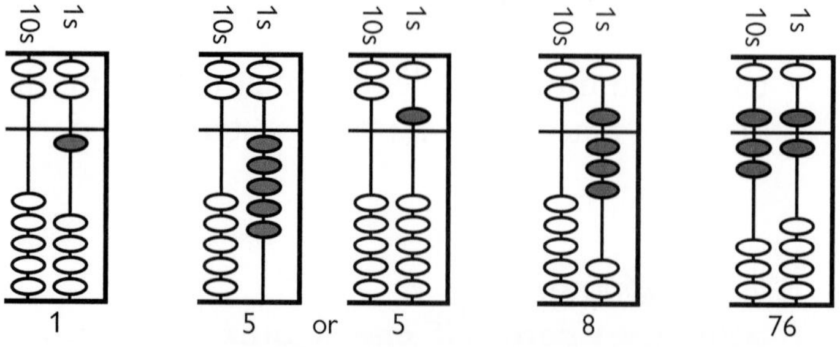

Investigate the Chinese Abacus

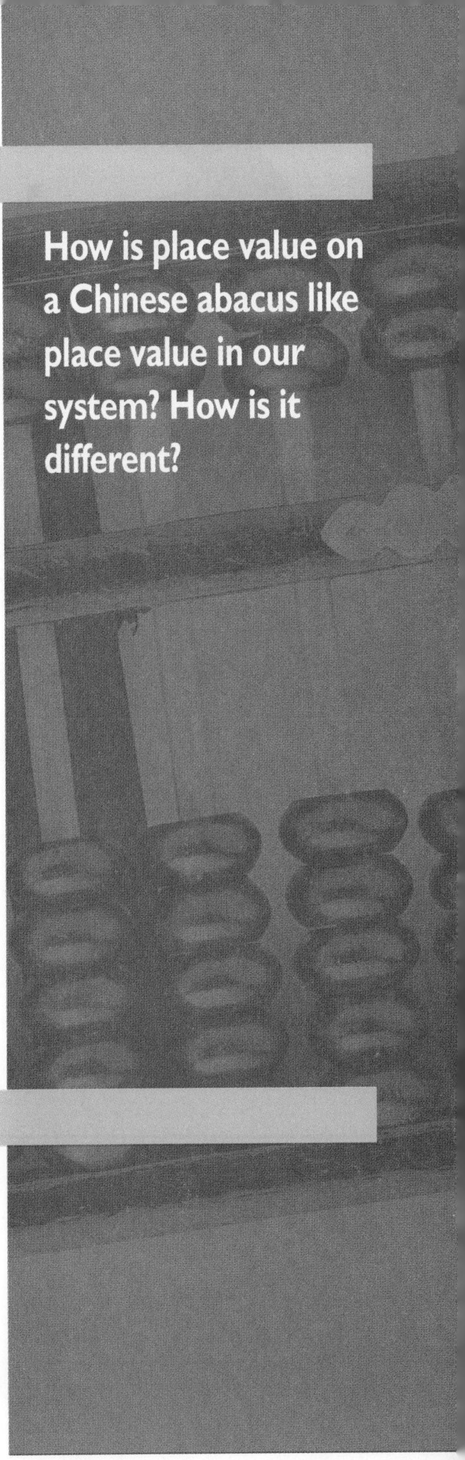

How is place value on a Chinese abacus like place value in our system? How is it different?

For each investigation below, explore different ways to make numbers on your abacus. Use both a drawing and an arithmetic notation to show how you made each number.

1 Make each of these numbers on the abacus in at least two different ways.

a. 25 **b.** 92 **c.** 1,342 **d.** 1,000,572

2 Use any 3 beads to find these numbers. You can use different beads for each number, but use exactly three beads.

a. the greatest number you can make

b. the least number you can make

3 Find some numbers we write in our system with 3 digits that can be made with exactly 3 beads.

a. Find at least five different numbers that you can make with 3 beads.

b. Make at least one number using only the first two columns of the abacus.

Define Place Value

Our number system uses place value. Each column has a value, and 0 is used as a place holder, so that 3 means 3 ones, 30 means 3 tens, 300 means 3 hundreds, and so on. How is the use of place value on the Chinese abacus like its use in our system? How is it different?

- Write a definition for place value that works for both the Chinese abacus and our system.

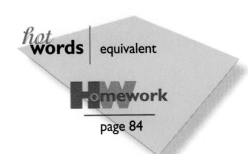

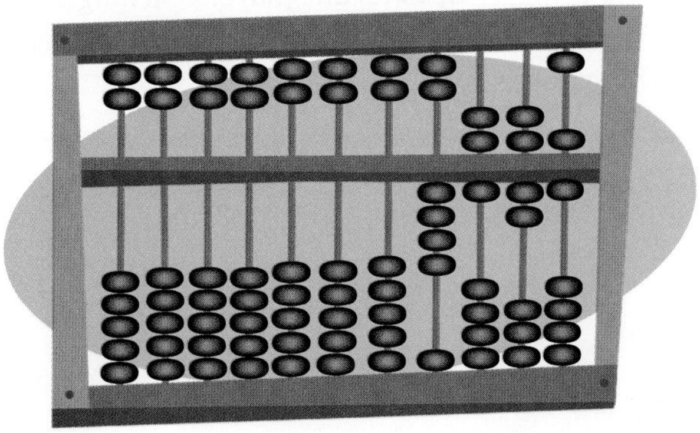

hot **words** | equivalent

Homework page 84

6 How Close Can You Get?

TRADING AND PLACE VALUE

How close can you get to a target number using a given number of beads? In this game you will explore trading among places on the Chinese abacus. You can use what you discover in playing the game to compare the Chinese abacus to our system.

Investigate Trading Relationships

How close can you get to 6,075 using exactly 14 beads?

Follow these steps to play the game How Close Can You Get. Try to get as close as possible to the target number using the given number of beads. If you can't make the number exactly, get as close as you can.

How Close Can You Get? Game Rules

1. One player picks a target number between 1,000 and 9,999.
2. Another player picks a number of beads, from 7 to 16. You must use the exact number of beads selected.
3. All players write down the group's challenge for the round: How close can you get to ____ using exactly ____ beads?
4. When all players have made a number, compare answers. The player or players who come closest to the target number score one point.
5. Continue playing, with different players picking the target number and number of beads. When someone reaches 10 points, the game is over.

When do we use trading in our number system?

Solve the Mystery Number Puzzles

Here are four mystery number puzzles. To solve each puzzle, you need to figure out which part of the abacus might be shown and give at least one number that the beads might make. Use drawings and expanded notation to show your answers. Beware! One of the mystery number puzzles is impossible to solve, and some puzzles can be solved in more than one way.

What trades can you make on the Chinese abacus to make both 5,225 and 5,225,000 with only 12 beads?

Puzzle A

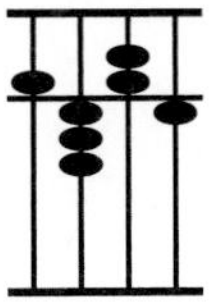

What number might be shown?

Clues:
- All the beads used to make the number are shown.
- One of the columns is the 10,000s column.
- The 10s column is not shown at all.

Puzzle B

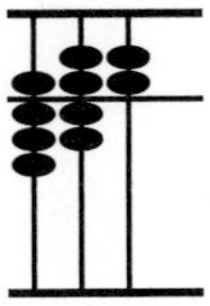

What number might be shown?

Clues:
- All the beads used to make the number are shown.
- One of the columns is the 100s column.
- The number is between 100,000 and 10,000,000.

Puzzle C

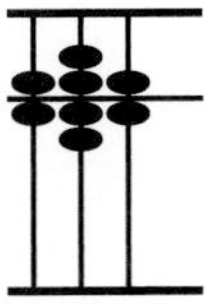

What number might be shown when you add the missing bead?

Clues:
- All the columns used to make the number are shown.
- One bead is missing from the figure.
- When we write the target number in our system, there is a 1 in the 1,000,000s place.

Puzzle D

What number might be shown when you add the missing beads?

Clues:
- All the columns used to make the number are shown.
- There are two beads missing from the figure.
- Beads are used only in the top part of the abacus to make the number.
- When we write the target number in our system, the only 5 used is in the 1,000s place.

hot words | equivalent expressions

Homework
page 85

7 Additive Systems

EXAMINING A DIFFERENT KIND OF SYSTEM

An additive system does not use place value. You simply add together the values of individual symbols to find the value of the number. For example, if △ equals 1 and □ equals 7, then □ □ △ equals 7 + 7 + 1 = 15. Do you think a number system like this would be easier or harder to use than our system?

What if place value was not used at all in a number system?

Investigate How Additive Systems Work

Figure out how your additive system works by making the following numbers. Record how you made each number on a chart. Write arithmetic expressions for the three greatest numbers.

1. Make the numbers 1 through 15.
2. Make the greatest number possible with three symbols.
3. Make five other numbers greater than 100.

Number in our System	Number in Additive System	Arithmetic Expression (for three greatest numbers only)
10	□△△△	7+1+1+1=10

What are the patterns in your system?

Compare the Three Systems

In our system, three-digit numbers are always greater than two-digit numbers. For example, 113 has three digits. It is greater than 99, which has two digits. In the system you are investigating, are numbers that have three symbols always greater than numbers that have two symbols? Explain why or why not.

Improve the Number Systems

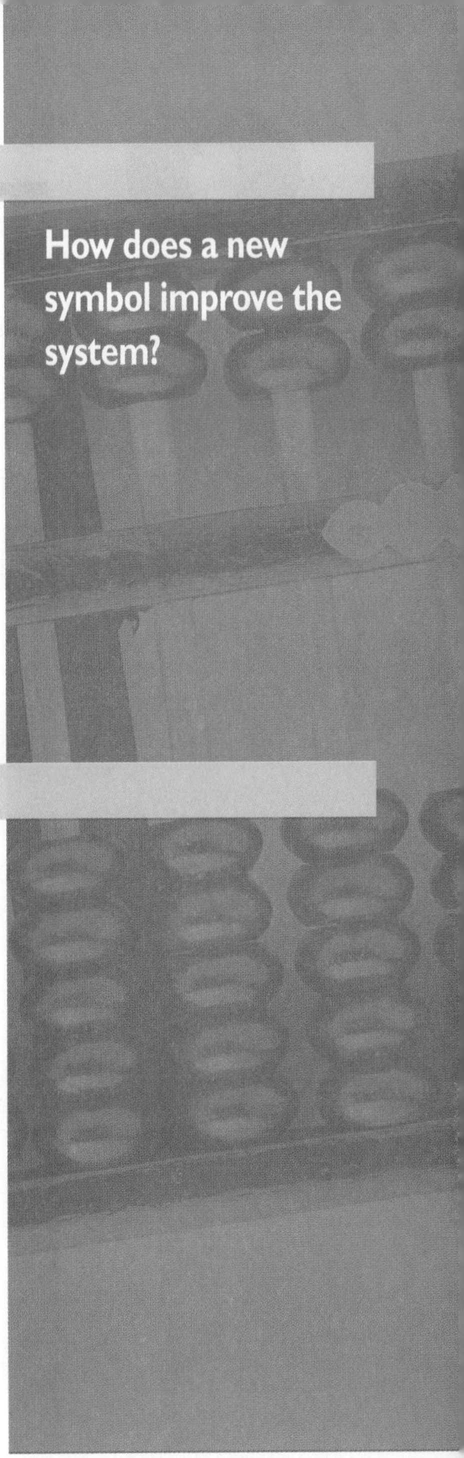
How does a new symbol improve the system?

Improve your additive number system by making up a new symbol. The new symbol should make the system easier to use or improve it in some other way.

1. Give the new symbol a value different from the other symbols in the system.
2. Make at least five numbers with your improved system. Write them on your recording sheet.

□□□□△△☆   □□△☆

Analyze the Improved Additive Systems

Work with a partner who investigated a different system. See if you can figure out how each other's improved systems work. Compare the largest numbers you can make with three symbols and with four symbols. Answer the following questions.

- Which system lets you make greater numbers more easily? Why?
- Which patterns of multiples are easy to recognize in each system? Are they the same? Why or why not?
- Can you find other ways in which the two systems are alike?
- Can you find other ways in which they are different?
- In what ways are these number systems like our system?

hot words | additive system
Roman numerals

HW Homework
page 86

8 The MD System

IDENTIFYING THE PROPERTIES OF NUMBER SYSTEMS

It's time to get out your Mystery Device and learn a new number system: the MD system. As you learn to use the MD system, you will investigate making numbers in more than one way. To do this, you will play How Close Can You Get.

Decode the MD Number System

How would you make 25 using the MD system?

This illustration shows how much the beads are worth in the MD system. The Mystery Device here shows 0—all of the beads are away from the center. To make numbers, you push beads toward the center of the Mystery Device. Always keep the short arms inside the hoop.

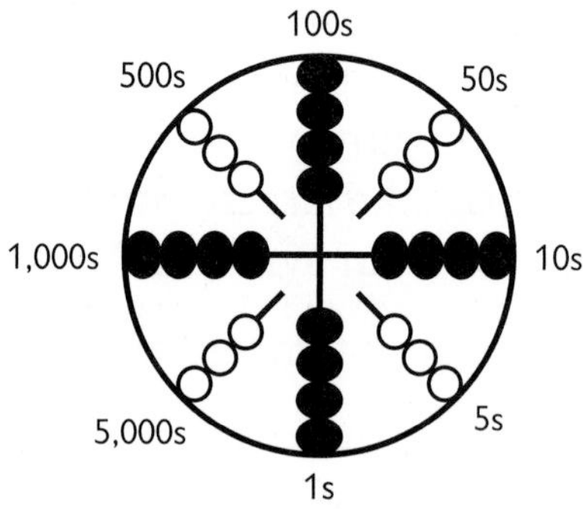

Create Multiple Representations of Numbers

Can you make numbers in more than one way in the MD system?

Use the MD system to make each of the following numbers on the Mystery Device.

1. Make a 4-digit number with 0 in one of the places.
2. Make a 3-digit number that can be made in at least two ways.
3. Make a number that can be shown in only one way.

Investigate Trading Relationships in the MD System

Jackie is playing How Close Can You Get? and she needs help. She has made 6,103 with 6 beads, but she doesn't know what to do next to get to 12 beads. Write a hint to Jackie using words, drawings, or arithmetic notation. Be careful not to give away the answer.

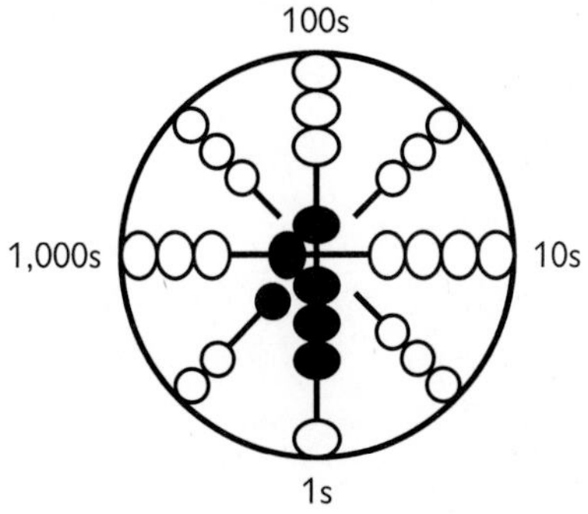

How close can you get to 6,103 with exactly 12 beads using the MD number system?

Compare the MD System to Other Systems

Make a chart like the one shown. Fill in the chart to compare the MD number system to your Mystery Device system, the abacus system, the additive system you investigated in Lesson 7, and our own number system.

How is the MD number system similar to each of the other systems?

Comparison Questions	MD System	My Mystery Device System	Chinese Abacus	Additive System Investigated	Our Number System
What are the kinds of symbols used to make numbers?					
What are the building blocks?					
Is there a limit on the highest number possible?					
Is there more than one way to show a number?					
How does the system use place value?					
How does the system use trading?					
How does the system use zero?					
What are the patterns in the system?					
How is the system additive?					

hot words | place value

Homework
page 87

PHASE THREE

Our system for representing numbers was developed over thousands of years. People from cultures all over the world have had a part in making it such a powerful tool for working with numbers.

Imagine that you have been asked to investigate different number systems to help improve our number system. How would you create a "new and improved" number system?

Number Power

WHAT'S THE MATH?

Investigations in this section focus on:

PROPERTIES of NUMBER SYSTEMS

- Understanding that number systems are efficient if every number can be represented in just one way and only a few symbols are used
- Describing in detail the features of our number system
- Identifying and describing a mathematically significant improvement to a number system

NUMBER COMPOSITION

- Writing an arithmetic expression using exponents
- Learning how to evaluate terms with exponents, including the use of 0 as an exponent
- Developing number sense with exponents

MathScape Online
mathscape1.com/self_check_quiz

9 Stacks and Flats

WRITING ARITHMETIC EXPRESSIONS USING EXPONENTS

Our number system is a base 10 place-value system. You can better understand our system by exploring how numbers are shown in other bases. As you explore other bases, you will learn to use exponents to record the numbers you make.

Make Numbers in the Base 2 System

How can you show a number using the fewest base 2 pieces?

Make a set of pieces in a base 2 system. Use your base 2 pieces to build the numbers 15, 16, 17, 26, and 31. Write an arithmetic expression using exponents for each number. You may need to make more pieces to build some numbers.

1. Fill in your Stacks and Flats Recording Sheet to tell how many pieces you used to make each number. Write 0 for the pieces you do not use.

2. Write an arithmetic expression using exponents for each number you make.

How does the pattern of exponents in base 2 compare to the pattern of exponents in base 10?

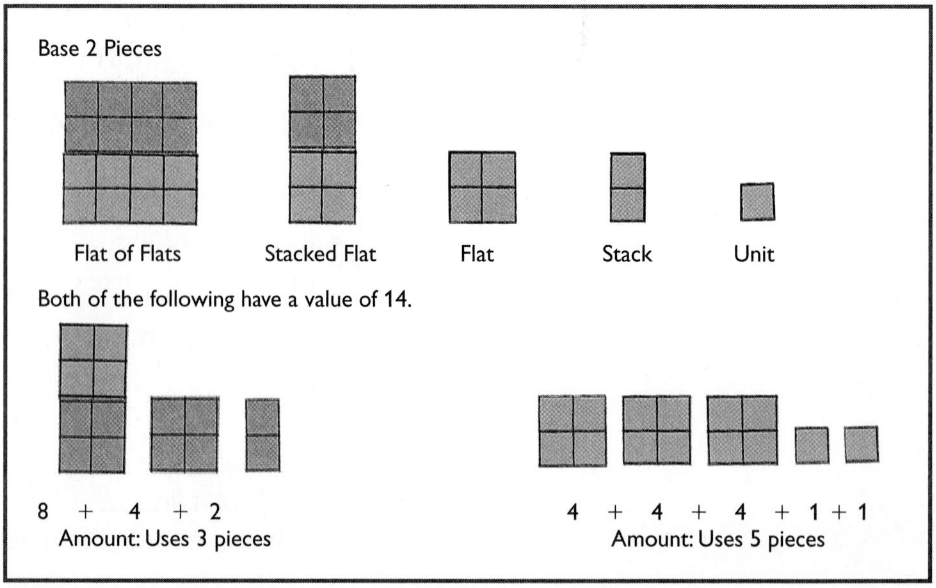

Investigate a Base 3 or Base 4 Number System

Choose whether you will work in base 3 or base 4. Make a new set of pieces for the base you have chosen. Your set should include at least 5 units, 5 stacks, 5 flats, 3 stacked flats, and 3 flats of flats.

Make at least four different numbers in your base, using as few pieces as possible for each.

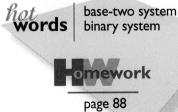

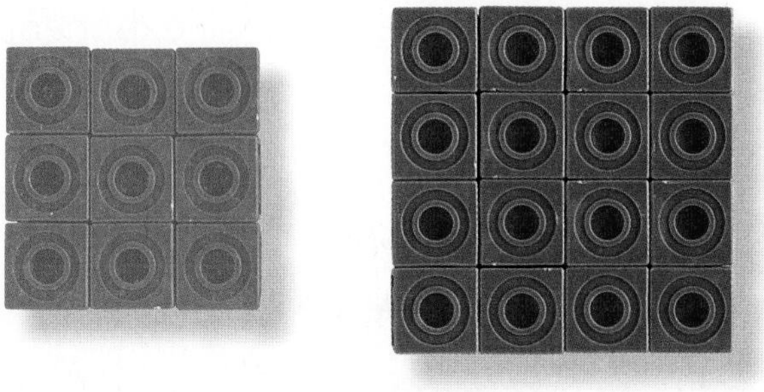

How can you use what you know about patterns of exponents to create a set of pieces for a different base?

Write a Report About the Different Base

Once you have created and investigated your set of base 3 or base 4 pieces, write a report about your set. Attach to your report one of each kind of piece from the base you chose.

1. Describe the patterns you find in your set, including what your base number to the zero power equals.
2. On a new Stacks and Flats Recording Sheet, write arithmetic expressions for, and record how you made, the numbers 11, 12, 35, and 36.
3. Figure out the next number after 36 that would have a 0 in the units column. Describe how you figured this out.

hot **words** | base-two system
binary system

Homework
page 88

10 The Power Up Game

EVALUATING EXPRESSIONS WITH EXPONENTS

Do you think switching the base and exponent will result in the same number? Can a small number with a large exponent be greater than a large number with a small exponent? Play the Power Up game and find out.

How can you use 3 digits to make the greatest possible expression using exponents?

Explore Exponents with the Power Up Game

Play the Power Up game with a partner. Make a chart modeled after the diagram below to record the numbers you roll, the expression you write, what the expression equals, and whether you score a point on the turn.

The Power Up Game Rules

1. Each player rolls a number cube four times and records the digit rolled each time. If the same digit is rolled more than two times, the player rolls again.
2. Each player chooses three of the four digits rolled to fill in the boxes in this arithmetic expression: $(\square + \square)^{\square}$
3. Players then evaluate their expressions. The player whose expression equals the greater number gets one point.

What did you learn about exponents from the Power Up game?

Evaluate Arithmetic Expressions with Exponents

Choose three of the four letters to Dr. Math that you want to answer. Write answers using words, drawings, and arithmetic expressions. Make sure that you describe how you solved each problem. Do not just give the answer.

Can you describe the effect of exponents on numbers in different ways?

Dear Dr. Math,

We're studying exponents in math class. I was asked to draw a picture that showed what 4^3 means. I drew this:

I drew 3 sets of 4 because 4 gets multiplied 3 times. But I know that $4 \times 4 \times 4 = 64$, so I don't understand why my picture shows 12. Why doesn't it show 64, even though it shows 4^3? Can you explain what's happening with my picture? How would I draw a picture of 4^3?

Exasperated with Exponents

Dear Dr. Math,

Isn't it true that $8 \times 3 = 3 \times 8$? I know I learned this! And aren't exponents a way of showing multiplication? But 8^3 does not $= 3^8$! This is really confusing. Can you tell me why this doesn't work? Is there ever a time when it does work to switch the two digits?

Bambfoozled in Boston

Dear Dr. Math,

Something is wrong with my calculator! I think that 3^6 should be much smaller than 5^4, because after all, 3 is smaller than 5. And 6 is not that much bigger than 4. So I don't understand why my calculator tells me that the answer for 3^6 is bigger than the answer for 5^4. I think it needs a new battery; what do you think? Why can't I tell which of two numbers is bigger by comparing the base numbers?

Crummy Calculator

Dear Dr. Math,

One of the problems I had to do for homework last night was to figure out what 10^0 was equal to. I called my friend to ask him, and he thought it was 0. He said it means that 10 is multiplied by itself 0 times, so you have nothing. I thought it was 1, but I don't remember why that works. Which of us is right? And can you please explain to me why?

Zeroing In

hot words exponent, power

Homework page 89

11 Efficient Number Systems

COMPARING THE FEATURES OF DIFFERENT SYSTEMS

Some number systems use a base, and others do not. Here you will decode different place-value systems. You will see that some systems work better than others. Your decoding work will help you think about the features that make different systems work and that make a number system easy to use.

Decode Three Place-Value Systems

How are bases used in a place-value system?

Each of the number systems shown uses a different place-value system. See if you can use the numbers on this page and on the Decoding Chart to decode each system.

1 Figure out what goes in the place-heading boxes (☐) to decode the system.

2 Choose two numbers that are not on the chart. Write an arithmetic expression that shows how the numbers are made in each system. Make sure you label each expression with the name of the system.

Our System	Hand System Place Values				
30		1	0	1	0
35		1	0	2	2
40		1	1	1	1
50		1	2	1	2
60		2	0	2	0
101	1	0	2	0	2

Our System	Crazy Places Place Values					
30			1	0	0	0
32			1	0	0	2
40		1	0	0	0	0
47		1	0	0	0	7
53	1	0	0	0	0	3

Our System	Milo's System Place Values				
30		●	★	★	★
34		●	★	●	●
58		●	◆	◆	◆
105	●	★	★	●	◆

Compare the Features of Many Number Systems

Use the features from class discussion to make a chart that compares some of the systems you have learned in this unit. Include at least one system that is additive, one that uses a base, and one that uses place value.

1 Make a list of at least six different features of a number system.

2 Choose at least six different number systems and describe how they use each feature.

3 Create your own chart format. Leave an empty column so you can add our system to your chart later.

Feature	MD	Chinese Abacus	Milo's System
Place Value	Yes, large beads are worth 10. Small beads = 1.	5 is on top. 1 is on bottom. Yes.	No. because the system use symbols.
base system	Yes. 1, 10, 100.	Base of 5. 5, 50, 500, 5,000 Base of 1. 1, 10, 100, 1,000	Yes, it have base. 1, 3, 10, 30, 100
#s represented in more than one way	Yes, you can use 10 small bead or 1 big bead to make 10.	Yes, you can use 5 ones or 1 5's to make 5.	Yes, you can use U and make 2.

What features make a number system efficient?

Check Whether Our System Is Efficient

What makes some number systems more efficient than others? Use your chart to check whether our system is efficient or not. Explain your reasoning in writing and use your chart as an example.

hot words
base (number)
place value

Homework
page 90

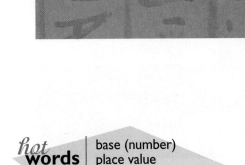

12 A New Number System

FINAL PROJECT

You have decoded many different types of number systems and looked at their features. Now it is time to improve one of the systems by bringing together the best of each. You will start by taking a look at the ancient Egyptian system.

Decode the Egyptian Number System

How does the ancient Egyptian system work?

Use what you have learned about decoding systems to figure out the value of each symbol on your Ancient Egyptian System Reference Sheet. On a separate sheet of paper, write the value of each symbol.

1. Choose four numbers that are not on the chart. Write each number in the ancient system.
2. Write arithmetic expressions to show how each of the four numbers is made.

Analyze the Ancient Number System

Describe the features of the Egyptian system. What are its disadvantages? Find at least one way to improve the ancient system.

What are the ways in which the ancient system is like our system?

Revise a Number System

What features would a number system need to be efficient?

Choose one of the following number systems. Find a way to make it more efficient. Present your revised system clearly with words, drawings, and arithmetic expressions, so that others could use it.

1 Revise either Alisha's (Lesson 4), Yumi's (Lesson 7), Milo's (Lesson 11), or your own Mystery Device system.

2 Find a more efficient way to make all the numbers between 0 and 120, using words or symbols. Show how at least four numbers are made using arithmetic expressions.

3 Describe how your system uses the following features:

- a. place value
- b. base system
- c. symbols
- d. rules
- e. a way to show zero
- f. trading
- g. range (greatest and least number)
- h. making a number in more than one way

4 Compare your system to our system and describe the differences and similarities.

Evaluate the Efficiency of an Improved System

A good way to evaluate a number system is to ask questions about the different properties of the system. What are the building blocks of the system? Does it use base or place value? Can a number be made in more than one way? Come up with at least three more questions and use them to explain why your partner's system is or is not efficient. Make sure your explanation talks about the mathematical features of the system.

hot words
base-ten system
place-value system

Homework
page 91

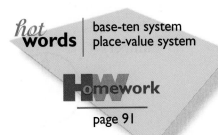

Homework 1

Inventing a Mystery Device System

Applying Skills

1. $3(7) + 5(2) = ?$
2. $6(5) + 3(2) + 7(8) = ?$
3. $4(25) + 6(15) + 2(10) = ?$
4. $9(100) + 4(10) + 4(1) = ?$
5. $5(1,000) + 3(100) + 2(10) + 9(1) = ?$
6. $6(1,000) + 3(10) + 4(1) = ?$

George's Mystery Device System
- Large beads = 10
- Small beads with arms pointing in = 1
- Small beads with arms pointing out = 5

Show each number in George's system. Draw only the beads you need for each number. Remember to draw the diagonal arms either in or out. Use the fewest beads you can.

7. Draw 128 in George's system.
8. Draw 73 in George's system.
9. Draw 13 in George's system.
10. What is this number?

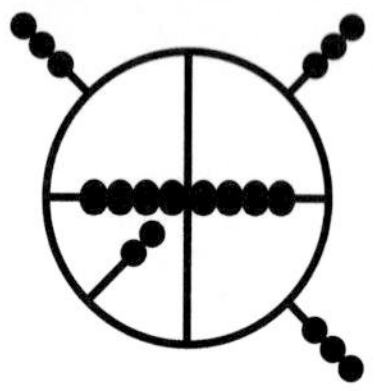

11. What is this number?

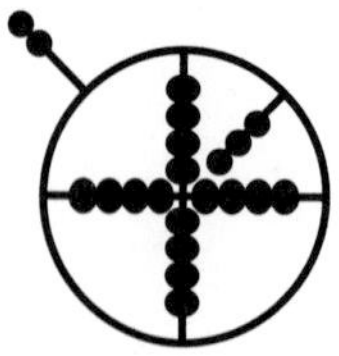

12. What is this number?

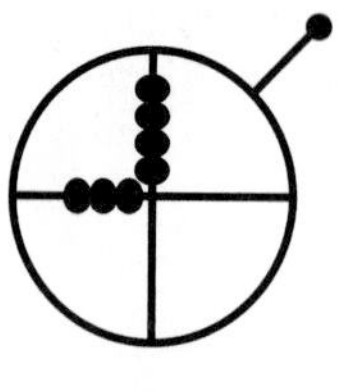

Extending Concepts

13. What is the greatest number you can make with this system? Explain how you know.
14. Can you find a number you can make in more than one way? Can you find a number that can be made in more than two ways?
15. Is there a number you can make in only one way in George's system? What is the arithmetic expression? What would the arithmetic expression be for that same number written in our system?

Writing

16. Answer the letter to Dr. Math.

Dear Dr. Math,
In my Mystery Device system, for numbers larger than 100, large beads mean 100, small beads with the arms pointing out are 20, and small beads with the arms pointing in are 10. To make numbers less than 100, large beads are 10, small beads with the arms pointing in are 5, and small beads with the arms pointing out are 1. My friends get confused using my system. How should I change it?
B. D. Wrong

HW 2 Homework

Comparing Mystery Device Systems

Applying Skills

Write each number in arithmetic expressions.

1. 6,782

2. 9,015

3. 609

4. 37,126

5. 132,056

6. 905,003

What number do these arithmetic expressions represent?

7. 8(100,000) + 7(10,000) + 6(1,000) + 3(100) + 2(10) + 1(1)

8. 3(10,000) + 1(1,000) + 6(100) + 3(1)

9. 5(100,000) + 2(1,000) + 3(100)

10. 7(1,000,000) + 5(100,000) + 6(10,000) + 2(1,000) + 4(100) +3(10)

Remember George's system from Lesson 1.

George's Mystery Device System

- Large beads = 10
- Small beads with arms pointing in = 1
- Small beads with arms pointing out = 5

11. Using only 3 beads, draw at least seven numbers in George's system. Only draw the beads that count.

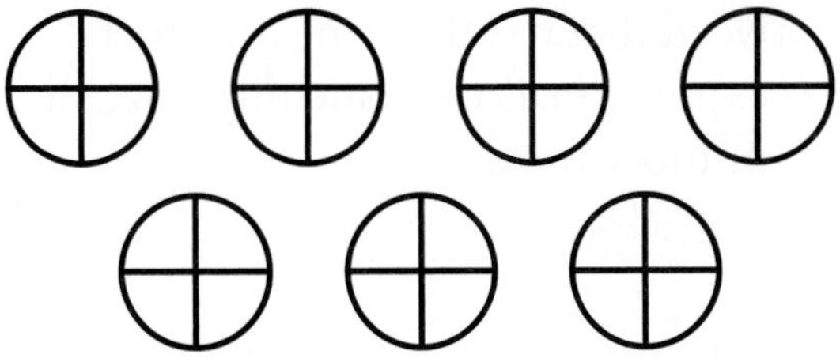

Extending Concepts

12. Using George's system, draw 32 in at least five different ways. Only draw the beads that count.

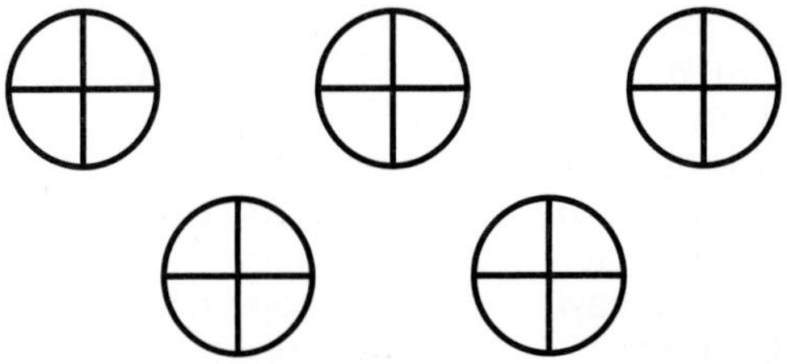

13. Can you think of an explanation for why you can make numbers so many ways in this system but not in our own?

Making Connections

14. Look at the way you solved items **11** and **12.** Did you just keep thinking of different solutions or did you try to use a pattern? Describe the pattern you used or one you might try next time.

Homework 3

Number Words in Many Languages

Applying Skills

Number Words in Fulfulde

1 go'o	11 sappo e go'o	30 chappan e tati
2 ɗiɗi	12 sappo e ɗiɗi	40 chappan e nayi
3 tati	13 sappo e tati	50 chappan e joyi
4 nayi	14 sappo e nayi	60 chappan e joyi e go'o
5 joyi	15 sappo e joyi	70 chappan e joyi e ɗiɗi
6 joyi e go'o	16 sappo e joyi e go'o	80 chappan e joyi e tati
7 joyi e ɗiɗi	17 sappo e joyi e ɗiɗi	90 chappan e joyi e nayi
8 joyi e tati	18 sappo e joyi e tati	100 teemerre
9 joyi e nayi	19 sappo e joyi e nayi	
10 sappo	20 noogas	

Number	Fulfulde Word	Arithmetic Expression	English Word	Arithmetic Expression
25	a.	b.	c.	d.
34	e.	f.	g.	h.
79	i.	j.	k.	l.
103	m.	n.	o.	p.

1. Copy and complete the chart above.

2. What building blocks does Fulfulde use that English also uses?

3. What building blocks does Fulfulde use that English does not use?

4. Write the Fulfulde number words that match these arithmetic expressions, and tell what each number equals:

a. $(10)(5 + 3) + 1(1)$

b. $(10)(4) + 1(5) + 3(1)$

c. $1(100) + 1(20) + 1(5) + 4(1)$

5. What are the arithmetic expressions for the English number words used in item **4**?

Extending Concepts

6. In Fulfulde, the arithmetic expression for chappan e joyi is 5(10). In English, the arithmetic expression for 50 is also 5(10). Find another Fulfulde number word that has the same arithmetic expression as the matching English number word.

Writing

7. How are the Fulfulde number words and arithmetic expressions like the English number words and arithmetic expressions? How are they different?

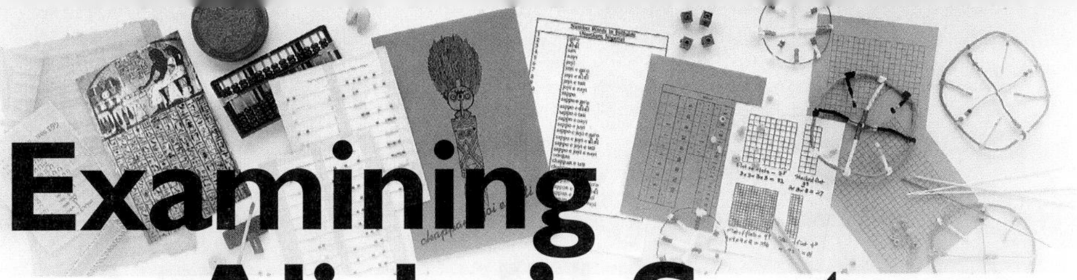

Homework 4: Examining Alisha's System

Applying Skills

Alisha's Mystery Device System

- Large beads = 4
- Small beads with arms pointing in = 1
- Small beads with arms pointing straight up = 20

Make a chart like the one below and use Alisha's system to show each number on the Mystery Device. Remember to draw in the diagonal arms in the correct position. Next, write the number in Alisha's number language. Then write the arithmetic expression for the number.

Number	Sketch	Number Word	Arithmetic Expression
65	**1.** 	**2.**	**3.**
143	**4.** 	**5.**	**6.**
180	**7.** 	**8.**	**9.**
31	**10.** 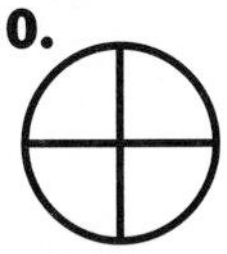	**11.**	**12.**

13. How would you commonly write the number word for 36? What is the arithmetic expression for this number?

Alisha's Number Language	
1 en	15 sim-vinta, sim
2 set	16 vinta-vinta
3 sim	17 vinta-vinta, en
4 vinta	18 vinta-vinta, set
5 vintaen	19 vinta-vinta, sim
6 vintaset	20 soma
7 vintasim	30 soma, set-vinta, set
8 set-vinta	40 set-soma
9 set-vinta, en	50 set-soma, set-vinta, set
10 set-vinta, set	60 sim-soma
11 set-vinta, sim	70 sim-soma, set-vinta, set
12 sim-vinta	80 vinta-soma
13 sim-vinta, en	90 vinta-soma, set-vinta, set
14 sim-vinta, set	100 vintaen-soma

Extending Concepts

14. What is the greatest number in the system that you can make in more than one way? Find all the different ways you can make the number and write an arithmetic expression for each one.

15. What is the least number you can make in more than one way? Make a different arithmetic expression for each way to show how the number is made.

Homework 5: Exploring the Chinese Abacus

Applying Skills

Abacus Rules

The beads above the crossbar are worth five times the value of the column if pushed toward the crossbar. Each bead below the crossbar is worth one times the value of the column if pushed toward the crossbar. The value of the column is zero when all beads are pushed away from the crossbar.

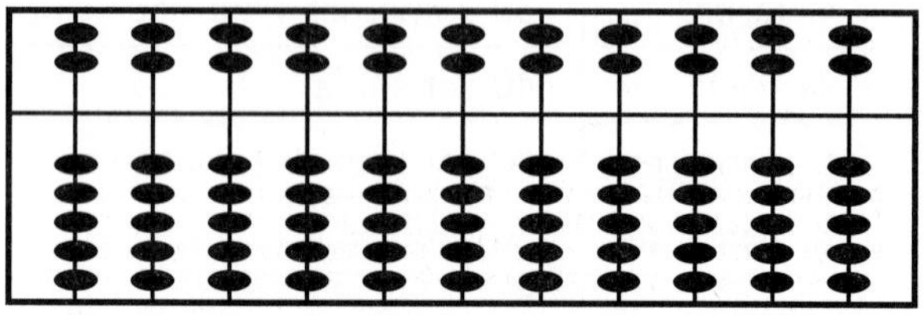

1. 2. 3. 4. 5. 6. 7. 8. 9. 10. 11.

Write the place value using words and numbers for each row marked on the abacus. For example, row 11 = 1 or *the ones place.*

Show each number on the abacus using the fewest beads. Show only the necessary beads. Write the arithmetic notation for each solution.

Number	Sketch	Arithmetic Expression
6,050	**12.** 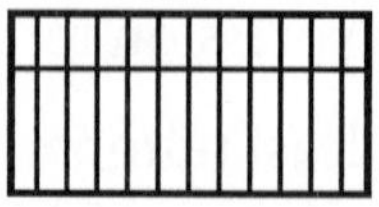	**13.**
28,362	**14.** 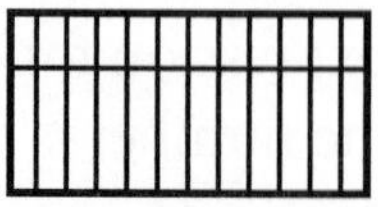	**15.**
4,035,269	**16.** 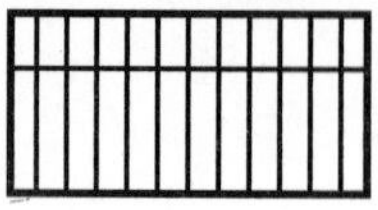	**17.**

Extending Concepts

18. How is zero shown on the abacus? Why is the zero important? How is zero on the abacus like or unlike zero in our system?

19. Find a number smaller than 100 that you could make on the abacus in more than one way. What is the arithmetic expression for each way you can make the number?

Writing

20. Answer the letter to Dr. Math.

Dear Dr. Math,
It seems like the Chinese abacus system has two different values for each column. Do the columns have the different values or do the beads?
Out O'Place

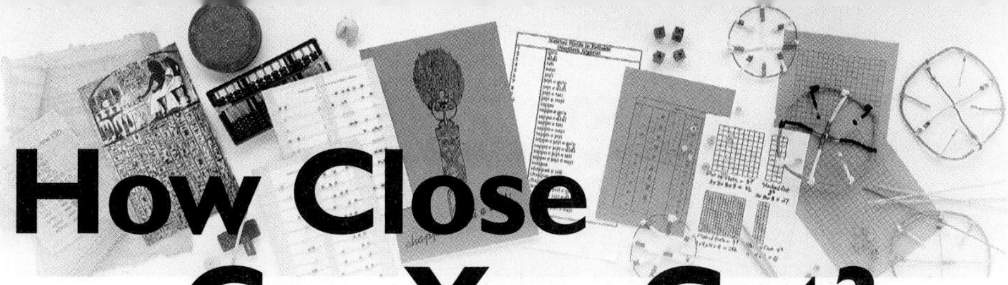

Homework 6: How Close Can You Get?

Applying Skills

Show different ways you can make this number on the abacus. Write the arithmetic expression for each solution.

Number	Sketch	Arithmetic Expression
852	**1.** 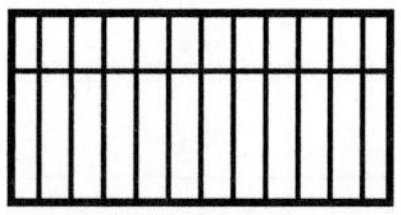	**2.**
852	**3.** 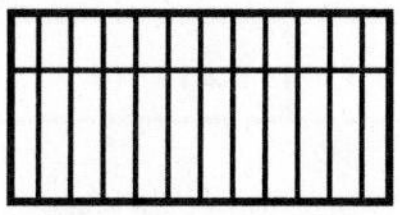	**4.**
852	**5.** 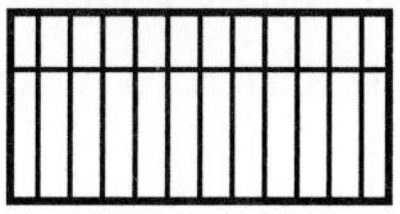	**6.**

What are some different ways you can make 555? How many beads do you use each time? Make a chart putting the number of beads you used in order from least to greatest.

	Number of Beads Used	Arithmetic Expression
7.		
8.		
9.		
10.		

Extending Concepts

11. What pattern do you see in the number of beads used to make 852 and 555? Why does this pattern work this way? Don't forget to explain how you used trading to make the different numbers.

Writing

12. Answer the letter to Dr. Math.

Dear Dr. Math,
To make the number 500, I can use five 100-beads from the bottom, or one 500-bead from on top. Or I can use four 100-beads from the bottom AND two 50-beads from on top. When would it make sense to use a different way to make the number?
Clu

Homework 7: Additive Systems

Applying Skills

Judy's System Judy invented a new additive system. ◊ = 1 □ = 9 ! = 81		
Judy's System	**Our Number System**	**Arithmetic Expression**
!!□◊◊◊	**1.**	**2.**
!!!!!□□□◊◊◊◊	**3.**	**4.**
□□□◊◊	**5.**	**6.**
!!!!!!!□□□□◊◊◊◊◊◊◊◊◊◊◊◊◊◊	**7.**	**8.**
9.	222	**10.**
11.	98	**12.**

What do these numbers in Judy's system represent in our system? How would you make numbers using Judy's system? Complete the chart and write an arithmetic expression for each number.

Extending Concepts

13. Write 1,776 in Judy's system. What number would you add to Judy's system to make writing larger numbers easier? Make sure your number fits the pattern. Write 1,776 using your added symbol.

14. How did you choose your added number?

Writing

15. Answer the letter to Dr. Math.

Dear Dr. Math,

When my teacher asked us to add a new symbol and value to the additive number system we had been using, I added ☆ to represent 0. But when I tried making numbers with it, things didn't turn out the way I planned. When I tried to use it to make the number 90, everyone thought the number was 9. Why didn't people understand? Here's what I did:

□☆

Z. Roe

Homework 8

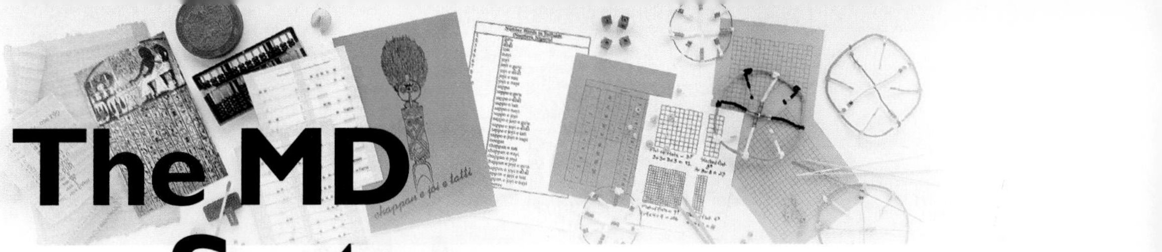

The MD System

Applying Skills

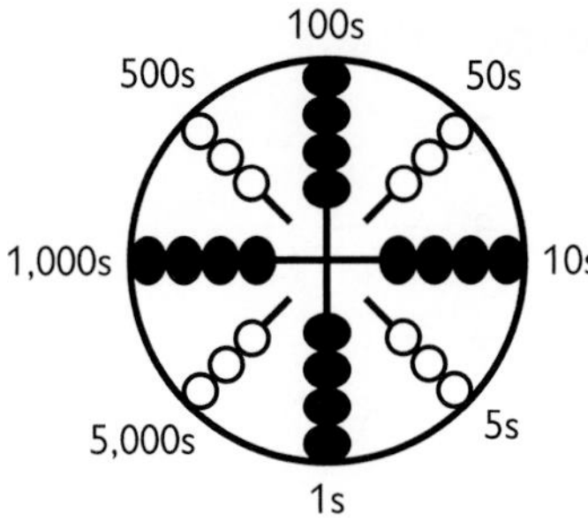

Make the numbers below using the MD system. Use the above illustration of the MD system to help you.

1. 9 **2.** 72 **3.** 665

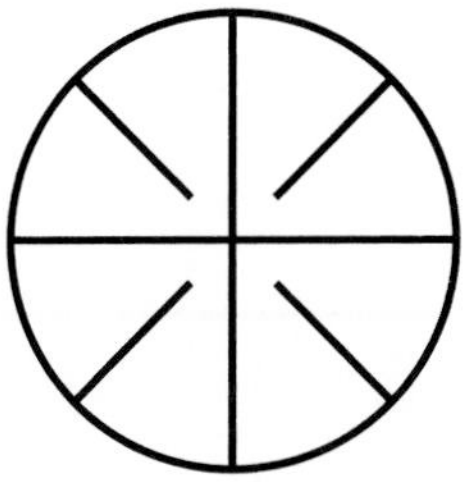

4. Show 7,957 in the MD system using the fewest beads you can.

5. Make the same number with the most beads you can.

Write the arithmetic notation for how you made each number using the MD system.

6. 9 **7.** 72 **8.** 665

9. 7,957 **10.** 7,957

Extending Concepts

11. What is the greatest number you can represent in the MD system?

12. How do you know there are not any higher numbers?

13. Can you make all the numbers in order up to that number?

14. What is one change you could make to the system, so that you could make some higher numbers?

15. What kind of trades could you make in the MD system?

16. How do you show 0 in the MD system? Is this like having a 0 in our system or is it different? Why?

Writing

17. The MD system and the Abacus system use the same place values. What other similarities do they have? differences? Which do you find easier to use? Explain.

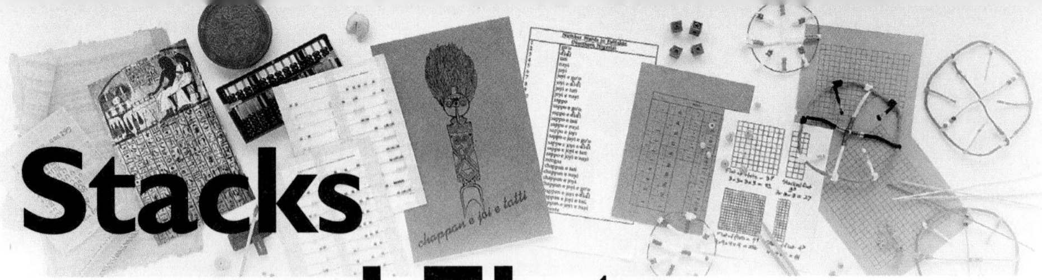

Homework 9

Stacks and Flats

Applying Skills

Make a list of powers through 4 for each number and write the value.

1. $2^0 =$	$2^1 =$	$2^2 =$	$2^3 =$	$2^4 =$
2. $3^0 =$	$3^1 =$	$3^2 =$	$3^3 =$	$3^4 =$
3. $4^0 =$	$4^1 =$	$4^2 =$	$4^3 =$	$4^4 =$

Figure out which base is used in each problem below.

4. $36 = 1(?^3) + 1(?^2)$

5. $58 = 3(?^2) + 2(?^1) + 2(?^0)$

6. $41 = 2(?^4) + 1(?^3) + 1(?^0)$

7. $99 = 1(?^4) + 1(?^2) + 3(?^1)$

Write the arithmetic expression for each number.

8. 25 base 2 **9.** 25 base 3 **10.** 25 base 4 **11.** 78 base 2 **12.** 78 base 3 **13.** 78 base 4

Extending Concepts

14. Describe how you figured out the arithmetic expressions above. Did you use a power greater than 4? Explain why.

15. In the base 2 system, how does the pattern continue after 1, 2, 4, 8, 16, ...? How is this pattern different from the pattern 2, 4, 6, 8, 10, 12, ...?

16. Look at the patterns in the chart below. Fill in the missing numbers for each pattern. Then answer the three questions at the bottom of the chart.

Number	Patterns of Powers	Patterns of Multiples
2	1, 2, 4, 8, 16, ___, ___, ___	2, 4, 6, 8, 10, 12, ___, ___, ___
3	1, 3, 9, 27, ___, ___, ___	3, 6, 9, 12, 15, 18, ___, ___, ___
4	1, 4, 16, 64, ___, ___, ___	4, 8, 12, 16, 20, ___, ___, ___

a. How can you use multiplication to explain the patterns of powers?

b. How can you use addition to explain the patterns of multiples?

c. Do you have another way to explain either group of patterns?

10 Homework

Power Up Game

Applying Skills

What do these expressions equal?

1. $1(2^4) + 2(2^3) + 2(2^2) + 1(2^1) + 2(2^0)$
2. $2(3^4) + 1(3^3)$
3. $1(4^3) + 2(4^2) + 2(4^0)$
4. $2(3^3) + 2(3^2) + 2(3^1) + 2(3^0)$
5. $2(2^3) + 1(2^1)$

Arrange each set of numbers to make the greatest and least values for each expression. The last blank represents an exponent.

6. 3, 6, 5, 2 greatest (____ + ____)—
   least (____ + ____)—
7. 7, 6, 6, 5 greatest (____ + ____)—
   least (____ + ____)—
8. 5, 2, 3, 4 greatest (____ + ____)—
   least (____ + ____)—
9. 1, 2, 3, 4 greatest (____ + ____)—
   least (____ + ____)—
10. **Powers Puzzle** Figure out each missing value. The sixteen numbers in the shaded area add to 11,104 when you have finished.

Number	To the 2nd Power	To the 5th Power	To the ___ Power	To the ___ Power
3			27	
	16			
6				1,296
		32		

Extending Concepts

11. What conclusion did you reach about the number that goes in the exponents place when you want a large or a small number? Find 2 digits where the larger digit raised to the smaller digit is bigger than the smaller digit raised to the larger digit. Find 2 digits where the larger digit raised to the smaller digit is equal to the smaller digit raised to the larger digit.

Making Connections

12. Earthquakes are rated on a Richter scale from 1 to 10. They are rated to one decimal place; the most powerful earthquake in North America was in Alaska and was rated 8.5. The power of an earthquake increases 10 times from one whole number to the next. An 8.5 earthquake is 10 times more powerful than a 7.5 earthquake. How many times more powerful is a 6.2 earthquake than a 4.2 earthquake?

11 Homework

Efficient Number Systems

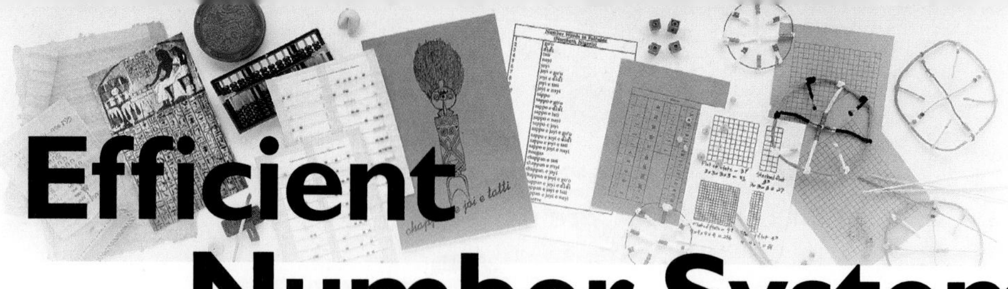

Applying Skills

Show how you would write each of these numbers in these systems.

1. Zany Places

	50	40	20	10	5	1
43						
72						
25						
17						

2. Maria's System ★ = 1 ● = 2 ◆ = 4

	100	20	10	2	1
31					
67					
183					
118					

Write an arithmetic expression for each number.

3. 43, Zany Places

4. 72, Zany Places

5. 25, Zany Places

6. 17, Zany Places

7. 31, Maria's system

8. 67, Maria's system

9. 183, Maria's system

10. 118, Maria's system

Extending Concepts

11. What is the greatest single digit in our base 10 system? What is the greatest single digit in the base 2 system? base 3? base 4? base 9?

12. How does the base of a place-value system affect the number of digits/symbols it contains?

13. What numbers cannot be made in Maria's system? Why?

14. Why can you make any number in Zany Places, but not in Maria's system?

15. Which system do you think is easier for writing numbers, Zany Places or Maria's system? Why?

Making Connections

16. The Metric system uses base 10 for measurement. Write the powers of 10 that are used in the measurements below. How is the English system (inches, feet, yards) different from the Metric system? Can you use exponents to describe the English system?

Metric	Number of Units	Exponents
deka	ten	$10^?$
kilo	thousand	$10^?$
giga	billion	$10^?$

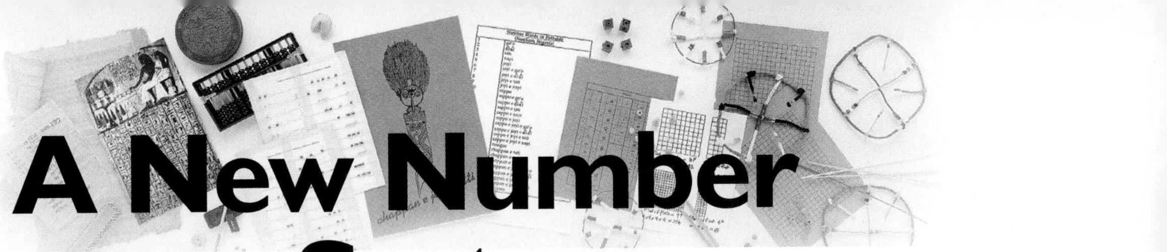

HW 12 Homework: A New Number System

Applying Skills

Give one example of a system you've learned about that uses each property, and explain how it works in that system.

1. place value **2.** base **3.** zero

4. trading **5.** building blocks

George's Mystery Device System

- Large beads = 10
- Small beads with arms pointing in = 1
- Small beads with arms pointing out = 5

George's New and Improved Mystery Device System

George decided to add place value to his system. He made each bead on the horizontal arms worth 100 and each bead on the vertical arms worth 10. He didn't change the value for the smaller beads.

- Large beads on vertical arm = 1 ten
- Large beads on horizontal arm = 1 hundred

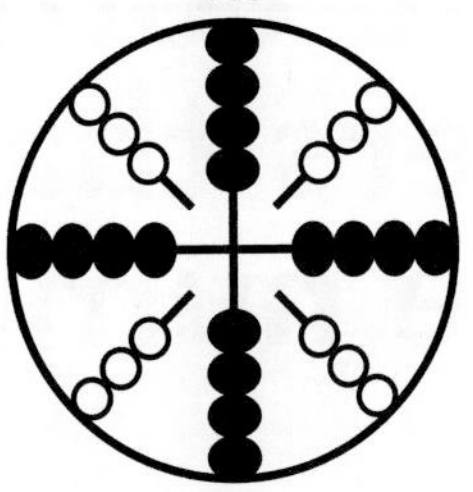

- Small beads with arms pointing in = 1
- Small beads with arms pointing out = 5

Draw each number in George's original system and in his revised system. Beware—some numbers cannot be made.

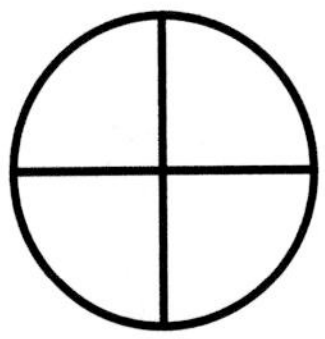
Original

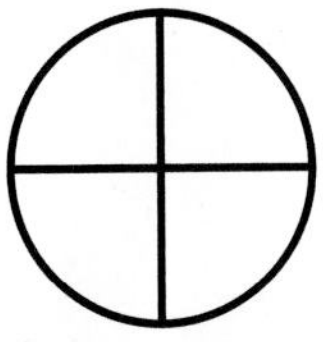
New and improved

6. 32 **7.** 97 **8.** 156 **9.** 371

Extending Concepts

Write the arithmetic expression for each of the numbers you made in both George's original system and his new and improved system.

10. 32 **11.** 97

12. 156 **13.** 371

14. What is the greatest number that George is able to make in his revised system? Are there any numbers less than this number that George cannot make? Tell why or why not.

15. In what way is George's revised system better than his old system? In what way is George's revised system not as good as his old system?

16. What are some ways that George's revised system is different from our system? Name at least 3 differences.

17. Make a list of all the different properties of our own number system that make it easy to use. For each item on your list, write one sentence explaining why.

PHASE**ONE**

The Whole of It

To work with fractions, it helps to be at ease with whole numbers. In this phase, you will focus on whole numbers. As you build models, figure out computational methods, and play fast-paced games, you will make connections among factors, multiples, and prime numbers.

How can you compute with numbers that are not whole numbers?

FROM WHOLES TO PARTS

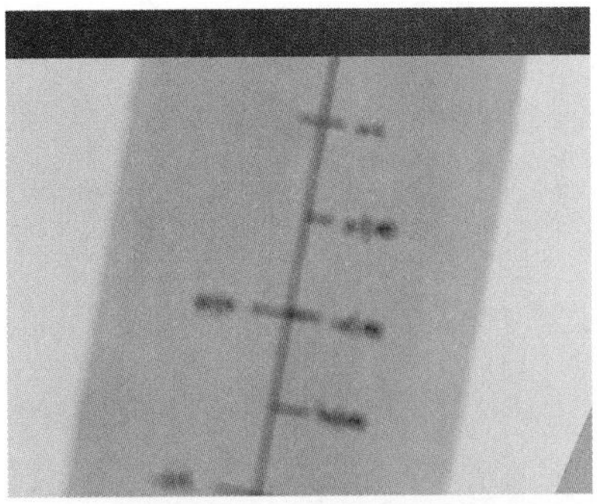

PHASE**TWO**

Between the Whole Numbers

Now it is time to focus on the parts of the whole. Making designs will help you to see how fractional parts relate to each other and the whole. By using number lines, you will discover ways to compare fractions. Games with cards and number cubes will make it lots of fun to try your new skills.

PHASE**THREE**

Adding Parts and Taking Them Away

With all that you have learned, you will be a natural at adding and subtracting fractions. You will have the chance to use area models, number lines, and computation. At the end of the phase, you will help a defective robot get out of a room and teach an extra-terrestrial visitor how to add and subtract fractions.

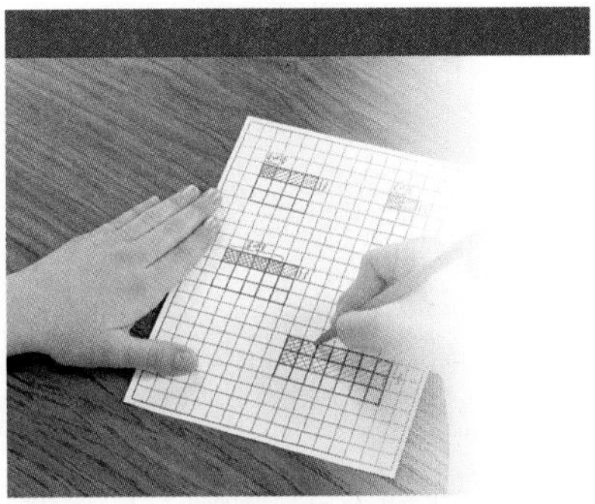

PHASE**FOUR**

Fractions in Groups

When you multiply and divide fractions, the answers are easy to compute. But, what is really happening? You will model fractions to sort out the confusing truth – sometimes multiplying results in a lesser answer and sometimes dividing results in a greater answer. How does this happen? You will figure out the hows and whys in this phase.

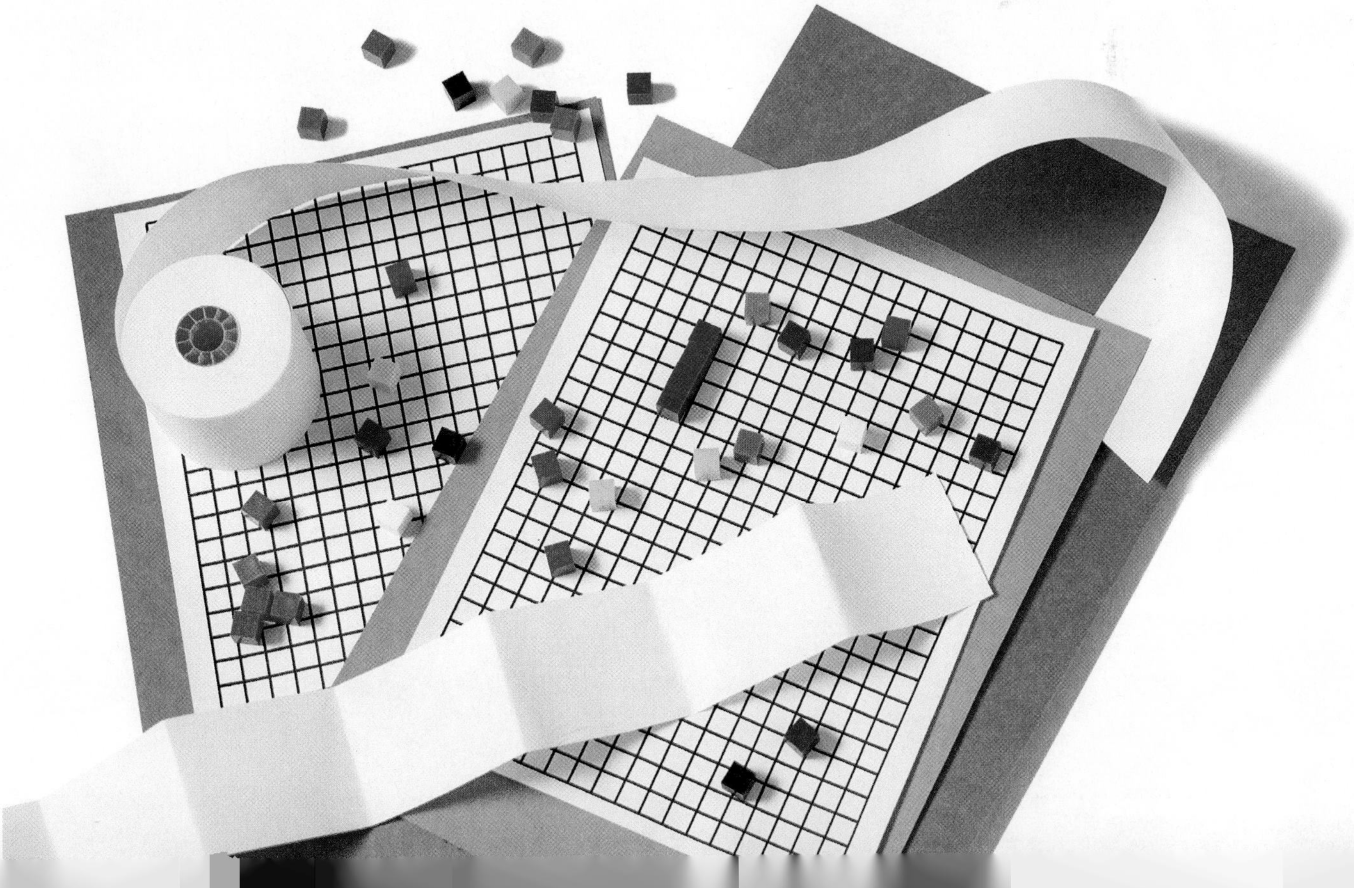

PHASE **ONE**

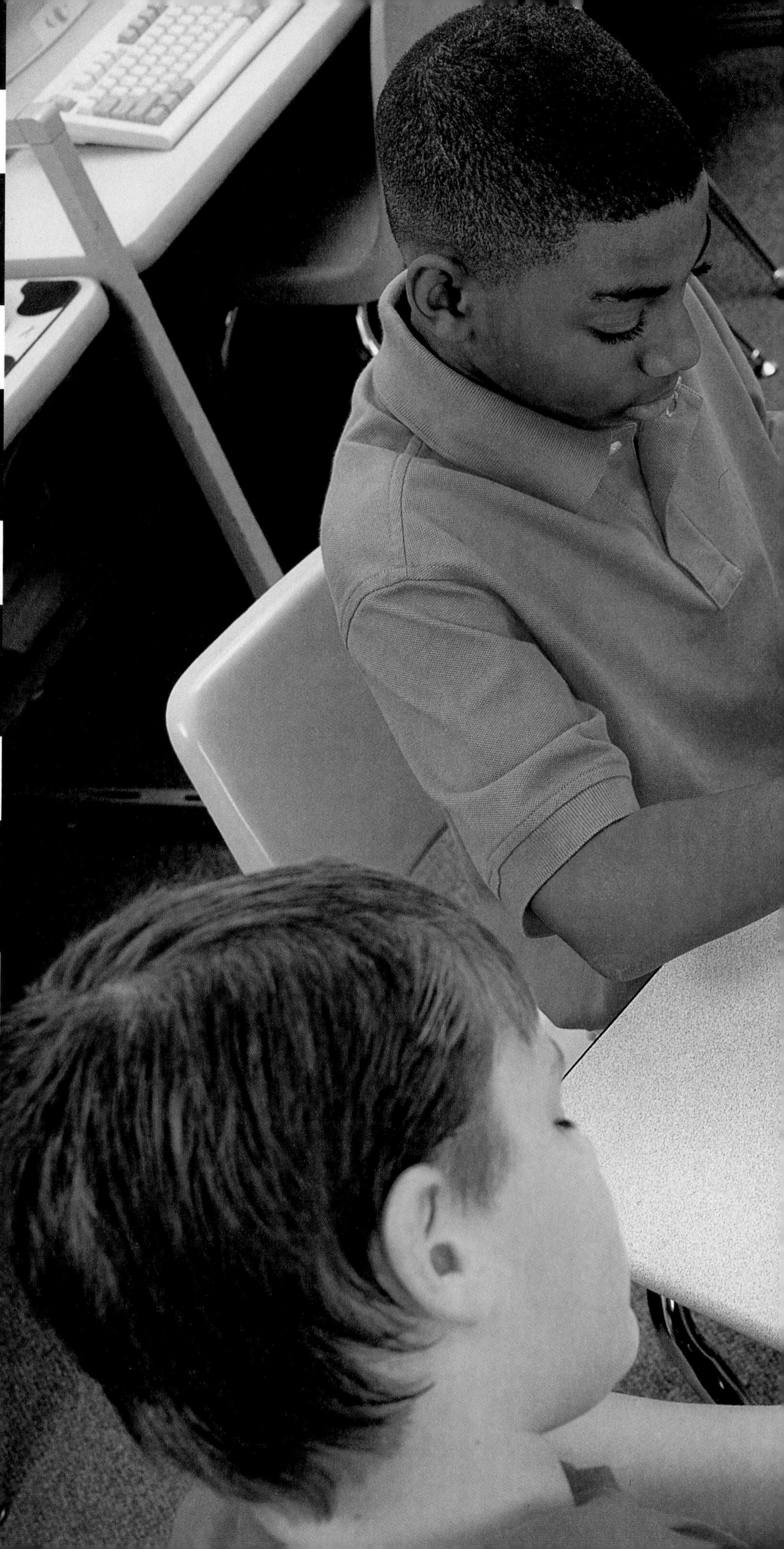

You will not see any fractions in this phase, but you will learn things that make working with fractions easier. Modeling shapes will help you find prime factors, which you can use to find greatest common factors. You will also find a way to identify least common multiples. You will look at the rules of order for solving problems and discover ways you can and cannot rearrange numbers in a problem to make it easier to solve.

The Whole of It

WHAT'S THE MATH?

Investigations in this section focus on:

NUMBER and OPERATIONS

- Understanding factors and prime factors
- Finding factors of whole numbers
- Finding the prime factorization of a whole number
- Finding the greatest common factor of two or more whole numbers
- Generating multiples of a number
- Finding the least common multiple of two or more numbers
- Performing operations in the correct order

MathScape Online
mathscape1.com/self_check_quiz

1 Shapes and Factors

MODELING NUMBERS TO IDENTIFY FACTORS

What can rectangles tell you about factors? To start this phase about whole numbers, you will do some model building. Then, you will take your modeling to a third dimension using cubes.

What are the possible side lengths of a rectangle made from a specific number of unit squares?

Find Factors Using Rectangles

Elaine uses square tiles to build tables with mosaic tops. Your class is going to find all the rectangular tabletops Elaine can make using specific numbers of tiles. For each number of tiles, you will make solid rectangles, in one layer, using all of the tiles.

Experiment until you are sure you have all the possible rectangles for each number you are assigned. Then, write the side lengths for each rectangle on the class chart.

1 Which numbers of tiles can produce more than one rectangle?

2 Which numbers of tiles can produce only one rectangle?

Dimensions and Factors

You can count the number of tiles on each side of a rectangle to find the lengths or **dimensions** of the sides.

Each dimension is a **factor** of the total number of tiles in the rectangle. A factor is a number that is multiplied by another number to yield a product. For example, 8 and 2 are factors of 16.

Find Factors Using Cubes

What can three-dimensional modeling tell you about factors?

Since rectangles have two dimensions, each tile tabletop you modeled showed you two factors. Now, you will use cubes to build three-dimensional shapes called rectangular prisms.

1 With your group, try to determine what numbers from 1 to 30 can be modeled as rectangular prisms without using 1 as a dimension. When you find a number for which you can make such a prism, see whether you can make a different prism with the same number of cubes. Your group should make a chart that shows:

- the total number of cubes, and
- the dimensions of each prism you made.

The number 8 can be modeled as a rectangular prism without using 1 as a dimension.

2 Compare your group's chart to the chart the class made for tile tabletops. How can you use the tabletop chart to predict the prism chart?

3 For which number could you make the most prisms? How many different prisms could you make for that number? What allows that number to be modeled in so many ways?

4 What factors do the following numbers have in common?

a. 8 and 15

b. 9 and 24

c. 12 and 16

d. 15 and 18

e. 20 and 24

f. 24 and 30

Write a Definition

On your copy of My Math Dictionary, write your own definitions and give examples of the following terms:

- factor
- prime factor
- common factor

hot words
prime number
factor
common factor

Homework
page 142

2 The Great Factor Hunt

FINDING PRIME AND COMMON FACTORS

As you saw when you built rectangles and prisms, numbers can be written as products of two, three, or more factors. How do you know when you have found the most possible factors for a number?

Find All the Factors

What is the longest string of factors you can multiply to get a given number?

One way to express the number 12 is to show it as a product of two factors: $12 = 2 \times 6$. How would you express the number 12 as the product of more than two factors? Without using 1 as a factor, what is the longest string of factors you can use to express the number 12?

Product of two factors: $12 = 2 \times 6$

Longest string of factors: $12 = 2 \times 2 \times 3$

The longest string of factors, excluding the factor 1, is called the *prime factorization* of the number.

Work with your classmates to find the longest factor string for each whole number from 1 to 30. When the class chart is finished, answer the following questions.

1. What kind of numbers cannot be part of these factor strings? What kind of numbers can be part of these factor strings?
2. Find three numbers that have a prime factor in common. List the numbers and the factor they share.
3. Find two pairs of numbers that have a string of more than one factor in common. List the number pairs and the factors they share.

$16 = 2 \times 8$
$= 2 \times 2 \times 4$
$= 2 \times 2 \times 2 \times 2$

$18 = 2 \times 9$
$= 2 \times 3 \times 3$

$27 = 3 \times 9$
$= 3 \times 3 \times 3$

Find the Greatest Common Factor

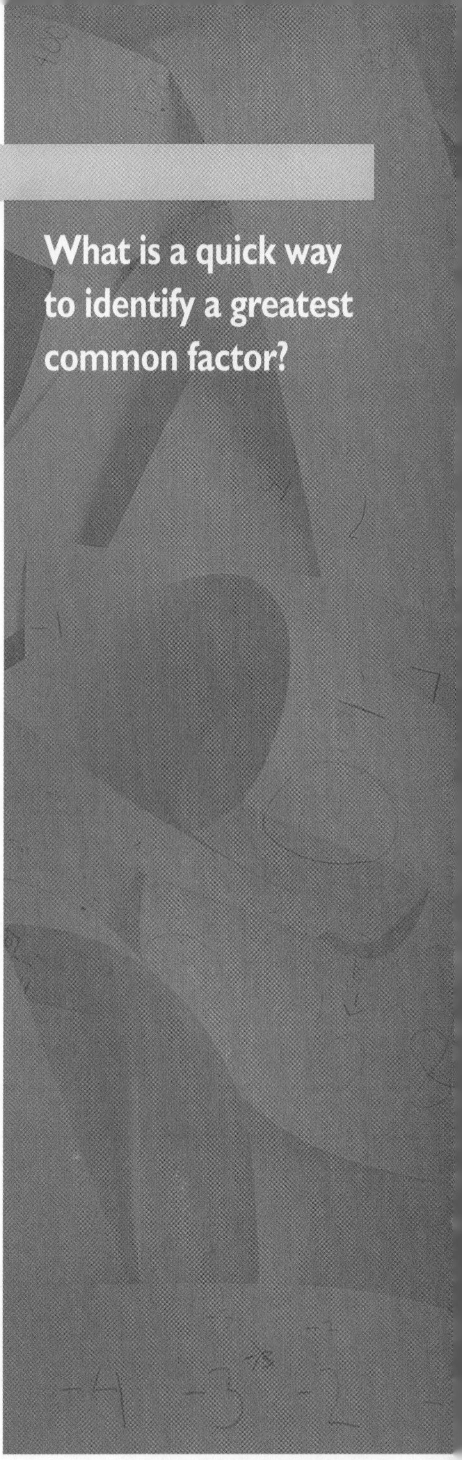

What is a quick way to identify a greatest common factor?

Look at the prime factorization chart that your class compiled. Compare the prime factor strings for 12 and 30.

$$12 = 2 \times \mathbf{2} \times \mathbf{3}$$

$$30 = \mathbf{2} \times \mathbf{3} \times 5$$

The numbers 12 and 30 have 2×3 in common. We will call "2×3" a "common string."

1 Write the longest "common string" of factors for each pair of numbers.

a. 9 and 27 **b.** 8 and 24

c. 24 and 30 **d.** 45 and 60

2 Use each pair's "common string" to find the greatest factor they share. For example, the longest "common string" of 12 and 30 is 2×3. The greatest factor they share is 2×3 or 6. This number is called the *greatest common factor* of 12 and 30.

3 Write each number in each pair as a product of their greatest common factor and another factor. For example, $12 = 6 \times 2$ and $30 = 6 \times 5$.

$12 = 2 \times (2 \times 3)$

$30 = (2 \times 3) \times 5$

$2 \times 3 = 6$ 6 is the greatest common factor of 12 and 30.

$12 = 6 \times 2$

$30 = 6 \times 5$

Write a Definition

On your copy of My Math Dictionary, write your own definitions and give examples of the following terms:

- prime factorization
- greatest common factor

hot words
prime factorization
greatest common factor

Homework
page 143

3 Multiple Approaches

FINDING COMMON MULTIPLES

When working with fractions, you often need to use common multiples of two or more numbers. In this lesson, you will use playing cards to find common multiples. Then, you will learn to find the least common multiple of any pair of numbers.

How can you use playing cards to determine common multiples?

Play "It's in the Cards"

You can use a standard deck of cards to study common multiples.

1. With a partner, try the example game of "It's in the Cards" shown below. What common multiples do the cards show?
2. Play the game three more times using the rates 3 and 4, 2 and 5, and 3 and 6. Make a table showing each pair of numbers and the multiples you find for each pair.
3. For each number pair, circle the least multiple that you found.
4. Using your own words, define the terms *common multiple* and *least common multiple.*

It's in the Cards

- Each player has a stack of 13 playing cards, representing the numbers 1 (ace) to 13 (king), in numerical order.
- Each player lays each of his or her cards facedown in a line.
- Each player turns over certain cards according to his or her rate. In the example, Player A had a rate of 2, so every second card is turned. Player B had a rate of 3, so every third card is turned.

The numbers that are faceup in both rows are common multiples of the rate numbers.

Find the Least Common Multiple

You may have noticed that the product of any two numbers is always a multiple of both numbers. For example, 54 is a common multiple of 6 and 9. But how can you find the least common multiple?

How can prime factors help you find least common multiples?

1 The table below shows steps for finding the least common multiple of various number pairs. Try to determine what happens in each step. Consider these questions.

- What prime factors do the given numbers have in common?
- How does the number in step 2 relate to the given numbers?
- Each given number is written as a product in step 3. What factors are used?
- Which numbers from step 3 are used in step 4?

2 When you think you know a method for finding the least common multiple of two numbers, use the method to find the least common multiple of 12 and 18, 14 and 49, and several other pairs of numbers. Does your method work? Make a table showing each pair of numbers, the steps you used, and the least common multiple of the numbers.

3 Describe your method for finding the least common multiple of two numbers.

4 How can you use your method to find the least common multiple of three numbers?

Steps to Find the Least Common Multiple

Given Numbers	Step 1	Step 2	Step 3	Step 4	Least Common Multiple
8 12	$2 \times 2 \times 2$ $2 \times 2 \times 3$	4	$8 = 4 \times 2$ $12 = 4 \times 3$	$4 \times 2 \times 3$	24
15 70	3×5 $2 \times 5 \times 7$	5	$15 = 5 \times 3$ $70 = 5 \times 14$	$5 \times 3 \times 14$	210
30 45	$2 \times 3 \times 5$ $3 \times 3 \times 5$	15	$30 = 15 \times 2$ $45 = 15 \times 3$	$15 \times 2 \times 3$	90

hot words
multiple
least common multiple

Homework
page 144

4 First Things First

FOLLOWING THE ORDER OF OPERATIONS

How much can you do with 1, 2, and 3? If you know the rules about the order of operations, you can make three numbers do a lot. In this lesson, you will use the rules to expand the possibilities.

Use the Order of Operations

In what order should an equation's operations be computed?

Work on your own to solve each of the following equations.

$1 + 3 \times 2 = n$ $\qquad 3 + 1 \times 2 = n$ $\qquad 4 + 6 \div 2 = n$

What answers did you get? Did everyone in your class get the same answers? Although everyone's computations may appear to be correct, why might some people get a different answer than you did?

Each of these problems has just one correct answer. To get the correct answer, you must follow the order of operations.

Your teacher will give you a handout. See if you can place the numbers 1, 2, and 3 in all ten equations to get a result of every whole number from 1 to 10. Make sure you follow the order of operations.

Order of Operations

- First, do any computations that are within **grouping symbols** such as parentheses.
- Then, evaluate any **exponents.**
- Next, **multiply** and/or **divide** in order from left to right.
- Finally, **add** and/or **subtract** in order from left to right.

Play "Hit the Target"

In this game, you will make equations using three numbers to equal a target number. Your team will have a better chance of winning if you know the order of operations.

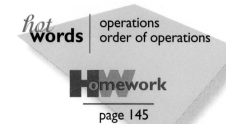

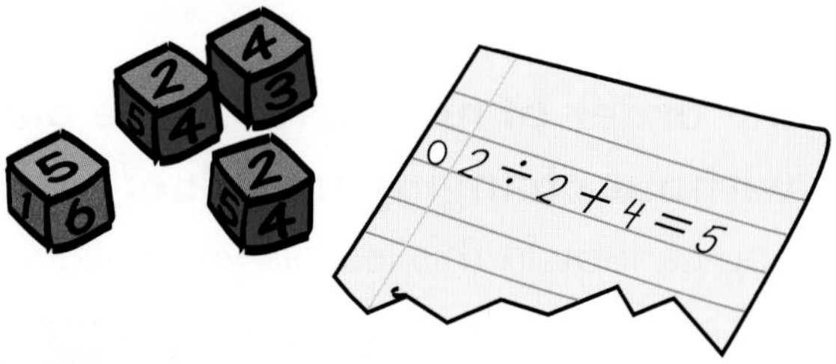

Can you get the target value using the order of operations?

Hit the Target

For each round of the game, your teacher will roll four number cubes. One is the target number. The other three are the building numbers. Your goal is to use all three building numbers and two operations to get as close to the target number as possible. Building numbers may be used in any order, but each number may only be used once.

Each team earns points as follows.

- One point for getting the target number.
- An additional point if you get the target number without using parentheses.
- An additional point if you get the target number using a combination of addition or subtraction and multiplication or division.

All teams that get the target number earn points. If no team gets the target number, the team with the number closest to the target number earns a point.

Write a Memory Helper

As you played "Hit the Target," you had to think fast and keep the order of operations in mind. What was your method of remembering the order of operations? Can you think of something that would help you remember? Write down your method and share it with the class.

hot words operations, order of operations

Homework
page 145

5 Putting It All Together

SOLVING PROBLEMS WITH WHOLE NUMBERS

Can you change the order of numbers or the parentheses in an expression without changing its value? Sometimes you can and sometimes you cannot. In this lesson, you will learn when you can change them. Then, you will solve "guess my number" problems.

Organize Your Math

What changes can you make in an expression to make it easier to find its value?

Look at the following pairs of expressions carefully. For each pair, try to predict whether the value of **A** will be the same as the value of **B**. If expression **A** can be changed to expression **B** without changing the value of **A**, explain why. If expression **A** cannot be changed to expression **B** without changing the value of **A**, explain why not.

1 **A** $3 + 34 + 27$
B $3 + 27 + 34$

2 **A** $524 - 412$
B $412 - 524$

3 **A** $(27 - 7) - 3$
B $27 - (7 - 3)$

4 **A** $4 \times (8 \times 2)$
B $(4 \times 8) \times 2$

5 **A** $(24 \div 4) \div 2$
B $24 \div (4 \div 2)$

6 **A** 18×6
B $10 \times 6 + 8 \times 6$

7 **A** $24 \div 8 + 2$
B $24 \div 2 + 8$

8 **A** 208×4
B $(200 \times 4) + (8 \times 4)$

9 **A** $5 + 5 \times 14$
B $(5 + 5) \times 14$

10 **A** $4 \times 6 + 5$
B $6 \times 4 + 5$

11 Write an expression of your own that could be made easier by a change in order or grouping. Show what change you would make. Then, describe the change. Explain:

- why it makes finding the value easier, and
- why it does not change the value.

Guess My Number

Can you find the mystery numbers?

Now, you will use what you have learned about factors and multiples to find mystery numbers.

1. Two numbers have a sum of 60. Both are multiples of 12. Neither number is greater than 40. What are the numbers?
2. Three numbers are each less than 20. They are all odd. One is the least common multiple of the other two numbers. What are the numbers?
3. The greatest common factor of two numbers is 11. One number is twice the other. Their least common multiple is one of the numbers. What are the numbers?
4. Sam and Susanna are brother and sister. The difference between their ages is a factor of each of their ages. Their combined age is 15. What are their possible ages?
5. Luis is 6 years older than his dog. His age and his dog's age are both factors of 24, but neither is a prime factor. What are their ages?
6. The number of pennies saved by Brian is a multiple of the number of pennies saved by his sister Carmen. Together, they have saved 125 pennies. Carmen has saved more than 10 pennies. Both numbers of pennies are multiples of 5. How many pennies does each have?

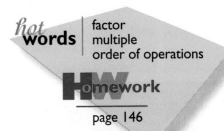

Write Your Own Problem

Create a problem of your own that involves factors or multiples. When you have finished, give your problem to another student and see if she or he can guess your numbers.

hot words
factor
multiple
order of operations

Homework
page 146

PHASE TWO

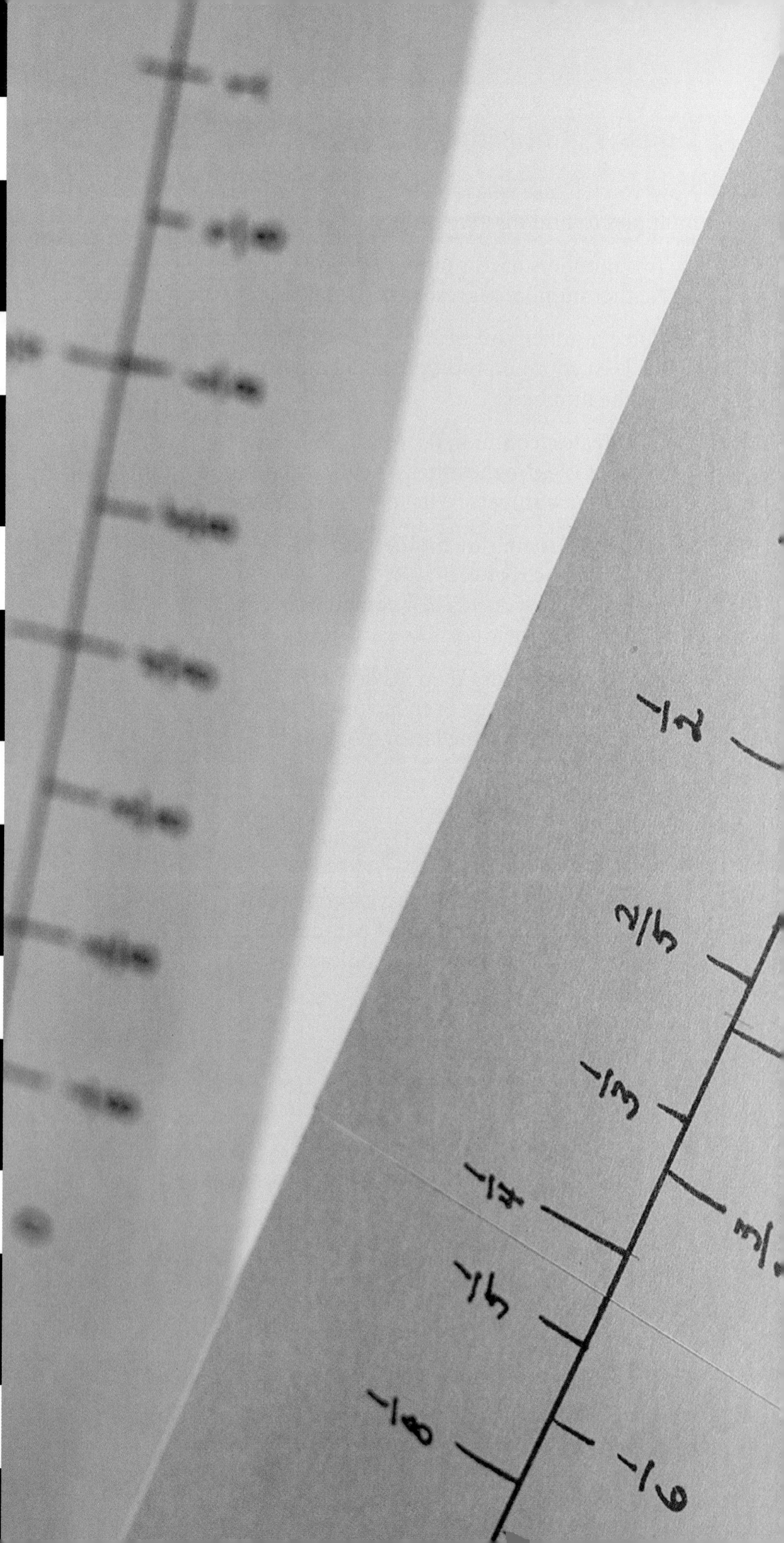

In daily life, you deal with fractions – a quarter pound, two and a half hours, six tenths of a mile – without thinking much about it. The experiences you have in this phase will give you a fuller understanding of fractions. You will make area models to show how different parts of a whole relate to each other. You will use number lines not only to order and compare fractions, but to see that every fraction has many names.

Between the Whole Numbers

WHAT'S THE MATH?

Investigations in this section focus on:

NUMBER and OPERATIONS

- Representing fractions as parts of a whole
- Writing an equivalent fraction for a given fraction
- Comparing and ordering fractions
- Graphing fractions on a number line
- Understanding improper fractions
- Finding a common denominator for two fractions

MathScape Online
mathscape1.com/self_check_quiz

6 Designer Fractions

DIVIDING WHOLE AREAS INTO FRACTIONAL PARTS

Can you be a math-savvy designer? In this lesson, you will divide whole rectangles into equal parts. Then, you will make designs with parts that are not the same size, but still represent one whole.

How many different ways can you divide a rectangle into four equal parts?

Make Designs to Show Fractions

Fractions can be represented as parts of a rectangle.

1 On centimeter grid paper, make a rectangle that is 8 centimeters by 10 centimeters and divide it into four equal parts. Each person in your group should make a different design using the following guidelines.

- The parts may look exactly alike or they may be different.
- You may divide parts with diagonal lines.

When you have finished, discuss with your group how you know that the four parts in your design are equal.

2 On centimeter grid paper, make four rectangles that are 6 centimeters by 8 centimeters. Your group will be assigned one of the following pairs of fractions.

$\frac{1}{3}$ and $\frac{1}{6}$ $\frac{1}{6}$ and $\frac{1}{12}$ $\frac{1}{3}$ and $\frac{1}{12}$

Your group should make two different designs for each of your two fractions. Then, answer these questions about your designs.

a. How do you know which of the fractional parts in the different designs are equal to each other?

b. Compare the sizes of the fractional parts for each of your two fractions. What do you notice?

c. What other fractions would be easy to make using the same rectangle?

Use Fractions to Make a Whole

You can make designs where each region represents a given fraction.

What designs can you make with different fractions that together make a whole?

1 Make one design for each group of fractions. First, outline the rectangle on grid paper using the dimensions given. Then, divide the rectangle into the fractional parts. Color each part and label it with the correct fraction.

a. On a 2-by-4-centimeter rectangle, show $\frac{1}{2}$, $\frac{1}{4}$, $\frac{1}{8}$, and $\frac{1}{8}$.

b. On a 3-by-4-centimeter rectangle, show $\frac{1}{3}$, $\frac{1}{3}$, $\frac{1}{6}$, and $\frac{1}{6}$.

c. On a 3-by-4-centimeter rectangle, show $\frac{1}{4}$, $\frac{1}{4}$, $\frac{1}{6}$, $\frac{1}{6}$, and $\frac{1}{6}$.

d. On a 4-by-6-centimeter rectangle, show $\frac{1}{4}$, $\frac{1}{8}$, $\frac{1}{8}$, $\frac{1}{3}$, and $\frac{1}{6}$.

2 Make a design to help determine the missing fraction in each statement. Label each part with the correct fraction.

a. Use a 3-by-4-centimeter rectangle to show $\frac{2}{3} + \frac{1}{4} + \underline{?} = 1$.

b. Use a 3-by-4-centimeter rectangle to show $\frac{1}{3} + \frac{1}{4} + \frac{1}{4} + \underline{?} = 1$.

Make Your Own Design

Create a design of your own that shows several fractions whose sum is one. Use the following guidelines.

- Use a different group of fractions than those listed above. Think about what size of rectangle would make sense for the group of fractions you choose.
- Divide your rectangle into at least four parts.
- Use at least four different fractions in your design.
- Label each part with the correct fraction.

hot words | fraction

Homework
page 147

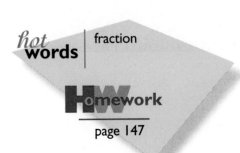

7 Area Models and Equivalent Fractions

USING AREA MODELS TO COMPARE FRACTIONS

How can you compare fractions? In this lesson, you will use area models and grid sketches.

Use Area Models to Compare Fractions

Which fraction is greater?

It's easy to compare two fractions like $\frac{1}{3}$ and $\frac{2}{3}$ because they have the same denominators. But what about fractions with different denominators? You can compare these fractions by using area models.

Use area models to compare each set of fractions.

1 $\frac{4}{5}$ and $\frac{2}{3}$ **2** $\frac{2}{7}$ and $\frac{1}{3}$ **3** $\frac{4}{5}$ and $\frac{3}{4}$

Using Area Models to Compare Fractions

To compare $\frac{1}{3}$ and $\frac{2}{5}$, make two 5-by-3 rectangles on grid paper. To show $\frac{1}{3}$, circle one of the columns. To show $\frac{2}{5}$, circle two rows.

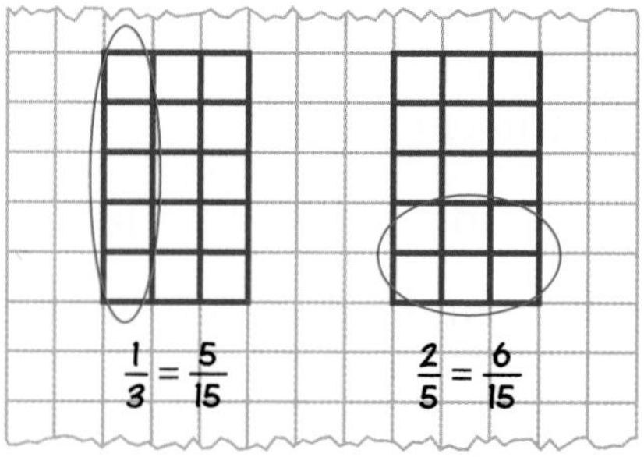

Since 5 of the 15 squares cover the same area as $\frac{1}{3}$, $\frac{5}{15}$ is equal to $\frac{1}{3}$. So $\frac{5}{15}$ and $\frac{1}{3}$ are **equivalent fractions.** Also, $\frac{6}{15}$ is **equivalent** to $\frac{2}{5}$. It is easier to compare $\frac{5}{15}$ and $\frac{6}{15}$ than it is to compare $\frac{1}{3}$ and $\frac{2}{5}$. So, $\frac{1}{3} < \frac{2}{5}$.

Use Grid Sketches to Compare Fractions

How can a grid sketch help you to compare fractions?

A quick sketch on one grid can be helpful for comparing fractions.

This sketch shows the comparison between $\frac{2}{5}$ and $\frac{1}{3}$.

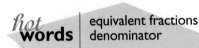

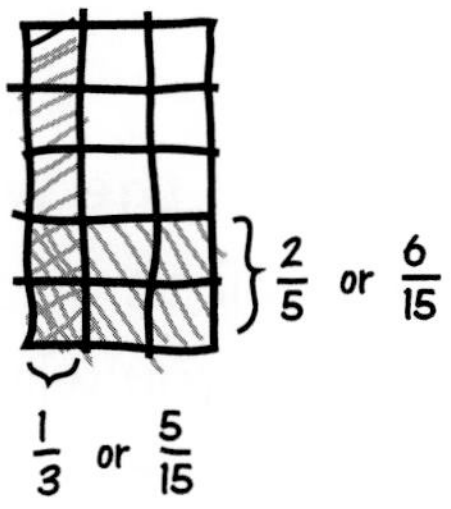

$\frac{6}{15}$ is greater than $\frac{5}{15}$.

1 Make a grid sketch for each pair of fractions. Label each sketch with the given fractions and their equivalent fractions. Compare the fractions.

a. $\frac{2}{3}$ and $\frac{3}{5}$ **b.** $\frac{2}{8}$ and $\frac{1}{5}$

c. $\frac{1}{3}$ and $\frac{2}{6}$ **d.** $\frac{6}{7}$ and $\frac{4}{6}$

2 Write a few sentences explaining how a sketch helps you decide which fraction is greater.

Practice Fraction Comparisons

How can you use common denominators to compare fractions?

Two fractions with the same denominator have a *common* denominator.

1 For each pair of fractions, find equivalent fractions that have a common denominator. Determine which fraction is greater. Write your answer using the original fractions.

a. $\frac{5}{6}$ and $\frac{7}{10}$ **b.** $\frac{4}{6}$ and $\frac{3}{4}$

c. $\frac{7}{8}$ and $\frac{4}{5}$ **d.** $\frac{3}{4}$ and $\frac{7}{10}$

2 Describe how you prefer to compare fractions. Why does your method work? If you prefer one method for some fractions and a different method for others, explain.

3 In your copy of My Math Dictionary, write a definition of *equivalent fractions*. Use a drawing to illustrate your definition.

hot words
equivalent fractions
denominator

Homework
page 148

8 Fraction Lineup

PLOTTING NUMBERS ON THE NUMBER LINE

What numbers belong between the whole numbers on a number line? In this lesson, you will be placing fractions in appropriate places on a number line. You will start with separate lines for different fractions and then combine them all onto one number line.

What are the positions of fractions on a number line?

Place Fractions on a Number Line

The number line below shows only whole numbers.

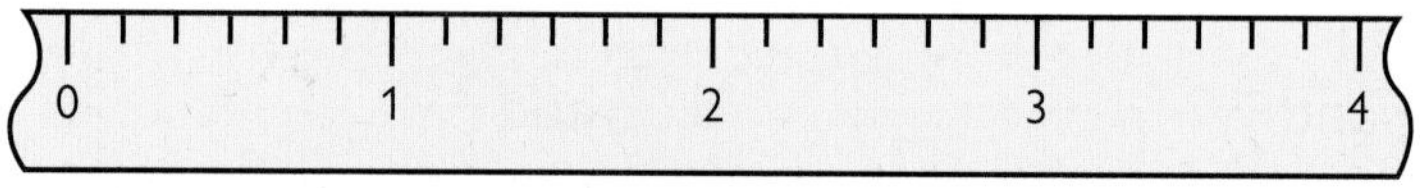

You can place fractions between the whole numbers.

1 Mark fractions on separate number lines as indicated.

a. On number line **A**, mark all the halves.

b. On number line **B**, mark all the thirds.

c. On number line **C**, mark all the fourths.

d. On number line **D**, mark all the sixths.

e. On number line **E**, write all the fractions that you wrote on lines **A** through **D**. Where two or more fractions have the same position, write each fraction.

2 Use your number lines to compare each pair of fractions. Write an expression using $<$, $>$, or $=$ to show the comparison.

a. $\frac{1}{4}$ and $\frac{2}{6}$ b. $\frac{4}{3}$ and $\frac{8}{6}$ c. $\frac{5}{3}$ and $\frac{10}{6}$

d. $\frac{5}{6}$ and $\frac{3}{4}$ e. $\frac{10}{6}$ and $\frac{7}{4}$ f. $\frac{3}{2}$ and $\frac{8}{6}$

3 Explain how you can use a number line to compare two fractions.

Use the Number Line to Find Equivalent Fractions

How can you use a number line to identify equivalent fractions?

In Lesson 7, you learned that equivalent fractions cover the same area on an area model. Equivalent fractions are also located at the same place on a number line.

1 Find an equivalent fraction for each fraction.

a. $\frac{3}{2}$ **b.** $\frac{5}{3}$ **c.** $\frac{6}{3}$

d. $\frac{4}{6}$ **e.** $\frac{6}{6}$ **f.** $\frac{4}{3}$

2 Imagine that you were to label all the twelfths ($\frac{0}{12}$, $\frac{1}{12}$, $\frac{2}{12}$, and so on) on your number line **E**. Which fraction(s) that you had already labeled are equivalent to $\frac{3}{12}$? to $\frac{10}{12}$? to $\frac{18}{12}$?

3 The number line below is marked in ninths. Which of these fractions would be equivalent to fractions that you labeled on number line **E**?

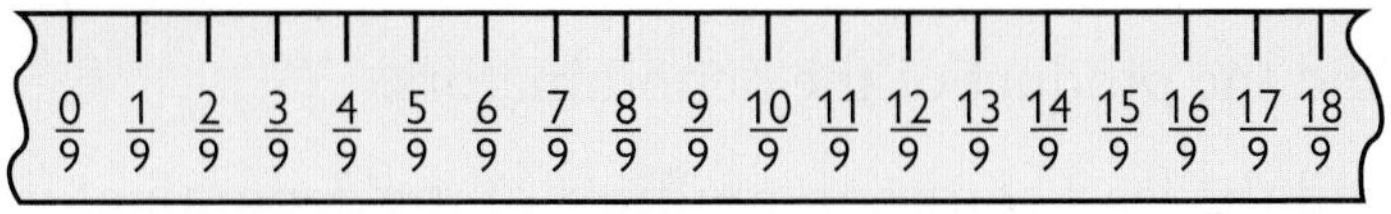

Write about Equivalent Fractions

Write two fractions that have a common denominator. Make one equivalent to $\frac{3}{4}$ and the other equivalent to $\frac{5}{6}$. Explain why they are equivalent. Include both an area model and a number line with your answer.

hot words
equivalent fractions
improper fraction

Homework
page 149

9 Focus on Denominators

COMPARING FRACTIONS USING COMMON DENOMINATORS

Writing equivalent fractions with common denominators makes them easier to compare. However, when you use the product of the denominators as the common denominator, this number may be larger than you need.

Look for the Least Common Denominator

How can you be sure that you have found the least common denominator?

You studied least common multiples in Lesson 3. You can use least common multiples to find least common denominators.

1 For each pair of fractions, complete the following:

- Find the product of the denominators.
- Find the least common multiple of the denominators.
- Write equivalent fractions with common denominators using the lesser of the two numbers, if they are different.
- Compare the fractions using the symbols <, >, or =.

a. $\frac{5}{6}$ and $\frac{7}{12}$ **b.** $\frac{3}{10}$ and $\frac{1}{6}$ **c.** $\frac{2}{3}$ and $\frac{3}{4}$

d. $\frac{4}{5}$ and $\frac{7}{10}$ **e.** $\frac{2}{7}$ and $\frac{1}{4}$ **f.** $\frac{2}{3}$ and $\frac{7}{9}$

2 In step **1**, which method of finding a common denominator gave the lesser common denominator for the fraction pairs?

3 Using the number line below, find equivalent fractions with the least common denominator for $\frac{5}{10}$ and $\frac{4}{6}$.

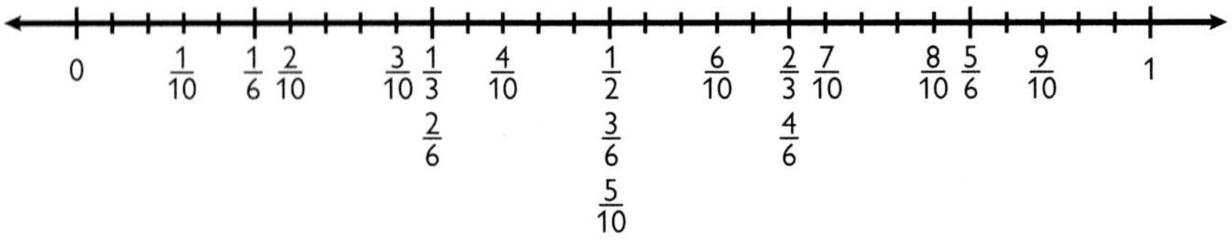

4 Explain why the least common multiple of the denominators of $\frac{5}{10}$ and $\frac{4}{6}$ is not their least common denominator. What should you do before finding the least common denominator?

Play "Get the Cards in Order"

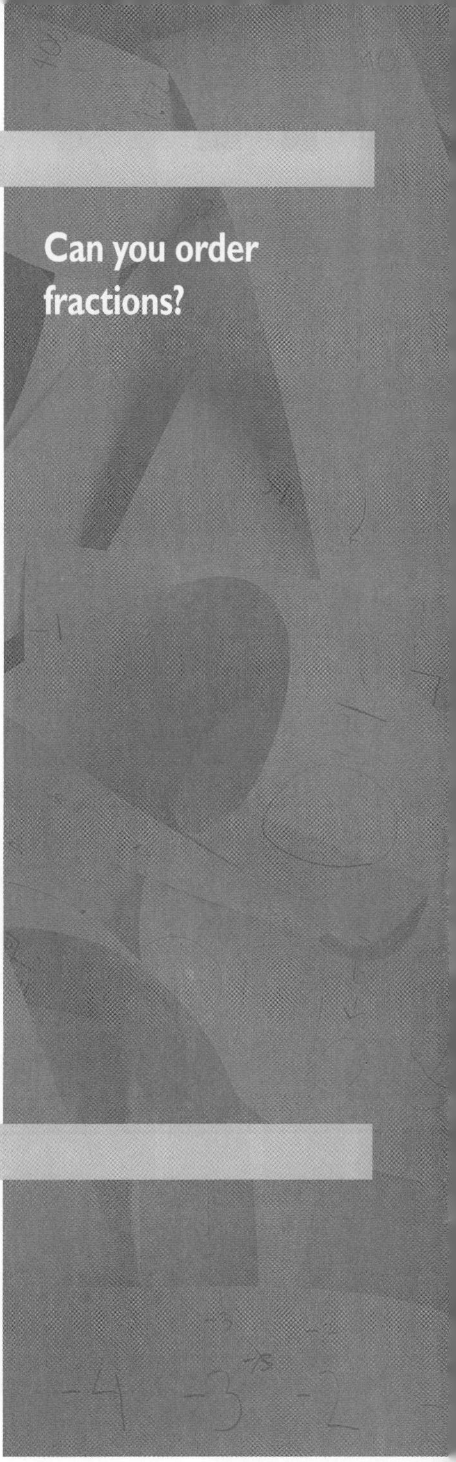

Can you order fractions?

Play "Get the Cards in Order" with a partner.

Get the Cards in Order

The goal of the game is get the longest line of fractions listed in order from least to greatest using the following rules:

- All 20 fraction cards are placed faceup in any order. This is the selection pot.
- Players take turns selecting one card from the pot. This card is placed in the player's own line of cards so that the cards are in order from least to greatest.
- If a player disagrees with the placement of an opponent's card, he or she can challenge the placement. Together, the players should decide the correct placement. If the original placement was incorrect, the challenging player can take the card and put it in his or her line or return it to the selection pot.
- The game ends when all cards in the pot have been played. The winner is the player with the greater number of cards in his or her line.

Write About Comparing Fractions

Write an explanation of how to compare two fractions. Give examples to support your explanation.

hot words

least common multiple
least common denominator
numerator

Homework

page 150

PHASE THREE

Now that you are familiar with so many aspects of fractions, it is time to add and subtract. You will start by using number lines. When you put fraction strips on the number line, the sums and differences make sense. Soon you will be adding and subtracting fractions without models. As you play games and solve unusual problems, you will become familiar with fractions greater than one.

Adding Parts and Taking Them Away

WHAT'S THE MATH?

Investigations in this section focus on:

NUMBERS and OPERATIONS

- Using a number line to model addition and subtraction of fractions
- Adding and subtracting fractions without using a model
- Converting between mixed numbers and improper fractions
- Adding and subtracting mixed numbers and improper fractions

MathScape Online
mathscape1.com/self_check_quiz

10 Sums and Differences on the Line

ADDING AND SUBTRACTING FRACTIONS ON THE NUMBER LINE

You have already learned to add and subtract whole numbers using a number line. In this lesson, you will use number lines to add and subtract fractions.

Use a Number Line to Add Fractions

How can you add fractions on a number line?

Each number line you use today is divided into twelfths.

1 Use number lines to find each sum. Then write a fraction that represents the sum.

a. $\frac{1}{3} + \frac{1}{4}$ b. $\frac{2}{3} + \frac{1}{4}$ c. $\frac{1}{4} + \frac{5}{12}$

d. $\frac{7}{12} + \frac{1}{2}$ e. $\frac{5}{6} + \frac{3}{4}$ f. $\frac{1}{6} + \frac{3}{4}$

2 Choose one of your addition problems. How did you determine the denominator for your answer?

Adding Fractions Using Number Lines

To show $\frac{1}{2}$, make a paper strip that fills $\frac{1}{2}$ of the space between 0 and 1.

To show $\frac{1}{3}$, make a paper strip that fills $\frac{1}{3}$ of the space between 0 and 1.

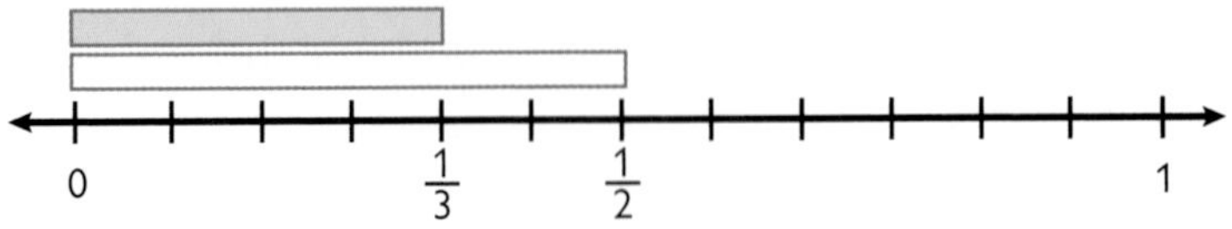

To find the sum of $\frac{1}{2} + \frac{1}{3}$, place the strips end to end with zero as the starting point. Find their total length. What fraction represents the sum?

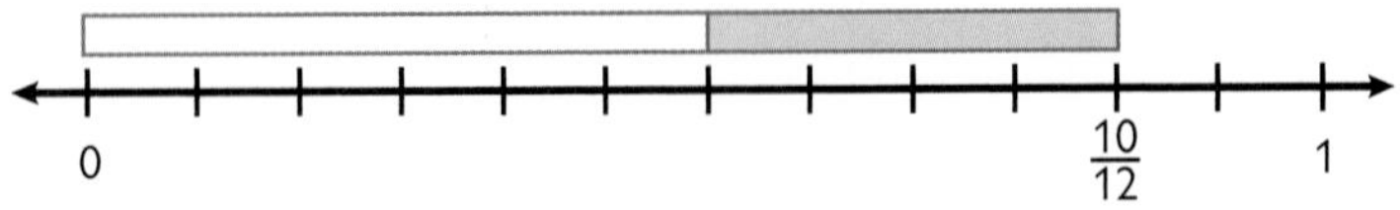

$$\frac{1}{2} + \frac{1}{3} = \frac{10}{12}$$

Use a Number Line to Subtract Fractions

How can you subtract fractions on a number line?

Now that you can add fractions on the number line, try subtracting them. To subtract, find how much greater one fraction strip is than the other.

1 Use number lines to find each difference. Write a fraction that represents each difference.

a. $\frac{5}{6} - \frac{1}{4}$ **b.** $\frac{3}{4} - \frac{1}{3}$

c. $\frac{1}{2} - \frac{1}{6}$ **d.** $\frac{11}{12} - \frac{1}{6}$

e. $\frac{2}{3} - \frac{7}{12}$ **f.** $\frac{1}{2} - \frac{1}{3}$

2 Choose one of your subtraction problems. Explain how you found the difference. How did you determine the denominator for your answer?

3 Find each sum or difference.

a. $\frac{1}{3} - \frac{1}{6}$ **b.** $\frac{3}{4} - \frac{1}{6}$

c. $\frac{1}{2} + \frac{3}{4}$ **d.** $\frac{5}{6} + \frac{1}{3}$

e. $\frac{11}{12} - \frac{2}{3}$ **f.** $\frac{5}{6} - \frac{1}{4}$

g. $\frac{2}{3} + \frac{5}{6}$ **h.** $\frac{1}{3} + \frac{1}{4}$

i. $1 - \frac{5}{6}$ **j.** $\frac{7}{12} - \frac{1}{2}$

Write a Definition

In your copy of My Math Dictionary, write a definition of *common denominator*, and give an example.

hot words
numerator
denominator
common denominator

Homework
page 151

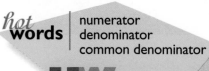

11 Numbers Only

ADDING AND SUBTRACTING FRACTIONS WITHOUT MODELS

Using area models or number lines is not always the simplest way to add and subtract fractions. In this lesson, you will devise a way to find sums and differences of fractions without these tools.

Add and Subtract Fractions Without Models

How can you add and subtract fractions without using models?

Use what you have learned about models for adding and subtracting fractions to devise a method to add and subtract fractions without models.

1 Copy and complete the addition table. Find each sum and describe how you found the sum.

	Problem	Answer	Description of Method
a.	$\frac{2}{5} + \frac{1}{5}$		
b.	$\frac{1}{2} + \frac{1}{8}$		
c.	$\frac{3}{8} + \frac{1}{4}$		
d.	$\frac{7}{10} + \frac{3}{5}$		

2 What steps would you use to find the sum of any two fractions?

3 Copy and complete the subtraction table. Find each difference and describe how you found the difference.

	Problem	Answer	Description of Method
a.	$\frac{6}{7} - \frac{2}{7}$		
b.	$\frac{2}{3} - \frac{4}{9}$		
c.	$\frac{7}{12} - \frac{1}{4}$		
d.	$\frac{3}{5} - \frac{2}{15}$		

4 What steps would you use to find the difference between any two fractions?

Play "Race for the Wholes"

The game "Race for the Wholes" can be played with 2, 3, or 4 players. Play the game with some of your classmates.

Can you make a runner land on a whole number?

Race for the Wholes

To play this game, you will need a set of fraction cards, a number line, and the same number of game markers (racers) as players. The number line should be marked with the numbers from 0 to 3 with 24 spaces between whole numbers.

- Begin the game with all racers on 0.
- Each player is dealt 3 cards and the rest remain facedown in a pile.
- Players take turns. On each turn, the player moves any racer by using one of his or her fraction cards. The player can move the racer forward or backward the distance shown on the card.
- After playing a card, the player puts it in a discard pile and draws a new card to replace it.
- A player earns one point each time he or she can make a racer land on a whole number.
- The first player with 5 points is the winner.

Solve the Magic Square

In a magic square, the sum of the numbers in each column, row, and diagonal is the same. Arrange the numbers into the nine boxes so that they form a magic square. The sum of the fractions in each column, row, and diagonal will be 1.

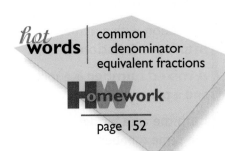

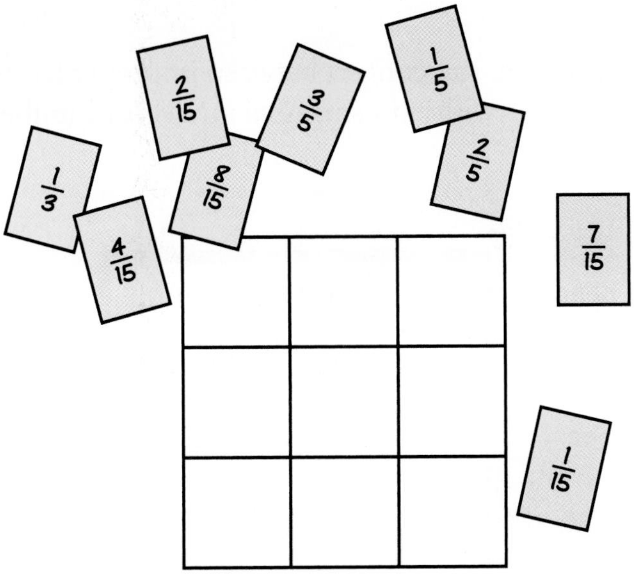

hot words
common denominator
equivalent fractions

Homework
page 152

12 Not Proper but Still Okay

ADDING MIXED NUMBERS AND IMPROPER FRACTIONS

What do you do with fractions that are greater than one? How you choose to write these numbers usually depends on what you are doing with them. But, do not be misled by the names. An improper fraction is not bad, and a mixed number is not confused!

How are mixed numbers and improper fractions related?

Write Mixed Numbers and Improper Fractions

Numbers that are not whole numbers and are greater than one can be written as mixed numbers or as improper fractions.

1 Draw an area model or number line that shows each improper fraction. Then, write the corresponding mixed number.

a. $\frac{4}{3}$ **b.** $\frac{7}{4}$ **c.** $\frac{12}{5}$ **d.** $\frac{7}{2}$

2 Draw an area model or number line that shows each mixed number. Then, write the corresponding improper fraction.

a. $2\frac{1}{3}$ **b.** $4\frac{1}{4}$ **c.** $3\frac{3}{5}$ **d.** $1\frac{5}{6}$

3 Write each number in a different form without using an area model or a number line.

a. $3\frac{4}{5}$ **b.** $\frac{15}{5}$ **c.** $\frac{10}{4}$ **d.** $6\frac{2}{3}$

4 Explain how you can convert between improper fractions and mixed numbers without using area models or number lines.

Improper Fractions and Mixed Numbers

A fraction with a numerator less than the denominator is a **proper fraction.**

A fraction with a numerator greater than or equal to the denominator is an **improper fraction.**

A **mixed number** is a mix of a whole number and a proper fraction.

Both the area model and the number line show the same number. It can be called $\frac{5}{3}$ or $1\frac{2}{3}$.

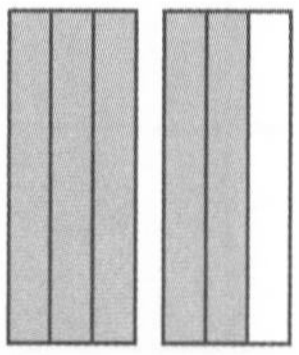

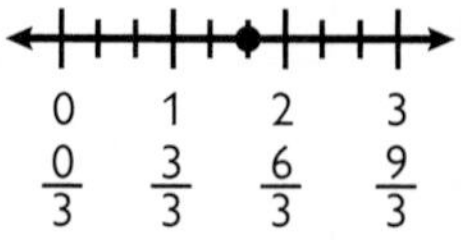

Add Fractions Greater than One

How can you add mixed numbers and improper fractions?

You can use your knowledge of adding fractions to help you add mixed numbers and improper fractions.

1 With a partner or group, find at least two ways to add the following numbers. If the problem is written with improper fractions, write the sum as an improper fraction. If the problem is written with mixed numbers, write the sum as a mixed number. If the problem has both types of numbers, write the sum as an improper fraction and as a mixed number.

a. $\frac{4}{3} + \frac{5}{6}$
b. $\frac{3}{2} + \frac{6}{5}$
c. $2\frac{1}{2} + 3\frac{3}{4}$
d. $4\frac{1}{3} + 1\frac{2}{5}$
e. $2\frac{1}{4} + \frac{5}{2}$
f. $\frac{4}{3} + 3\frac{1}{3}$

2 With a partner, discuss two different methods of adding mixed numbers and improper fractions. Is one method better or more efficient than the other? Explain your reasoning. Be ready to share your methods with the class.

Mixed Number Rule

Mixed numbers should not include an improper fraction. So, you sometimes need to rename the fraction.

For example, if you add $2\frac{3}{5}$ and $4\frac{4}{5}$, you might find that the sum is $6\frac{7}{5}$. But, you cannot have an improper fraction in a mixed number. So, your next step is to rename the improper fraction within the mixed number.

Since $\frac{7}{5} = 1\frac{2}{5}$, $6\frac{7}{5}$ is the same as $6 + 1 + \frac{2}{5}$.

So, you should write $6\frac{7}{5}$ as $7\frac{2}{5}$.

Write About Fractions

In your copy of My Math Dictionary, describe and give examples of *improper fraction* and *mixed number.*

hot words | improper fractions, mixed numbers

Homework
page 153

13 Sorting Out Subtraction

SUBTRACTING MIXED NUMBERS AND IMPROPER FRACTIONS

Now that you have added mixed numbers and improper fractions, it is time to consider subtraction. You can use what you know about adding mixed numbers and improper fractions to help you find ways to subtract these types of numbers.

How can you subtract mixed numbers and improper fractions?

Subtract Fractions Greater than One

The information below about regrouping in subtraction will help you subtract mixed numbers.

1 Find each difference.

a. $\frac{10}{9} - \frac{2}{9}$

b. $\frac{25}{12} - \frac{7}{6}$

c. $3 - 1\frac{2}{7}$

d. $5\frac{3}{4} - 2\frac{1}{7}$

e. $4\frac{11}{12} - 3\frac{3}{4}$

f. $5 - 2\frac{4}{5}$

g. $6\frac{2}{3} - \frac{9}{2}$

h. $\frac{21}{5} - 2\frac{3}{10}$

2 When you have found the answers to all eight problems, be ready to explain the steps you used to get each answer.

Regrouping in Subtraction

When you subtract mixed numbers, you sometimes need to regroup. That is, you need to take one whole from the whole number part and add it to the fraction part.

To find $5\frac{1}{12} - \frac{17}{12}$, use the following steps:

- First, regroup.

$5\frac{1}{12} = 4 + 1 + \frac{1}{12}$ *Take one whole from the 5.*

$= 4 + \frac{12}{12} + \frac{1}{12}$ *Rename the whole as $\frac{12}{12}$.*

$= 4 + \frac{13}{12}$ *Add $\frac{12}{12}$ and $\frac{1}{12}$.*

- Then, subtract.

$4\frac{13}{12} - 1\frac{7}{12} = 3\frac{6}{12}$ or $3\frac{1}{2}$

Get That Robot Out of Here!

Against the back wall of your room there is a large robot with an annoying bug in its programming. The robot has to travel 10 meters forward to get through the door, and you need to get it out as soon as possible. You can tell the robot to move toward the door using any of the given distances listed below. However, because of the bug, on every other move the robot will move in the wrong direction. So, if you tell the robot to go $\frac{1}{2}$ meter and then tell it to go $\frac{1}{3}$ meter, it will go forward $\frac{1}{2}$ meter and then back $\frac{1}{3}$ meter. The robot will only be $\frac{1}{6}$ meter closer to the door.

To get the robot out, you can choose any of the numbers below, but you can use each number only once. Create a table like the one below to record your progress. Notice the *distance from wall* plus *distance to door* must always equal 10 meters.

What is the least number of moves that you can use to get the robot out of the room?

Move Number	Add or Subtract	Distance from Wall	Distance to Door
1	$+2\frac{1}{3}$	$2\frac{1}{3}$	$7\frac{2}{3}$
2	$-\frac{1}{4}$	$2\frac{1}{12}$	$7\frac{11}{12}$

Possible Moves

Each move can be used only once. All distances are in meters.

$\frac{1}{2}$	$\frac{1}{4}$	$\frac{4}{3}$	$\frac{13}{6}$	$\frac{5}{6}$
$\frac{3}{2}$	$\frac{9}{4}$	$2\frac{1}{3}$	$\frac{1}{12}$	$\frac{11}{12}$
$\frac{1}{6}$	$\frac{7}{2}$	$\frac{3}{4}$	$\frac{5}{12}$	$1\frac{1}{6}$
$\frac{5}{3}$	$\frac{7}{3}$	$\frac{4}{4}$	$\frac{1}{3}$	$2\frac{1}{4}$

What is the least number of computations that will get the robot out of the room?

hot words improper fraction, mixed number

Homework
page 154

14 Calc and the Numbers

DESCRIBING AND USING RULES FOR FRACTION ADDITION AND SUBTRACTION

An extraterrestrial visitor named Calc has landed in your classroom. He likes math, but so far he can only work with whole numbers. You are going to teach him about fractions. Fortunately, he is excellent at following directions.

Write Directions for Adding Fractions

Can you write directions to explain how to add fractions?

Calc is from the planet Integer. He can only work with whole numbers. He follows directions perfectly, and he understands terms such as *numerator* and *denominator.*

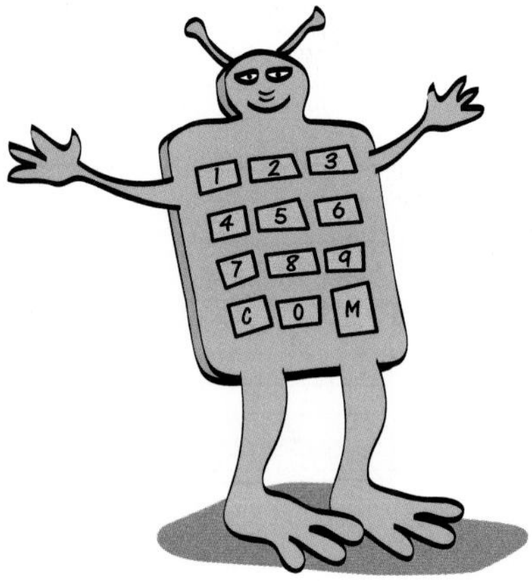

1 Write step-by-step instructions that tell Calc how to add two fractions. He needs to learn how to add:

a. fractions with like denominators,

b. fractions with unlike denominators,

c. improper fractions, and

d. mixed numbers.

2 Test your instructions on the problems below. Would Calc be able to find all of the answers using your instructions?

a. $\frac{3}{7} + \frac{2}{7}$ **b.** $\frac{1}{3} + \frac{1}{4}$ **c.** $\frac{9}{5} + \frac{11}{5}$

d. $\frac{5}{3} + \frac{14}{12}$ **e.** $7\frac{1}{2} + 4\frac{1}{3}$ **f.** $2\frac{3}{4} + 1\frac{5}{8}$

Use Fractions

Can you add and subtract fractions to solve problems?

Calc returned home and told his fellow Integerlings the news about fractions. Now, some Integerlings need to use fractions. See if you can answer each of their problems.

1 Maribeth is planning to carpet two rooms with matching carpet.

a. Both rooms are 12 feet wide. One room is $9\frac{3}{4}$ feet long. The other is $8\frac{2}{3}$ feet long. She finds some carpet that is 12 feet wide. How long of a piece of carpet does she need?

b. Maribeth decides she should have some extra carpet in case she makes a mistake installing it. She buys a 20-foot length. She does a perfect job installing the carpet. How much carpet is left over?

2 Ronita plans to replace the floor molding around the perimeter of a room.

a. The room measures $12\frac{1}{4}$ feet by 11 feet. It has one doorway, which is $3\frac{1}{4}$ feet wide. If Ronita does not put molding along the doorway, what is the total length of molding she needs?

b. Ronita's brother gives her $30\frac{1}{2}$ feet of molding. How much more molding does Ronita need?

3 John is a tailor and is making some suits.

a. He has $12\frac{1}{4}$ yards of fabric. He needs $4\frac{2}{3}$ yards for one customer and $5\frac{3}{4}$ yards for another. How much fabric does he need in all? Does he have enough fabric?

b. John's customers both decided to order vests, so he needs $\frac{3}{4}$ yard more of the fabric for each customer's suit. How much fabric does he need now? How much will he have leftover when he is done?

hot words
equivalent fractions
common denominator

Homework
page 155

PHASE FOUR

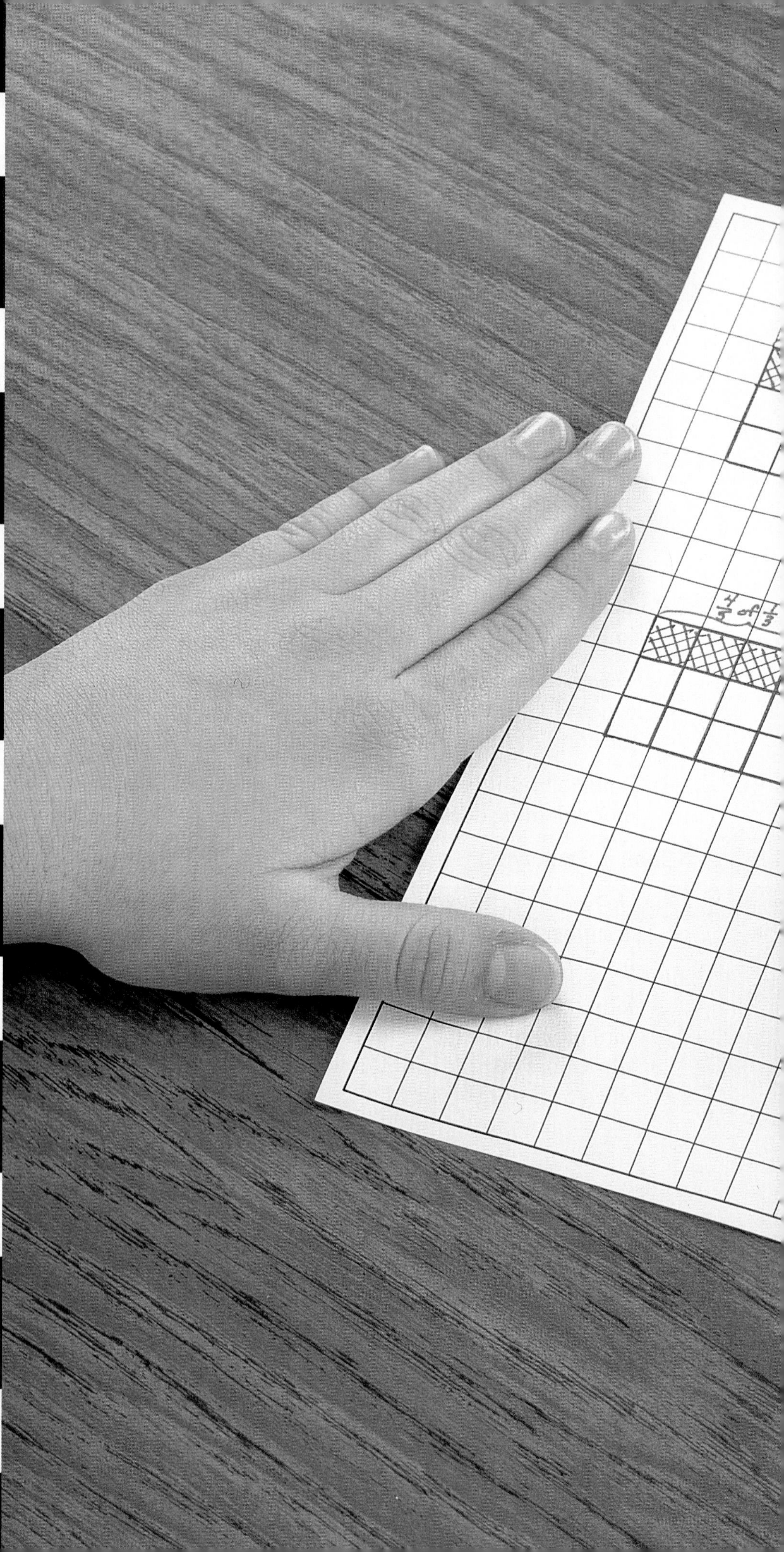

As you will see in this phase, fraction multiplication and division are easy to compute. But, why do they work the way they do and how do they relate to each other? Think of this phase as a study in group behavior – of fractions! By the end of the phase, you will be playing fraction multiplication and division games. With your new understanding of fraction groups, you will be able to estimate and make predictions about problems that used to seem mysterious.

Fractions in Groups

WHAT'S THE MATH?

Investigations in this section focus on:

NUMBERS and OPERATIONS

- Finding and applying methods to multiply a whole number by a fraction
- Using an area model to find a fraction of a fraction
- Finding and applying a method for multiplying fractions
- Estimating products
- Using models to divide by fractions
- Finding and applying a method for dividing fractions

MathScape Online
mathscape1.com/self_check_quiz

15 Picturing Fraction Multiplication

FINDING FRACTIONS OF WHOLES

How can you use drawings to multiply a fraction by a whole? In this lesson, you will learn several ways to multiply using drawings. Which drawings work best for you may depend on how you think about the problems.

How can you use a drawing to find a fraction of a whole number?

Find Fractions of Wholes

When a math class was asked to show $\frac{1}{3}$ of 3, the students made a variety of drawings. How does Maya's drawing show that $\frac{1}{3}$ of 3 is 1? How does David's drawing show that $\frac{1}{3}$ of 3 is 1?

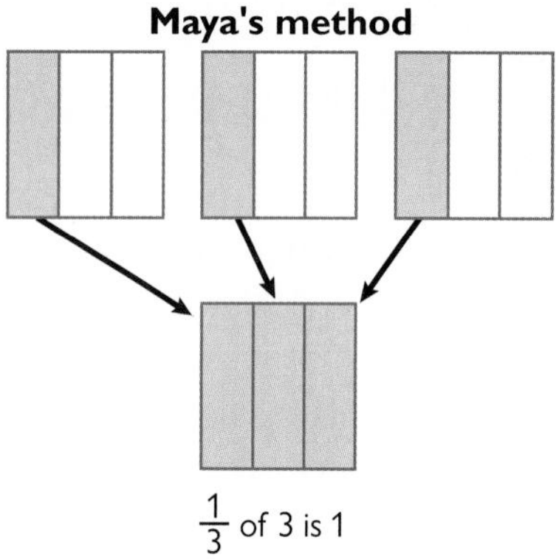

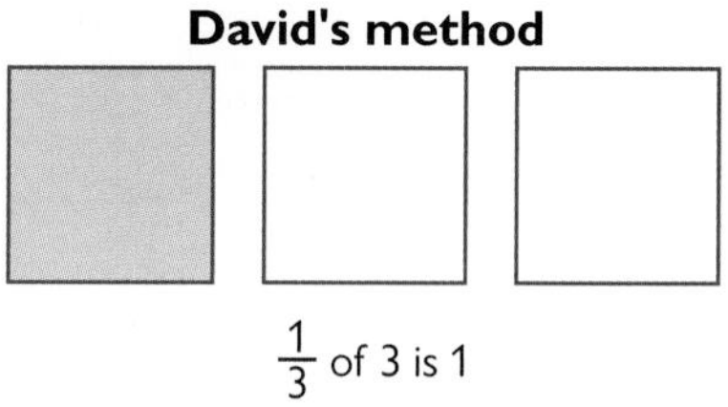

1 Solve each problem in the handout Reproducible R18.

2 Use Maya's method, David's method, or a method of your choice to solve each problem.

a. $\frac{5}{6}$ of 12 **b.** $\frac{2}{3}$ of 6

c. $\frac{3}{5}$ of 25 **d.** $\frac{1}{3}$ of 18

e. $\frac{5}{8}$ of 40 **f.** $\frac{5}{9}$ of 3

3 How did you solve each problem in part 2? Did you use the same method for every problem? Are some problems easier to solve using Maya's method than David's method?

Use Number Lines to Find Fractions of Wholes

How can you use a number line to find a fraction of a whole number?

To find $\frac{2}{3}$ of 6 on a number line, first draw a number line 6 units long. Then, divide the line into three equal sections. Each section is one third of the line. To find $\frac{2}{3}$ of 6, you will need two sections. So, $\frac{2}{3}$ of 6 is 4.

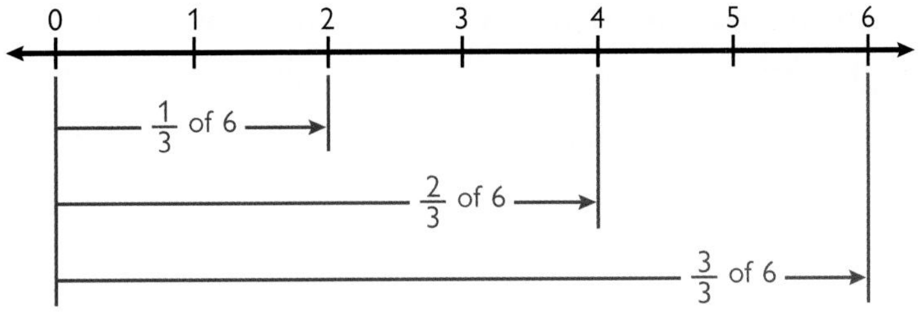

1 Use a number line to solve each problem.

a. $\frac{4}{5}$ of 20 **b.** $\frac{5}{6}$ of 36

c. $\frac{2}{9}$ of 18 **d.** $\frac{1}{6}$ of 24

e. $\frac{3}{4}$ of 16 **f.** $\frac{2}{5}$ of 10

2 Do you prefer to use drawings or number lines to find fractions of whole numbers? Explain.

Compute a Fraction of a Whole Number

After Kary and Jesse multiplied fractions using drawings and number lines, each of them devised a plan to multiply without using models.

Kary's Method	Jesse's Method
First, multiply the whole number by the numerator of the fraction. Then, divide the result by the denominator of the fraction.	First, divide the whole number by the denominator of the fraction. Then, multiply the result by the numerator of the fraction.

1 Use both Kary's method and Jesse's method to solve each problem from part **1** above using the number line.

2 Decide whether Kary, Jesse, or both are correct. If both are correct, which do you prefer?

3 Find $\frac{1}{6}$ of 15. Explain how you found your answer.

hot words
numerator
denominator
product

Homework
page 156

16 Fractions of Fractions

MULTIPLYING FRACTIONS

In this lesson, you will find fractions of fractions. When you find a fraction of a fraction, you are actually multiplying the numbers. Making rectangular models will help you find a method to multiply fractions.

Finding Fractions of Fractions

How can you find a fraction of a fraction?

You can use a rectangular model to multiply fractions.

1 For each problem, the rectangle represents one whole. Use the rectangle to find each solution. Write a multiplication statement for each problem.

a. $\frac{1}{4}$ of $\frac{1}{3}$

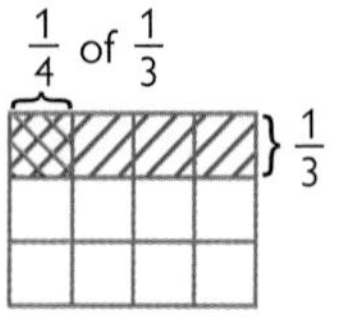

b. $\frac{1}{2}$ of $\frac{1}{4}$

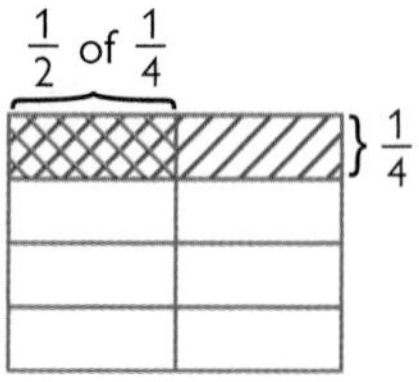

c. $\frac{4}{5}$ of $\frac{1}{3}$

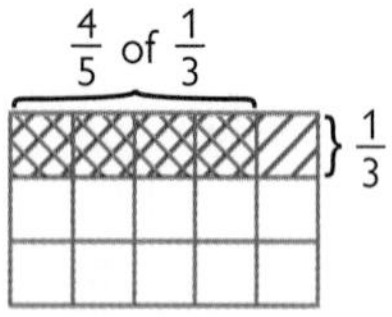

d. $\frac{2}{7}$ of $\frac{2}{3}$

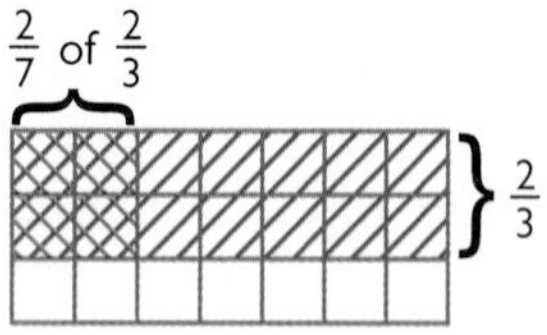

2 Study the numerators and denominators in the multiplication statements you wrote. Write a method for multiplying fractions.

Test a Hypothesis

Study the results of a variety of multiplication problems to determine how the first factor affects the product.

1 Make a table like the one below. Then, complete the table using the given problems. The factor type can be *whole number*, *improper fraction*, or *proper fraction*.

Problem	Answer	First Factor Type	Is Answer Greater or Less Than the Second Factor?
3 x 5	15	whole number	greater than
$\frac{3}{4}$ x 12	9	proper fraction	less than

a. $12 \times \frac{3}{4}$ **b.** $\frac{2}{3} \times 9$ **c.** $9 \times \frac{2}{3}$

d. $\frac{5}{4} \times 20$ **e.** $10 \times \frac{3}{2}$ **f.** $\frac{1}{4} \times \frac{1}{4}$

g. $\frac{4}{9} \times \frac{1}{3}$ **h.** $\frac{1}{3} \times \frac{4}{3}$ **i.** $\frac{3}{2} \times \frac{4}{3}$

j. $\frac{5}{4} \times \frac{12}{10}$ **k.** $0 \times \frac{2}{3}$ **l.** $6 \times \frac{5}{4}$

2 Highlight each row that has a *less than* in the last column. Use a different color to highlight each row that has a *greater than* in the last column.

3 Analyze your results. What type of factor(s) makes the product greater than the other number you were multiplying? What type of factor(s) makes the product less than the other number you were multiplying? Why do you think this is true?

When does a multiplication problem result in a product that is less than a factor?

Multiplying Fractions

To multiply two fractions use the following procedure.

$\frac{2}{3} \times \frac{3}{5}$

$\frac{2 \times 3}{3 \times 5} \rightarrow \frac{}{15}$ Multiply the denominators. This is like dividing each fifth into 3 pieces. Each piece is a fifteenth.

$\frac{2 \times 3}{3 \times 5} \rightarrow \frac{6}{15}$ Multiply the numerators. The numerators tell how many pieces you want. You have 3 fifths, and you want 2 of the 3 pieces in each fifth. You want a total of 6 pieces.

hot words
numerator
denominator
improper fraction

Homework
page 157

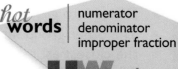

17 Estimation and Mixed Numbers

ESTIMATING RESULTS AND MULTIPLYING MIXED NUMBERS

Estimating is an important skill. In this lesson, you will estimate fraction products. As you get a better sense of how numbers work in fraction multiplication, you will be better at estimating fractions.

Estimate Fraction Products

Can you estimate fraction products?

For each multiplication problem, choose the estimate that you think will be the closest to the actual answer. Then, play "Best Estimate" with your classmates.

1 $\frac{10}{5} \times \frac{3}{4}$	**a.** $\frac{1}{2}$	**b.** 3	**c.** $1\frac{1}{2}$
2 $\frac{5}{4} \times 13$	**a.** 13	**b.** 15	**c.** 10
3 $\frac{1}{5} \times 26$	**a.** 100	**b.** $5\frac{4}{5}$	**c.** 5
4 $11 \times \frac{1}{5}$	**a.** 2	**b.** $2\frac{1}{2}$	**c.** 55
5 $\frac{3}{2} \times 48$	**a.** 16	**b.** 75	**c.** 148

Best Estimate

Get Ready

- Each student makes up 3 multiplication problems and solves the problems.
- Each student makes up 3 possible answers for each problem. One of the answers is close to the actual answer. The other two answers are not.
- Each student writes each problem plus its 3 answers on an index card.

Play the Game

- Divide into groups. Each student takes a turn presenting a problem.
- The other group members have 10 seconds to decide which of the answers is the closest estimate and write the choice on a piece of paper.
- All students who choose the best estimate receive a point.

Multiply Mixed Numbers

How can you multiply mixed numbers?

You know how to multiply fractions and whole numbers and how to multiply fractions and fractions. How can you multiply with mixed numbers?

1 Study the equations involving mixed numbers. Try to find a method or methods for multiplying mixed numbers. Be sure your methods work for all the examples.

$1\frac{1}{3} \times 2 = \frac{8}{3}$ or $2\frac{2}{3}$	$\frac{4}{5} \times 3\frac{2}{5} = \frac{68}{25}$ or $2\frac{18}{25}$
$2\frac{1}{2} \times 3 = \frac{15}{2}$ or $7\frac{1}{2}$	$\frac{1}{2} \times 1\frac{1}{2} = \frac{3}{4}$
$4 \times 1\frac{1}{8} = \frac{36}{8}$ or $4\frac{4}{8}$ or $4\frac{1}{2}$	$1\frac{1}{2} \times 1\frac{1}{2} = \frac{9}{4}$ or $2\frac{1}{4}$
$2\frac{1}{2} \times \frac{1}{2} = \frac{5}{4}$ or $1\frac{1}{4}$	$1\frac{1}{3} \times 1\frac{1}{4} = \frac{20}{12}$ or $1\frac{8}{12}$ or $1\frac{2}{3}$

2 Write a method or methods for multiplying mixed numbers. Be prepared to explain your methods to your teacher and classmates.

Write about Multiplying Mixed Numbers

What happens when you multiply a number by a mixed number? Is the product greater or less than the other number? Before writing your conclusion, make sure you try multiplying a mixed number by:

- proper fractions,
- improper fractions,
- whole numbers, and
- mixed numbers.

hot **words**
estimate
mixed number
improper fraction

Homework
page 158

18 Fraction Groups within Fractions

DIVIDING WITH FRACTIONS

You know how to multiply a number by a fraction. What happens when you divide a number by a fraction? What you have learned about multiplying fractions will be helpful when you divide fractions.

Find Fraction Groups in Whole Numbers

How can you divide a whole number by a fraction?

Here are two ways to model the problem $6 \div \frac{1}{2}$.

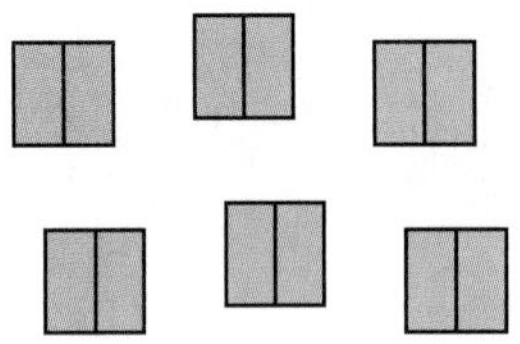

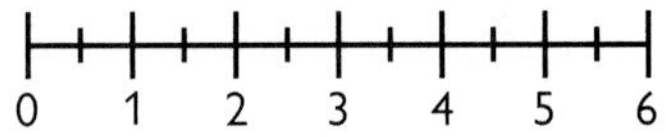

There are 12 halves in 6.

$$6 \div \frac{1}{2} = 12$$

1 Make either a drawing or a number line to find each quotient.

a. $6 \div \frac{1}{3}$ **b.** $6 \div \frac{1}{4}$ **c.** $6 \div \frac{1}{5}$

2 Look at the problems and answers for part **1**.

a. Describe a method that you could use to find the answers.

b. Use your method to find $8 \div \frac{1}{3}$ and $6 \div \frac{1}{6}$.

3 To find each quotient, refer to part **1**.

a. Use the drawing or number line for part **1a** to find $6 \div \frac{2}{3}$.

b. Use the drawing or number line for part **1b** to find $6 \div \frac{3}{4}$.

c. Use the drawing or number line for part **1c** to find $6 \div \frac{3}{5}$.

4 Look at the problems and answers for part **3**.

a. Describe a method that you could use to find the answers.

b. Use your method to find $8 \div \frac{2}{3}$ and $6 \div \frac{5}{6}$.

Divide Fractions and Mixed Numbers

Use what you have learned about fraction division to find each quotient.

1. $9 \div \frac{2}{3}$
2. $\frac{4}{5} \div 9$
3. $\frac{5}{6} \div \frac{8}{9}$
4. $\frac{3}{4} \div \frac{1}{2}$
5. $\frac{7}{10} \div \frac{4}{5}$
6. $\frac{1}{2} \div \frac{1}{4}$
7. $\frac{9}{5} \div \frac{2}{3}$
8. $\frac{5}{9} \div \frac{1}{4}$
9. $1\frac{1}{2} \div \frac{1}{8}$
10. $9 \div \frac{4}{3}$
11. $12 \div \frac{5}{4}$
12. $3\frac{3}{4} \div \frac{3}{8}$

How can you divide fractions?

Dividing Fractions

When you divide a number by a fraction, the denominator of the fraction breaks the number into more parts. So, you multiply by the denominator. Then, the numerator of the fraction tells you to regroup these parts. So, you divide by the numerator.

You can use this method to find $6 \div \frac{3}{4}$.

$6 \times 4 = 24$ This tells you there are a total of 24 parts.

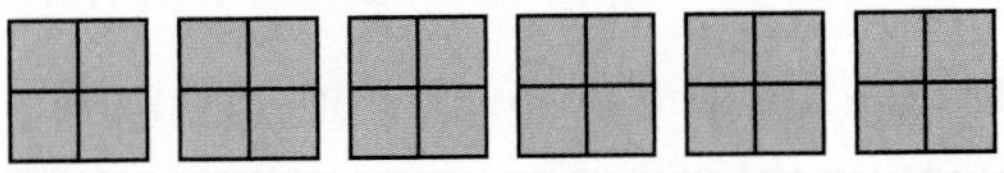

$24 \div 3 = 8$ There are 8 groups of 3 parts.

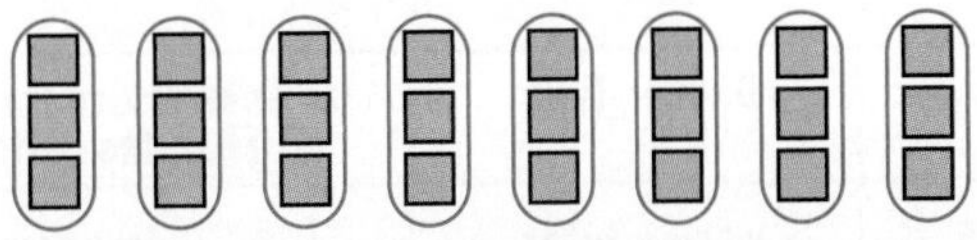

A shortcut for dividing with fractions is to rewrite the division problem as a number multiplied by the **reciprocal** of the divisor.

$$6 \div \frac{3}{4} = 6 \times \frac{4}{3} = \frac{6}{1} \times \frac{4}{3} = \frac{24}{3} \text{ or } 8$$

This shortcut works with any division problem.

$14 \div 7 = \frac{14}{1} \div \frac{7}{1} = \frac{14}{1} \times \frac{1}{7} = \frac{14}{7}$ or 2

$\frac{2}{3} \div \frac{3}{4} = \frac{2}{3} \times \frac{4}{3} = \frac{8}{9}$

$\frac{5}{6} \div 2 = \frac{5}{6} \div \frac{2}{1} = \frac{5}{6} \times \frac{1}{2} = \frac{5}{12}$

$1\frac{1}{2} \div \frac{6}{5} = \frac{3}{2} \div \frac{6}{5} = \frac{3}{2} \times \frac{5}{6} = \frac{15}{12}$ or $1\frac{1}{4}$

hot words | reciprocal

Homework page 159

19 Understanding Fraction Division

FINDING TRENDS AND ESTIMATING RESULTS

What trends can you discover in division with fractions? In this lesson, you will use a familiar investigation to learn more about division by fractions.

What happens when a number is divided by a number between 0 and 1?

Explore the Effects of Division

Sue said, "We have already learned that when you multiply a number by a number that is greater than one, the product is greater than the other number. When you multiply by a proper fraction, the product is less than the other number. I think the opposite is true for division."

Is Sue correct? The following investigation will help you decide.

1 Make a table like the one below. Then, complete the table using the given problems. The divisor type can be *whole number, improper fraction, proper fraction,* or *mixed number.*

Problem	Answer	Divisor Type	Is Answer Greater or Less Than the Dividend?
$10 \div 2$	5	whole number	less than
$10 \div \frac{1}{2}$	20	proper fraction	greater than

a. $\frac{3}{4} \div 4$ **b.** $\frac{5}{12} \div 6$ **c.** $\frac{1}{2} \div \frac{1}{4}$

d. $\frac{1}{8} \div \frac{1}{3}$ **e.** $\frac{4}{3} \div \frac{1}{3}$ **f.** $\frac{2}{9} \div \frac{5}{4}$

g. $\frac{5}{8} \div \frac{3}{2}$ **h.** $1\frac{1}{2} \div \frac{1}{8}$ **i.** $\frac{4}{5} \div 3\frac{1}{4}$

j. $2\frac{1}{4} \div 1\frac{1}{2}$ **k.** $5 \div 1\frac{1}{3}$ **l.** $\frac{5}{9} \div \frac{5}{6}$

2 Analyze your results. What type(s) of divisors make the quotient greater than the dividend? What type(s) of divisors make the quotient less than the dividend? Why do you think this is true?

Estimate the Quotients

Can you estimate fraction quotients?

For each division problem, choose the estimate that you think will be closer to the real answer. Then play "Better Estimate" with your classmates.

1. $4 \div \frac{1}{8}$ a. 32 b. $\frac{1}{2}$
2. $3\frac{1}{2} \div \frac{1}{3}$ a. 21 b. 10
3. $4 \div \frac{5}{4}$ a. $4\frac{1}{4}$ b. 3
4. $\frac{5}{9} \div \frac{1}{4}$ a. 2 b. $\frac{1}{2}$
5. $25 \div \frac{5}{6}$ a. 23 b. 30
6. $\frac{1}{2} \div \frac{25}{12}$ a. $\frac{1}{4}$ b. 20

Better Estimate

Get Ready

- Each student makes up three division problems and solves the problems. The problems should include whole numbers, proper fractions, improper fractions, and mixed numbers.
- Each student makes up two possible answers for each problem. One of the answers is close to the actual answer. The other answer is an inaccurate guess or poor estimate.
- Each student writes each problem plus the two answers on an index card or quarter-sheet of paper.

Play the Game

- Divide into groups. Each student takes a turn presenting a problem.
- The other group members have 15 seconds to decide which of the answers is the closer estimate and write the choice on a piece of paper.
- When time is up, everyone holds up his or her answer choice. All students who choose the better estimate receive a point.
- The player with the most points after all problems have been presented wins.

hot words
mixed number
improper fraction
estimate

HW Homework
page 160

20 Multiplication vs. Division

APPLYING FRACTION MULTIPLICATION AND DIVISION

Now, you are familiar with how to multiply and divide fractions. In this lesson, you will apply what you have learned about fractions in a game that combines both operations.

How can you multiply and divide fractions to get the least answer?

Play "Get Small"

Play a round of "Get Small" on your own or with a partner. Make a recording sheet like the one below to show your work.

First Fraction	Second Fraction	× or ÷	Resulting Fraction
$\frac{5}{4}$	$\frac{3}{4}$	×	$\frac{15}{16}$
$\frac{15}{16}$	$\frac{3}{2}$	÷	$\frac{30}{48}$ or $\frac{5}{8}$

Get Small

In this game, the player rolls a number cube to create fractions. These fractions are multiplied or divided to create the least possible number. Use the following rules to play the game.

- The player rolls the number cube and places the result in either the numerator or the denominator of the first fraction.
- Then, the player rolls the number cube to determine the other part of the first fraction.
- The player rolls two more times to create a second fraction.
- The player decides whether to multiply or divide the two fractions. The result becomes the first fraction for the next round.
- As before, the player rolls twice to create the second fraction for the next round. Then, he or she decides to multiply or divide.
- The player continues the game for 10 rounds. The object of the game is to end with the least possible number.

Find the Relationship between Multiplication and Division

How are division and multiplication of fractions related?

You and your partner will solve two similar sets of problems. One of you will work on multiplication and the other will work on division. When you have both finished, compare your work. What relationships do you see between the problems?

Student A

1. $\frac{1}{2} \times \frac{1}{3}$
2. $\frac{5}{6} \times \frac{3}{4}$
3. $\frac{4}{3} \times \frac{1}{3}$
4. $1\frac{1}{5} \times \frac{2}{5}$
5. $5 \times \frac{1}{4}$
6. $\frac{7}{5} \times 3$
7. $\frac{3}{8} \times \frac{1}{4}$
8. $\frac{5}{4} \times \frac{3}{2}$
9. $3\frac{1}{2} \times 4$
10. $\frac{4}{5} \times \frac{20}{3}$

Student B

1. $\frac{1}{6} \div \frac{1}{2}$
2. $\frac{5}{8} \div \frac{5}{6}$
3. $\frac{4}{9} \div \frac{4}{3}$
4. $\frac{12}{25} \div 1\frac{1}{5}$
5. $\frac{5}{4} \div \frac{1}{4}$
6. $\frac{21}{5} \div 3$
7. $\frac{3}{32} \div \frac{1}{4}$
8. $\frac{15}{8} \div \frac{5}{4}$
9. $14 \div 4$
10. $\frac{16}{3} \div \frac{20}{3}$

Solve Real-Life Problems

Solve each problem using multiplication or division.

1. There are $3\frac{3}{4}$ pies left after Julia's Fourth of July party. Her family has 4 members. How much pie will each family member get if they divide the remaining pies evenly?
2. Joe earns \$8.00 an hour mowing lawns. He has spent $5\frac{1}{4}$ hours mowing lawns this week. How much money did he earn?
3. Keisha reads 3 books each week. On average, how many books does she read in a day?
4. Claire has a CD that takes $\frac{3}{4}$ of an hour to play. How long will it take to play the CD 7 times?

hot words
numerator
denominator

Homework
page 161

HW Homework 1

Shapes and Factors

Applying Skills

Determine how many rectangles can be made from each number of tiles. Give the dimensions for each rectangle.

1. 14 tiles

2. 18 tiles

3. 23 tiles

4. 25 tiles

The statements in items 5 and 6 are about the rectangle below. Complete each statement.

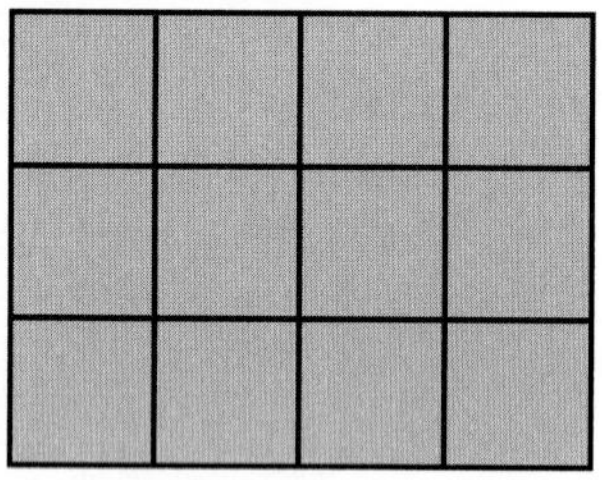

5. This rectangle's ___?___ are 3 centimeters by 4 centimeters.

6. The two ___?___ of 12 that this rectangle models are 3 and 4.

7. What two factors are common to every even number?

8. What factor is common to every number?

Suppose you are making solid rectangular prisms using 12, 15, 22, 27, 30, 40, and 48 cubes. Do not use 1 as a dimension.

9. Which numbers of cubes could form a rectangular solid that is 2 cubes high?

10. Which numbers of cubes could form a rectangular solid that is 3 cubes high?

Extending Concepts

Making a factor tree is one way to show the prime factorization of a number. Each branch of the tree shows the factors of the number above it. Two possible factor trees for the number 48 are shown below.

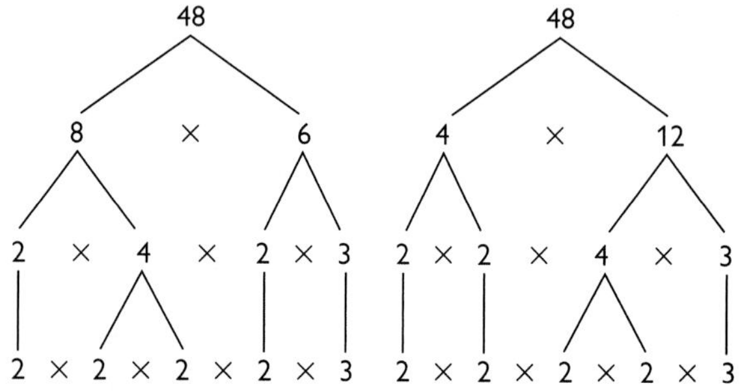

Make a factor tree for each number. Then write the number's prime factors.

11. 30

12. 81

13. 100

14. 66

15. 120

16. 250

Making Connections

17. Tamara wants to put a tiled patio in her backyard that is 8 feet by 12 feet. She plans to use 1-foot-square paving tiles. How many tiles does she need?

18. Masao built a raised planter box and needs to buy soil to put into it. The box is 4 feet wide and 8 feet long. He wants to fill it with soil to a depth of 2 feet. How many cubic feet of soil will he need?

Homework 2: The Great Factor Hunt

Applying Skills

Write each number as a product of its prime factors. Do not include the factor 1.

1. 12 **2.** 16 **3.** 21

4. 26 **5.** 30 **6.** 45

List all the factors other than 1 that each pair of numbers has in common.

7. 10 and 15 **8.** 8 and 19

9. 14 and 28 **10.** 6 and 24

11. 16 and 26 **12.** 40 and 50

Find the greatest common factor of each pair of numbers.

13. 4 and 6 **14.** 9 and 16

15. 12 and 24 **16.** 30 and 45

17. 20 and 50 **18.** 36 and 48

Extending Concepts

For items 19 to 24, use the number map below. Each time you move from one number to another, you multiply. You may move in any direction along the dotted line, but you may use each number only once. For example, the value of the highlighted path is 1 × 5 × 4 × 11 or 220.

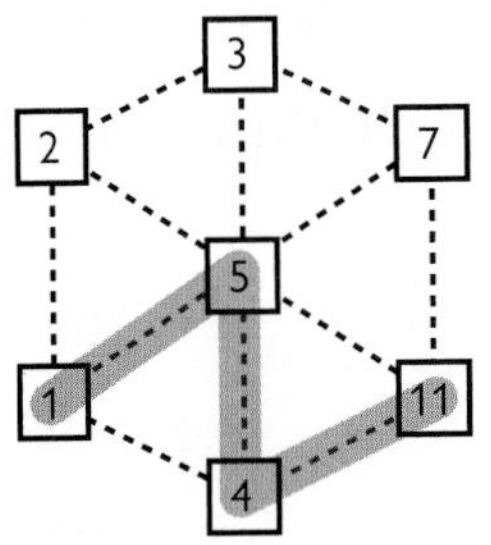

19. What is the longest path you can find? What is its value?

20. Can you find a path that is shorter than the one for item 19, but has the same value?

21. How can you make sure that a path's value will be an even number?

22. How can you make sure that a path's value will be an odd number?

23. Can you find a two-number path with a value that is prime? If possible, give an example.

24. Can you find a three-number path with a value that is prime? If possible, give an example.

Writing

25. Make up your own number map. In your map, include only prime numbers. Use the pattern below or make up your own. Write two questions to go with your map and give their answers.

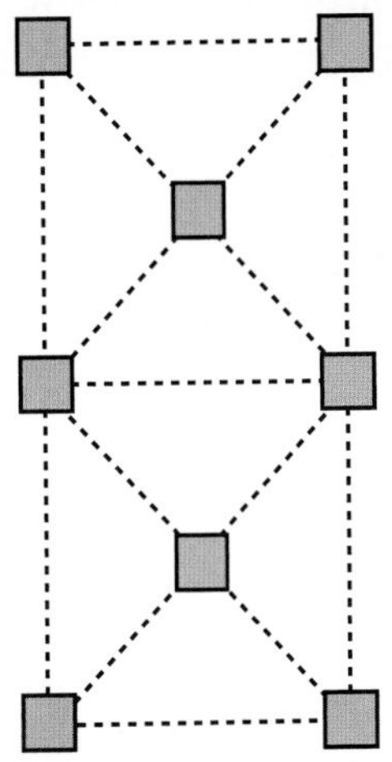

Homework 3

Multiple Approaches

Applying Skills

Write the first five multiples of each number.

1. 3

2. 8

3. 13

4. 25

5. 40

6. 100

7. Determine whether 150 is a multiple of 6. Explain.

8. List three common multiples of 2 and 5.

9. List three common multiples of 3 and 9.

Find the least common multiple of each pair of numbers.

10. 6 and 8

11. 5 and 7

12. 9 and 12

13. 12 and 18

14. 9 and 21

15. 45 and 60

16. 11 and 17

17. 16 and 42

18. When will the least common multiple of two numbers be one of the numbers?

19. Suppose the only factor two numbers have in common is 1. How can you find their least common multiple?

Extending Concepts

To find the least common multiple of three numbers, first find the least common multiple of two of the numbers. Then find the least common multiple of that number and the third number.

Consider the numbers 10, 25, and 75. The least common multiple of 10 and 25 is 50. The least common multiple of 50 and 75 is 150. Therefore, the least common multiple of 10, 25, and 75 is 150.

Find the least common multiple of each set of numbers.

20. 16, 24, and 26

21. 15, 30, and 60

22. 4, 22, and 60

23. 14, 35, and 63

24. 9, 12, and 15

25. 15, 25, and 35

Making Connections

Our calendar year is based on the length of time Earth takes to travel around the Sun. The solar year is not exactly 365 days long. Therefore, every fourth year is a leap year with 366 days. However, each solar year is not precisely $365\frac{1}{4}$ days. So, leap year is skipped every 400 years. Therefore, leap years are the years that are multiples of 4, but are not multiples of 400.

Determine which of the following years were leap years.

26. 1600

27. 1612

28. 1650

29. 1698

30. 1700

31. 1820

32. 1908

33. 2000

First Things First

Applying Skills

Solve for n in each equation.

1. $5 + 3 \times 7 = n$

2. $6 \div 3 + 4 = n$

3. $10 - 8 \div 2 + 3 = n$

4. $(6 - 3) \times 2 = n$

5. $3 \times 4 \div 6 + 2 = n$

6. $2 + 3 \times 4 = n$

7. $6 - 2 \times 3 = n$

8. $9 \div 3 + 4 = n$

9. $(3 \times 2)^2 - 10 = n$

10. $(3 + 6)^2 = n$

11. $5 \times (4 + 2)^2 = n$

12. $(5^2 + 3) \div 7 = n$

13. $13 + 6 - 4 + 12 = n$

14. $7 \times 9 - (4 + 3) = n$

15. $4 + 2(8 - 6) = n$

16. $3(4 + 7) - 5 \times 4 = n$

17. $19 + 45 \div 3 - 11 = n$

18. $9 + 12 - 6 \times 4 \div 2 = n$

19. $26 \div 13 + 9 \times 6 - 21 = n$

20. $27 - 8 \div 4 \times 3 = n$

Place each of the numbers 2, 5, and 6 into each equation so that the equation is true.

21. $\underline{\ ?\ } + \underline{\ ?\ } \times \underline{\ ?\ } = 32$

22. $\underline{\ ?\ } - \underline{\ ?\ } \div \underline{\ ?\ } = 2$

23. $\underline{\ ?\ } \times \underline{\ ?\ } \div \underline{\ ?\ } = 15$

24. $\underline{\ ?\ }^2 \div \underline{\ ?\ } - \underline{\ ?\ } = 13$

25. $\underline{\ ?\ }^2 \div \underline{\ ?\ }^2 + \underline{\ ?\ } = 14$

26. $(\underline{\ ?\ } + \underline{\ ?\ }) \times \underline{\ ?\ } = 22$

Extending Concepts

Imagine you are playing Hit the Target. For each set of building numbers, write an equation with the target number as the result.

27. building numbers: 1, 4, 5
target number: 2

28. building numbers: 2, 3, 6
target number: 1

29. building numbers: 3, 4, 6
target number: 2

30. building numbers: 3, 4, 6
target number: 6

Writing

31. Answer the letter to Dr. Math.

Dear Dr. Math:
My teacher gave our class a really easy problem to solve. Here is the problem.

$$16 - 8 \div 4 + 3 = n$$

Priscilla Pemdas got 17 for the answer—crazy, huh? Priscilla said my answer, which was 5 of course, was wrong. I know I added, subtracted, and divided correctly. But the teacher agreed with Priscilla! What is going on?
D. Finite Disorder

Homework 5

Putting It All Together

Applying Skills

For each pair, tell whether expression A has the same answer as expression B.

1. **A.** $5 \times 6 + 11$
B. $5 \times 11 + 6$

2. **A.** $48 \div 12 \div 2$
B. $48 \div (12 \div 2)$

3. **A.** $(30 - 11) - 2$
B. $30 - (11 - 2)$

4. **A.** $3 \times 5 \times 7$
B. $7 \times 3 \times 5$

5. **A.** $50 \div 2 + 25$
B. $50 \div 25 + 2$

6. **A.** 25×5
B. $20 \div 5 + 5 \times 5$

7. **A.** $3 + 3 \times 10$
B. $(3 + 3) \times 10$

Solve each problem.

8. Angie is thinking of two numbers. Their sum is 50. Each of the numbers has 10 as a factor. Both numbers are greater than 10. What are the numbers?

9. Julio has read two books so far this month. One book has twice as many pages as the other. All together, the two books have 300 pages. How many pages does each book have?

10. Mr. Robinson has two children. One child is 3 times as old as the other. The sum of their ages is 16. What are the ages of the children?

11. Atepa has two cats of different ages. The least common multiple of their ages is 24. The greatest common factor of their ages is 4. How old are his cats?

12. Marcus is thinking of three numbers. Their greatest common factor is 6. Their sum is 36. What are the numbers?

Extending Concepts

Debbie, Mariana, Eduardo, and Gary all love to read. For items 13–16, use the clues to determine how many pages each of them has read so far this month.

- **Debbie has read 300 pages.**
- **The number of pages read by Eduardo and Debbie share a common factor, 50.**
- **Gary has read twice as many pages as Debbie.**
- **All together, the four students have read 1,425 pages.**
- **The least common multiple of the number of pages read by Mariana and Eduardo is 350.**

13. Debbie

14. Mariana

15. Eduardo

16. Gary

Writing

17. Use the ages of some family members, friends, or fictitious people to create a "guess my number" problem. Your problem should include the terms *factor* and/or *multiple.* Write the problem and its solution.

Homework 6

Designer Fractions

Applying Skills

For items 1–6, use a grid.

1. Shade $\frac{1}{2}$ of a 4-by-4 rectangle. How many squares are shaded?

2. Shade $\frac{1}{4}$ of a 4-by-4 rectangle. How many squares are shaded?

3. Shade $\frac{1}{8}$ of a 4-by-4 rectangle. How many squares are shaded?

4. Shade $\frac{1}{3}$ of a 4-by-3 rectangle. How many squares are shaded?

5. Shade $\frac{1}{6}$ of a 4-by-3 rectangle. How many squares are shaded?

6. Shade $\frac{1}{12}$ of a 4-by-3 rectangle. How many squares are shaded?

Write a fraction that describes the part of the whole represented by each color.

7.

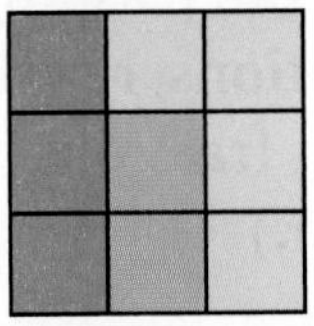

8.

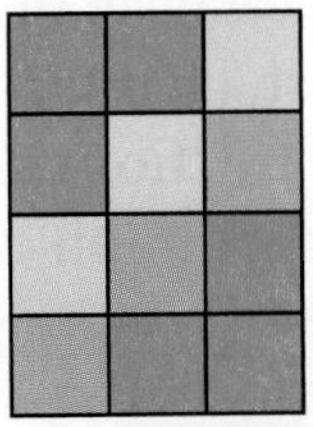

9.

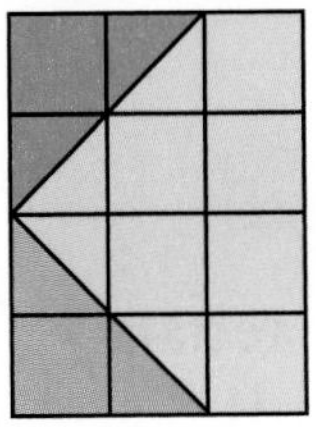

10.

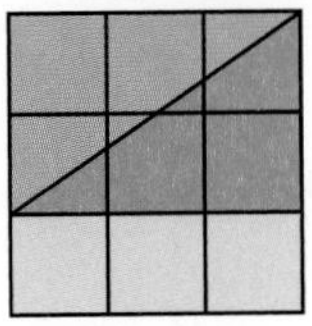

11.

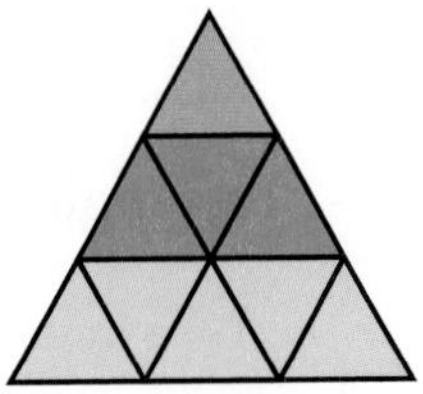

12.

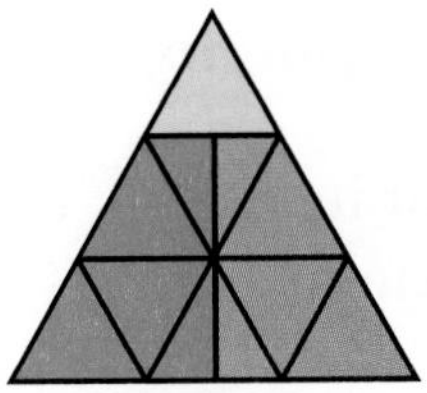

13.

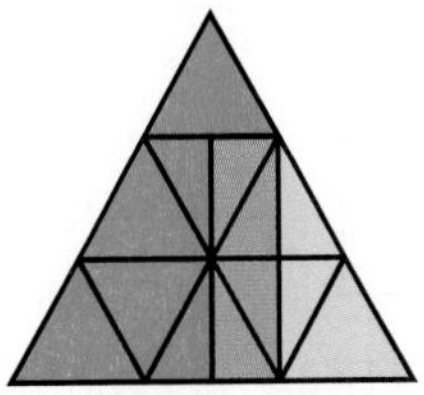

14.

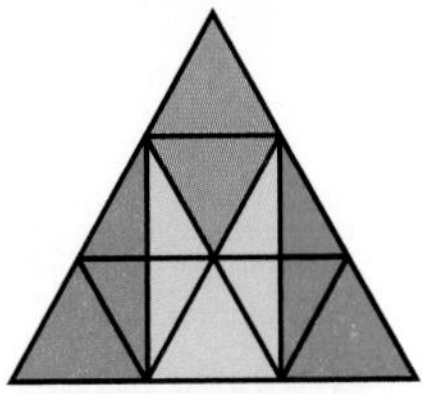

Extending Concepts

Answer each question.

15. If $\frac{1}{3}$ of a rectangle is shaded, what fraction of the rectangle is *not* shaded?

16. If $\frac{3}{7}$ of a rectangle is shaded, what fraction of the rectangle is *not* shaded?

Making Connections

17. The quilt pattern below is called Jacob's Ladder. Each block contains 9 same-sized squares. Each square is made of smaller squares or triangles. What fraction of the block is made of blue fabric? red fabric? white fabric?

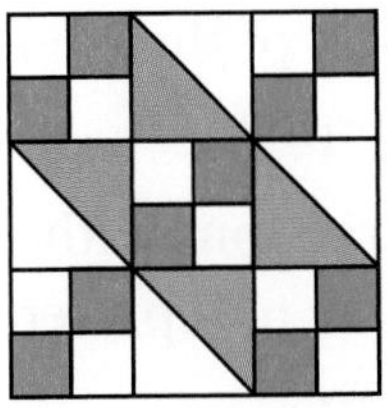

Area Models and Equivalent Fractions

Homework 7

Applying Skills

For items 1–4, use the 4-by-3 rectangle below.

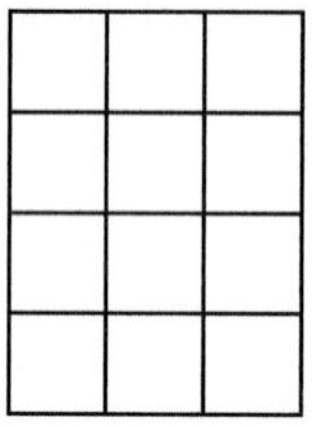

1. What fraction of the whole rectangle does one column represent?
2. What fraction of the whole rectangle do two rows represent?
3. Write two fractions with common denominators to represent one column and two rows.
4. Which fraction is greater?

For items 5–8, use the 2-by-7 rectangle below.

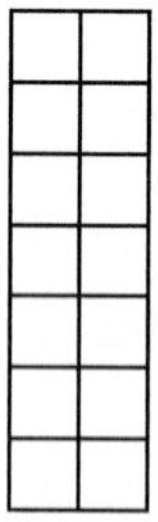

5. What fraction of the whole rectangle does one column represent?
6. What fraction of the whole rectangle do three rows represent?
7. Write two fractions with common denominators to represent one column and three rows.
8. Which fraction is greater?

Make a grid sketch for each pair of fractions. Label each sketch with the given fractions and their equivalent fractions. Compare the fractions.

9. $\frac{5}{8}$ and $\frac{3}{5}$
10. $\frac{5}{9}$ and $\frac{2}{3}$
11. $\frac{3}{4}$ and $\frac{6}{7}$
12. $\frac{3}{5}$ and $\frac{4}{7}$

For each pair of fractions, find equivalent fractions that have a common denominator. Compare the fractions.

13. $\frac{5}{12}$ and $\frac{1}{6}$
14. $\frac{2}{5}$ and $\frac{1}{4}$
15. $\frac{2}{7}$ and $\frac{3}{8}$
16. $\frac{1}{3}$ and $\frac{4}{7}$

Extending Concepts

Order each set of fractions from least to greatest. (*Hint:* To order fractions, rewrite all the fractions as equivalent fractions with a common denominator.)

17. $\frac{1}{2}, \frac{3}{5}, \frac{5}{6}$, and $\frac{2}{3}$
18. $\frac{1}{2}, \frac{4}{5}, \frac{2}{5}$, and $\frac{3}{4}$

Making Connections

19. Crystal's entire garden is a rectangle measuring 5 yards by 6 yards. She plans to plant a vegetable garden in $\frac{2}{3}$ of this area. How many square yards will she plant in vegetables? (Include a sketch with your solution.)

Homework 8: Fraction Lineup

Applying Skills

Make a number line like the one shown below. Place each fraction on the number line.

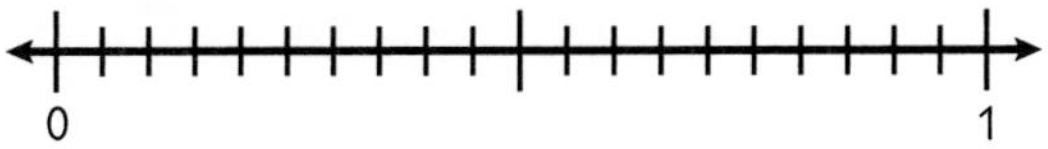

1. $\frac{1}{2}$ **2.** $\frac{3}{4}$

3. $\frac{4}{5}$ **4.** $\frac{3}{10}$

5. $\frac{14}{20}$ **6.** $\frac{3}{5}$

Write an equivalent fraction for each fraction.

7. $\frac{1}{2}$ **8.** $\frac{2}{10}$

9. $\frac{15}{20}$ **10.** $\frac{6}{10}$

11. $\frac{4}{5}$ **12.** $\frac{7}{10}$

Use a ruler to compare each pair of fractions. Write an expression using <, > or = to show the comparison.

13. $\frac{3}{8}$ and $\frac{1}{4}$ **14.** $\frac{8}{16}$ and $\frac{1}{2}$

15. $1\frac{1}{4}$ and $1\frac{3}{8}$ **16.** $2\frac{3}{4}$ and $2\frac{11}{16}$

17. $1\frac{1}{2}$ and $1\frac{9}{16}$ **18.** $\frac{3}{16}$ and $\frac{1}{8}$

Extending Concepts

19. Make a number line like the one shown below.

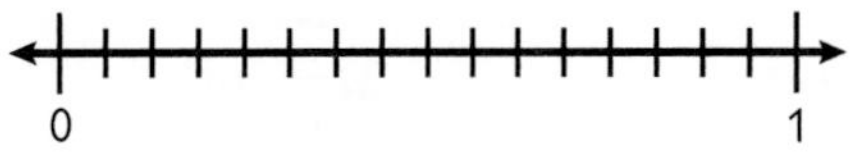

a. Label the number line with the fractions $\frac{1}{2}$, $\frac{1}{4}$, and $\frac{1}{8}$.

b. What would be the next fraction in the pattern? Label its position.

c. What would be the next two fractions in the pattern?

20. Make a number line like the one shown below. Then, locate each fraction.

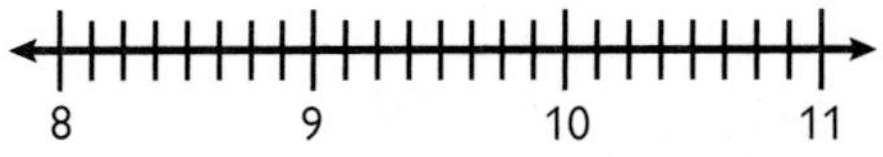

a. $8\frac{1}{2}$ **b.** $10\frac{3}{4}$

c. $9\frac{1}{8}$ **d.** $8\frac{7}{8}$

Making Connections

Americans use fractions in daily life when measuring lengths and distances. Complete each sentence. Use 12 inches = 1 foot and 3 feet = 1 yard.

21. $1\frac{2}{3}$ yards is the same as ___?___ feet.

22. 10 feet is the same as ___?___ yards.

23. $\frac{1}{4}$ foot is the same as ___?___ inches.

24. 18 inches is the same as ___?___ feet.

25. $1\frac{1}{4}$ yards is the same as ___?___ inches.

Homework 9

Focus on Denominators

Applying Skills

Find the least common denominator of each pair of fractions.

1. $\frac{2}{3}$ and $\frac{4}{15}$

2. $\frac{7}{8}$ and $\frac{3}{10}$

3. $\frac{7}{10}$ and $\frac{3}{5}$

4. $\frac{5}{7}$ and $\frac{2}{3}$

5. $\frac{7}{24}$ and $\frac{5}{12}$

6. $\frac{7}{8}$ and $\frac{5}{6}$

7. $\frac{3}{10}$ and $\frac{4}{15}$

8. $\frac{7}{12}$ and $\frac{5}{8}$

Compare each pair of fractions using $>, <$ or $=$.

9. $\frac{4}{7}$ and $\frac{5}{8}$

10. $\frac{1}{3}$ and $\frac{3}{15}$

11. $\frac{4}{5}$ and $\frac{6}{7}$

12. $\frac{3}{8}$ and $\frac{5}{12}$

13. $\frac{3}{4}$ and $\frac{5}{8}$

14. $\frac{16}{20}$ and $\frac{40}{50}$

15. $\frac{5}{6}$ and $\frac{7}{9}$

16. $\frac{3}{16}$ and $\frac{1}{6}$

Order each set of fractions from least to greatest.

17. $\frac{2}{3}, \frac{5}{12},$ and $\frac{5}{6}$

18. $\frac{5}{16}, \frac{1}{4},$ and $\frac{9}{32}$

19. $\frac{32}{16}, \frac{1}{2},$ and $\frac{15}{16}$

20. $1\frac{1}{8}, \frac{5}{4},$ and $\frac{17}{16}$

Extending Concepts

Use each fraction once to correctly complete each statement.

$$\frac{5}{16} \quad \frac{41}{48} \quad \frac{37}{48} \quad \frac{17}{24}$$

21. $\frac{5}{6} < \underline{\ ?\ } < \frac{7}{8}$

22. $\frac{2}{3} < \underline{\ ?\ } < \frac{3}{4}$

23. $\frac{1}{4} < \underline{\ ?\ } < \frac{3}{4}$

24. $\frac{9}{12} < \underline{\ ?\ } < \frac{19}{24}$

Place each of the nine numbers as a numerator or denominator so that all of the equations are true. Use each of the nine numbers only once.

1 2 3 4 5 6 7 8 9

25. $\frac{?}{?} = \frac{?}{10}$

26. $\frac{2}{?} = \frac{?}{14}$

27. $\frac{2}{?} = \frac{?}{12}$

28. $\frac{8}{12} = \frac{?}{?}$

Writing

29. Answer the letter to Dr. Math.

Dear Dr. Math:
We are doing a unit about fractions, but our first phase was not even about fractions! It was just about whole-number stuff like factors and multiples. What do greatest common factors and least common multiples have to do with fractions anyway?
Unclear N. Concept

Homework 10

Sums and Differences on the Line

Applying Skills

Find each sum or difference. Use a number line if you wish.

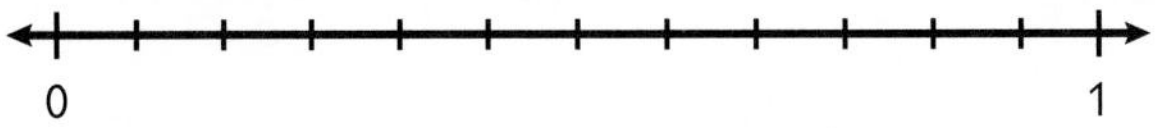

1. $\frac{2}{3} + \frac{1}{12}$

2. $\frac{1}{6} + \frac{1}{3}$

3. $\frac{3}{4} + \frac{1}{12}$

4. $\frac{5}{12} + \frac{1}{4}$

5. $\frac{1}{2} + \frac{1}{12}$

6. $\frac{1}{4} + \frac{2}{3}$

7. $\frac{5}{6} + \frac{1}{4}$

8. $1 - \frac{5}{6}$

9. $\frac{3}{4} - \frac{1}{3}$

10. $\frac{5}{6} - \frac{1}{2}$

11. $\frac{11}{12} - \frac{7}{12}$

12. $\frac{12}{12} - \frac{1}{3}$

13. $\frac{1}{4} - \frac{1}{6}$

14. $\frac{11}{12} - \frac{2}{3}$

Answer each question. Use a ruler if you wish.

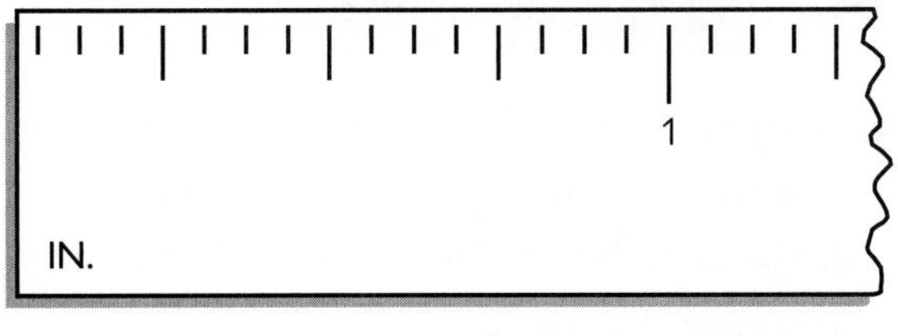

15. What is the sum of $\frac{5}{16}$ inch and $\frac{1}{4}$ inch?

16. What is the sum of $\frac{1}{2}$ inch and $\frac{3}{8}$ inch?

17. How much longer than $\frac{7}{8}$ inch is $\frac{3}{4}$ inch?

18. How much longer than $\frac{9}{16}$ inch is $\frac{7}{8}$ inch?

Extending Concepts

19. Find a path that has a sum equal to 2. You may enter any open gate at the top, but you must exit from the gate at the bottom.

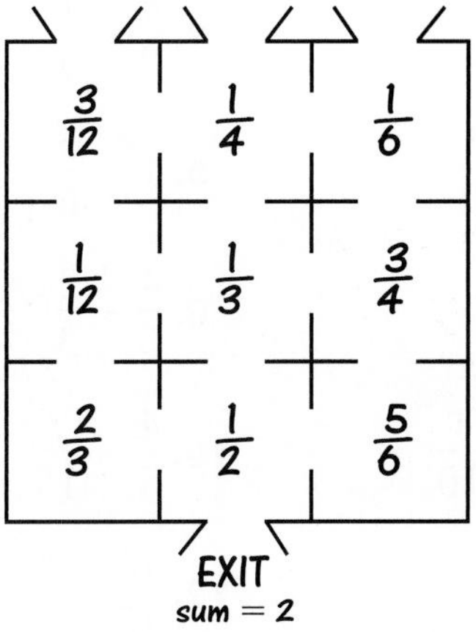

Making Connections

20. Theo has a growth chart in the doorway of his closet. He likes to check his height every three months. In the year between his tenth and eleventh birthdays, he grew $\frac{1}{2}$ inch, $\frac{3}{4}$ inch, $\frac{3}{8}$ inch, and $\frac{5}{8}$ inch. How much did he grow that year?

21. In the thirteenth century, people used a unit of measure less than an inch. It was called a barleycorn because it was about the length of a barley seed. Three barleycorns make an inch.

a. What is the length in inches of 1 barleycorn? 5 barleycorns? 9 barleycorns?

b. Find the sum in inches of 1 barleycorn, 5 barleycorns, and 9 barleycorns.

Homework 11

Numbers Only

Applying Skills

Find each sum or difference.

1. $\frac{4}{9}+\frac{1}{3}$

2. $1-\frac{3}{8}$

3. $\frac{5}{6}+\frac{5}{24}$

4. $\frac{8}{9}-\frac{1}{3}$

5. $\frac{5}{12}+\frac{1}{8}$

6. $1-\frac{5}{6}$

7. $\frac{5}{6}-\frac{1}{3}$

8. $\frac{4}{5}+\frac{2}{3}$

9. $\frac{9}{40}-\frac{1}{10}$

10. $\frac{4}{5}-\frac{2}{15}$

11. $\frac{5}{12}+\frac{4}{8}$

12. $\frac{14}{15}-\frac{1}{6}$

13. $\frac{3}{7}+\frac{9}{14}$

14. $\frac{7}{10}-\frac{1}{6}$

15. $\frac{5}{8}-\frac{1}{2}$

16. $\frac{3}{7}+\frac{4}{5}$

17. $\frac{4}{5}-\frac{1}{6}$

18. $\frac{5}{8}-\frac{7}{12}$

19. $\frac{5}{9}+\frac{5}{6}$

20. $\frac{3}{5}+\frac{1}{15}$

21. Find $\frac{5}{8}$ minus $\frac{5}{12}$.

22. Find the sum of $\frac{9}{10}$ and $\frac{4}{15}$.

23. How much more is $\frac{3}{4}$ than $\frac{2}{3}$?

24. What is the sum of $\frac{3}{8}$ and $\frac{5}{6}$?

25. Evaluate $\frac{5}{8}-\frac{1}{4}$.

26. What is the sum of $\frac{2}{3}$, $\frac{5}{8}$, and $\frac{7}{12}$?

Extending Concepts

Find each missing fraction.

27. $\frac{4}{5}+\frac{?}{\quad}=1$

28. $\frac{?}{\quad}+\frac{2}{3}=1$

29. $\frac{2}{3}-\frac{?}{\quad}=\frac{4}{9}$

30. $\frac{?}{\quad}-\frac{3}{8}=\frac{3}{16}$

31. $\frac{?}{\quad}-\frac{3}{4}=\frac{1}{12}$

32. $\frac{1}{5}+\frac{?}{\quad}=\frac{13}{20}$

33. Find the path that has a sum equal to 1. You may enter any open gate at the top, but you must exit from the gate at the bottom.

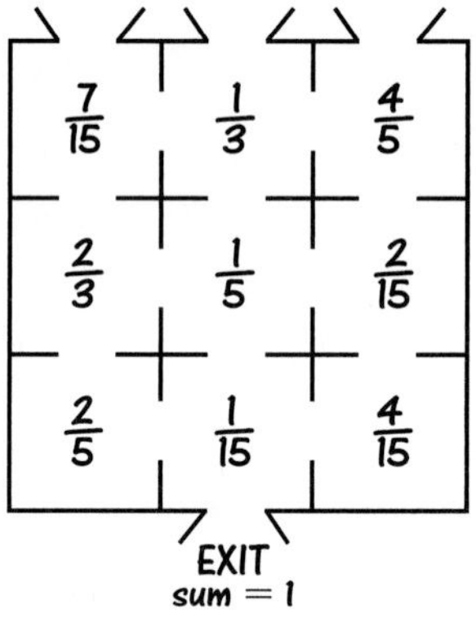

Making Connections

34. Benjamin Franklin liked to make up magic squares. His magic squares could be as large as 16 by 16. Here is a smaller one. Arrange the numbers,

$$\frac{1}{12}\ \frac{1}{6}\ \frac{1}{4}\ \frac{1}{3}\ \frac{5}{12}\ \frac{1}{2}\ \frac{7}{12}\ \frac{2}{3}\ \frac{3}{4}$$

into nine boxes so that the sum of each column, row, and diagonal is $1\frac{1}{4}$.

12 Homework

Not Proper but Still Okay

Applying Skills

Write each number in a different form: improper fraction, mixed number, or whole number.

1. $\frac{7}{3}$

2. $3\frac{5}{8}$

3. $\frac{55}{11}$

4. $5\frac{2}{5}$

5. $3\frac{7}{8}$

6. $\frac{24}{7}$

7. $\frac{35}{5}$

8. $11\frac{11}{12}$

9. $3\frac{9}{10}$

10. $\frac{44}{15}$

Find each sum. If the problem is written with improper fractions, write the sum as an improper fraction. If the problem is written with mixed numbers, write the sum as a mixed number. If the problem has both types of numbers, write the sum as an improper fraction and as a mixed number.

11. $2\frac{3}{4} + 3\frac{1}{2}$

12. $5\frac{1}{4} + 2\frac{5}{12}$

13. $2\frac{1}{2} + 5\frac{2}{3}$

14. $11\frac{3}{4} + 9\frac{3}{5}$

15. $\frac{5}{3} + \frac{11}{6}$

16. $\frac{22}{10} + \frac{57}{50}$

17. $\frac{7}{3} + \frac{12}{9}$

18. $\frac{7}{6} + \frac{9}{4}$

19. $4\frac{4}{5} + \frac{78}{15}$

20. $\frac{23}{7} + 3\frac{1}{4}$

21. $2\frac{3}{4} + \frac{19}{8}$

22. $\frac{7}{6} + 5\frac{5}{9}$

Extending Concepts

Find the path that has the given sum. You may enter any open gate at the top, but you must exit from the gate at the bottom.

23.

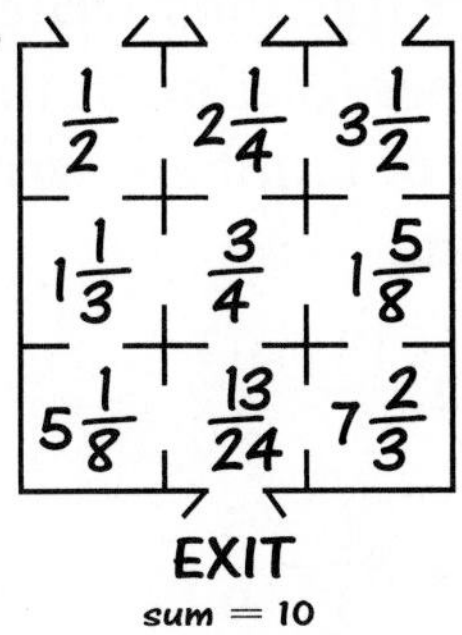

24.

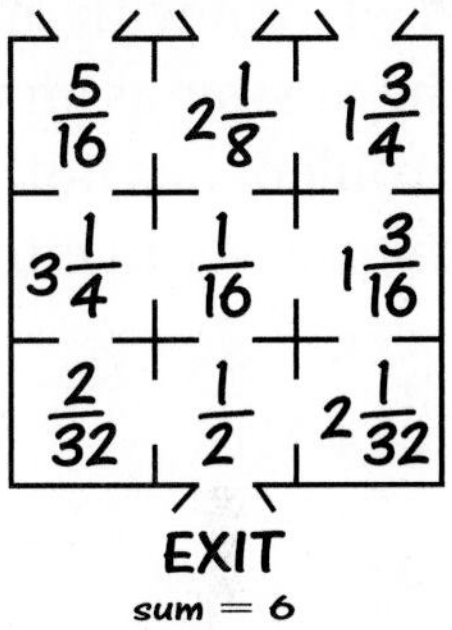

Writing

25. Answer the letter to Dr. Math.

Dear Dr. Math,

When I needed to solve the problem $2\frac{5}{16} + \frac{19}{4}$, a classmate said it would be easier if I changed $2\frac{5}{16}$ to an improper fraction. But, I do not want to do anything that is not proper! Do I have to solve the problem her way?

Yours truly,

Pollyanna DoRight

Homework 13

Sorting Out Subtraction

Applying Skills

Find each sum or difference. If a problem is written with mixed numbers, write the answer as a mixed number. If the problem is written with improper fractions, write the answer as an improper fraction. If the problem has both types of numbers, write the answer in either form.

1. $\frac{5}{4} - \frac{3}{8}$

2. $\frac{13}{5} - \frac{11}{10}$

3. $4 - 3\frac{5}{6}$

4. $8\frac{1}{3} - \frac{5}{9}$

5. $12\frac{3}{8} - 5\frac{3}{4}$

6. $\frac{33}{10} - 2\frac{3}{5}$

7. $\frac{5}{8} + \frac{3}{4}$

8. $2\frac{4}{5} - 1\frac{1}{6}$

9. $4\frac{9}{11} - \frac{35}{11}$

10. $1\frac{2}{3} + \frac{3}{4} + \frac{5}{4}$

11. $8\frac{8}{9} - 3\frac{1}{3}$

12. $2\frac{3}{13} + \frac{3}{2}$

13. $5\frac{1}{6} - 2\frac{1}{2}$

14. $7\frac{2}{5} - \frac{17}{10}$

15. $\frac{9}{4} + 5\frac{2}{3}$

16. $\frac{17}{6} - 1\frac{1}{9}$

17. $\frac{13}{4} - 2\frac{5}{6}$

18. $19 - \frac{12}{7}$

19. $4\frac{5}{9} + \frac{19}{6}$

20. $\frac{15}{4} + 19$

21. How much longer is $28\frac{1}{2}$ seconds than $23\frac{7}{10}$ seconds?

22. Find the sum of $4\frac{1}{5}$, $8\frac{7}{8}$, and $1\frac{7}{10}$.

23. What is the difference between $\frac{77}{9}$ and $5\frac{1}{3}$?

Extending Concepts

Complete each magic square. Remember that the sum of each column, row, and diagonal must be the same.

24.

$\frac{7}{2}$		$1\frac{1}{6}$
	$2\frac{11}{12}$	
		$\frac{7}{3}$

25.

5		$\frac{5}{3}$
$\frac{5}{6}$		
$6\frac{2}{3}$		

Making Connections

26. Kaylee is planning to bake a cake. She changed her recipe amounts to serve more people, and the resulting recipe involved many fractions. The first five ingredients are listed below.

Chocolate Cake

$2\frac{5}{8}$ cups flour

$1\frac{1}{8}$ cups cocoa powder

$1\frac{1}{2}$ cups sugar

$1\frac{1}{2}$ teaspoons baking soda

$\frac{3}{8}$ teaspoon salt

Kaylee has a bowl that holds 6 cups. Can she combine these ingredients in the bowl? Why or why not?

Homework 14

Calc and the Numbers

Applying Skills

Find each sum or difference. If a problem is written with mixed numbers, write the answer as a mixed number. If the problem is written with improper fractions, write the answer as an improper fraction. If the problem has both types of numbers, write the answer in either form.

1. $\frac{3}{5} + \frac{7}{8}$

2. $\frac{31}{32} - \frac{3}{16}$

3. $\frac{1}{30} + \frac{2}{30}$

4. $\frac{49}{50} - \frac{24}{25}$

5. $2\frac{3}{4} - 1\frac{1}{3}$

6. $\frac{8}{5} + 2\frac{2}{3}$

7. $\frac{15}{15} - \frac{3}{7}$

8. $1\frac{5}{8} + 3\frac{3}{4}$

9. $\frac{3}{7} + \frac{3}{4}$

10. $7\frac{1}{8} - 3\frac{3}{4}$

Extending Concepts

11. Three fourths of the students in Ms. Smith's class have standard backpacks. One eighth of the students have rolling backpacks. What fraction of the class has either a standard or a rolling backpack? What fraction of the class has neither?

12. Chapa received a check for her twelfth birthday. She spent $\frac{3}{8}$ of the money the day she got it. The following week, she spent $\frac{1}{4}$ of the total amount. What fraction of her money does she have left to spend?

13. The Hilliers are making pizza for dinner. Jill put anchovies on $\frac{1}{2}$ of the pizza. Then, Jack put sausage on $\frac{2}{3}$ of the pizza, including the half with the anchovies. What fraction of the pizza has sausage, but no anchovies?

14. At his graduation party, Gianni served $1\frac{5}{6}$ pounds of cheese and $2\frac{2}{3}$ pounds of peanuts. What is the total number of pounds of cheese and peanuts served?

15. Sophie has two watermelons. One weighs $3\frac{3}{4}$ pounds. The other weighs $5\frac{3}{8}$ pounds. What is the total weight of the watermelons? What is the difference between the weights of the two watermelons?

16. Analise spent $3\frac{1}{2}$ days visiting her grandmother. Then, she spent $\frac{5}{4}$ days at her uncle's house. How much time did she spend visiting these relatives? How much more time did she spend with her grandmother than her uncle?

Writing

17. Write a problem of your own, using a subject of your choice. Your problem should include at least two fractions, and its solution can be a fraction, mixed number, or a whole number.

15 Homework

Picturing Fraction Multiplication

Applying Skills

For items 1–2, use the drawing below.

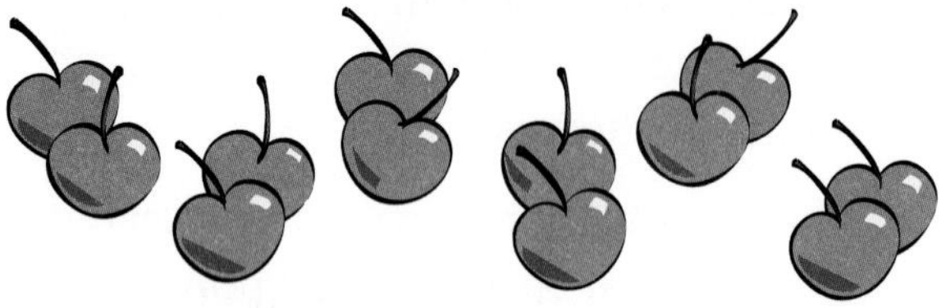

1. How many cherries would represent $\frac{2}{3}$ of the cherries?

2. How many cherries would represent $\frac{5}{6}$ of the cherries?

For items 3–4, use the drawing below.

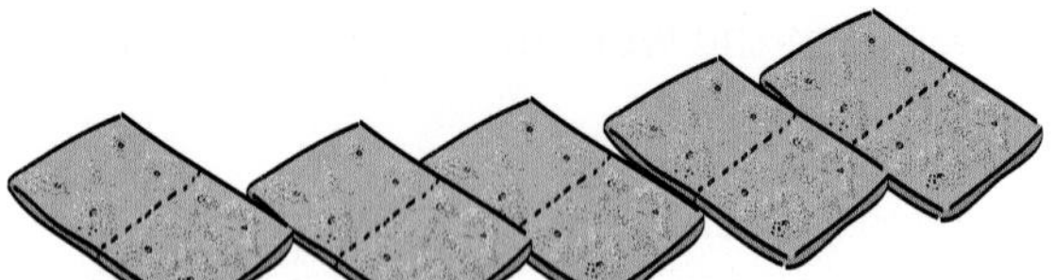

3. How many halves would represent $\frac{3}{10}$ of the crackers?

4. How many halves would represent $\frac{4}{5}$ of the crackers?

Solve each problem.

5. $\frac{4}{5}$ of 10

6. $\frac{3}{8}$ of 40

7. $\frac{2}{3}$ of 18

8. $\frac{5}{6}$ of 24

9. $\frac{3}{5}$ of 35

10. $\frac{3}{5}$ of 15

11. $\frac{1}{12}$ of 24

12. $\frac{3}{4}$ of 24

Extending Concepts

Solve each problem.

13. $\frac{2}{3}$ of 16

14. $\frac{3}{4}$ of 30

15. $\frac{3}{7}$ of 4

16. $\frac{7}{9}$ of 3

17. Danielle ate 11 of the orange wedges shown below. What fraction of the oranges did she eat?

18. Juanita served 3 giant submarine sandwiches at her party. Each sandwich was cut into 12 equal pieces. The guests ate $\frac{11}{12}$ of the sandwiches. How many pieces did they eat?

Making Connections

19. Myron manages a bakery. He has two pies that are exactly the same. However, the pies are cut into different size pieces and are priced differently. Which pie will earn more money?

$2 per piece.
Three pieces for $4.

Homework 16 Fractions of Fractions

Applying Skills

For each problem, the rectangle represents one whole. Use the rectangle to find each solution.

1. $\frac{2}{3}$ of $\frac{3}{4}$

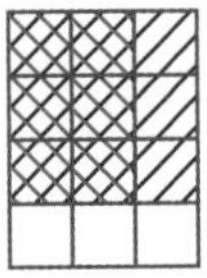

2. $\frac{3}{5}$ of $\frac{1}{6}$

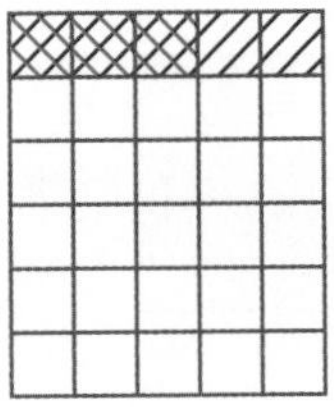

3. $\frac{1}{9}$ of $\frac{2}{3}$

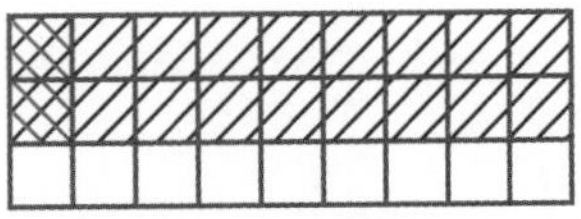

4. $\frac{5}{6}$ of $\frac{4}{5}$

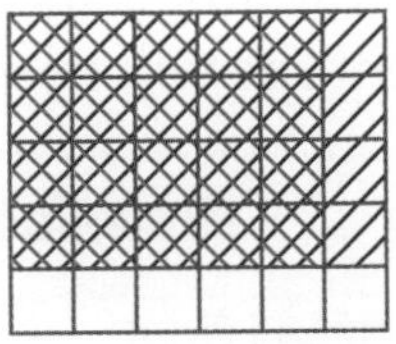

Find each product.

5. $\frac{3}{4} \times \frac{7}{8}$

6. $\frac{1}{2} \times \frac{4}{5}$

7. $\frac{2}{3} \times \frac{4}{7}$

8. $\frac{4}{9} \times \frac{3}{4}$

9. $\frac{2}{3} \times 60$

10. $\frac{3}{8} \times \frac{4}{9}$

11. $\frac{3}{5} \times \frac{10}{15}$

12. $\frac{1}{5} \times \frac{1}{5}$

13. $\frac{2}{3} \times \frac{1}{2}$

14. $\frac{1}{2} \times \frac{4}{9}$

Extending Concepts

15. Isabella spent $\frac{3}{4}$ of an hour doing homework. She spent $\frac{1}{2}$ of that time on math. What fraction of an hour did she spend on math?

16. Mia has a collection of baseball cards. Two-thirds of her collection represent National League players. One eighth of her National League cards represent the Giants. What fraction of her whole collection represents the Giants?

17. Ming has $\frac{1}{3}$ of a pizza left from last night's dinner. If he and his friend eat $\frac{1}{2}$ of the leftovers for an afternoon snack, how much of the whole pizza will they eat?

18. Jamil spends $\frac{1}{4}$ of his day at school. He spends $\frac{1}{10}$ of his school day at lunch. What fraction of his whole day does he spend at lunch?

Making Connections

19. About $\frac{3}{10}$ of Earth's surface is land.

a. At one time, $\frac{1}{2}$ of the land was forested. What fraction of Earth was forested?

b. Today about $\frac{1}{3}$ of Earth's land is forested. What fraction of Earth is forested today?

c. How much less of Earth is forested today?

Homework 17: Estimation and Mixed Numbers

Applying Skills

Choose the best estimate.

1. $\frac{15}{4} \times \frac{1}{2}$

a. 2 **b.** 3 **c.** 8

2. $\frac{1}{3} \times \frac{5}{9}$

a. 2 **b.** $\frac{1}{6}$ **c.** $\frac{1}{2}$

3. $2\frac{1}{4} \times \frac{1}{5}$

a. $10\frac{1}{4}$ **b.** $\frac{1}{2}$ **c.** 1

4. $\frac{6}{5} \times \frac{3}{4}$

a. $1\frac{1}{4}$ **b.** $\frac{18}{4}$ **c.** 1

Find each product.

5. $1\frac{1}{3} \times \frac{1}{2}$

6. $1\frac{4}{5} \times \frac{5}{9}$

7. $2\frac{3}{4} \times 3$

8. $1\frac{1}{5} \times \frac{1}{2}$

9. $\frac{2}{3} \times \frac{3}{4}$

10. $\frac{6}{7} \times \frac{9}{10}$

11. $\frac{3}{8} \times 2\frac{1}{4}$

12. $1\frac{1}{8} \times 1\frac{1}{8}$

13. $\frac{4}{3} \times 2\frac{1}{3}$

14. $\frac{3}{10} \times 20$

15. $3\frac{1}{4} \times 2\frac{2}{3}$

16. $2\frac{1}{2} \times \frac{5}{8}$

17. $5\frac{1}{3} \times \frac{4}{5}$

18. $2\frac{1}{2} \times 2\frac{2}{3}$

19. $3 \times 2\frac{1}{7}$

20. $3\frac{2}{3} \times 9$

Extending Concepts

The area of a rectangle equals the length times the width. Find the area of each rectangle.

21.

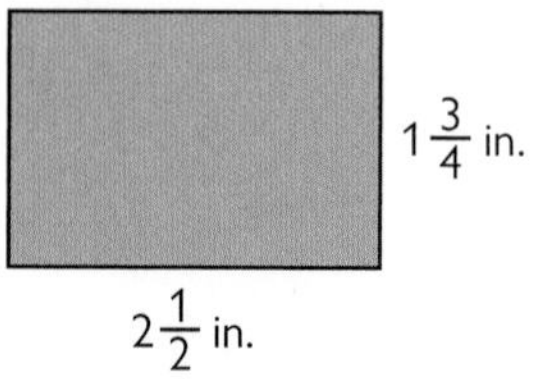

22.

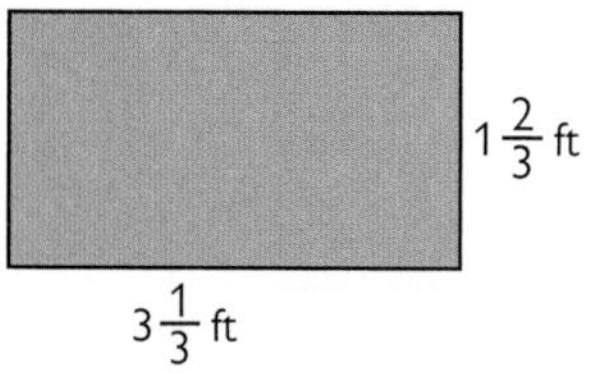

Making Connections

23. Tia is making brownies for a crowd. She wants to triple her usual recipe shown below. Rewrite her recipe showing the amounts of each ingredient she will need.

Brownies

- $\frac{2}{3}$ cup sifted flour
- $\frac{1}{2}$ teaspoon baking powder
- $\frac{1}{4}$ teaspoon salt
- $\frac{1}{3}$ cup butter
- 2 squares bitter chocolate
- 1 cup sugar
- 2 well beaten eggs
- $\frac{1}{2}$ cup chopped nuts
- 1 teaspoon vanilla

18 Homework

Fraction Groups within Fractions

Applying Skills

For items 1–3, use the following drawing to find each quotient.

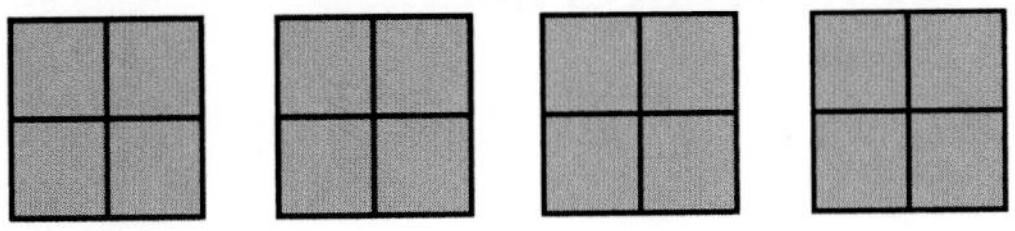

1. $4 \div \frac{1}{4}$

2. $4 \div \frac{1}{2}$

3. $4 \div \frac{1}{8}$

For items 4–6 use the following drawing to find each quotient.

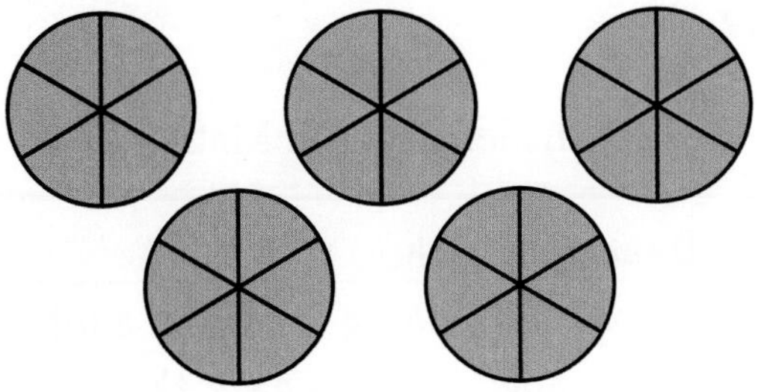

4. $5 \div \frac{1}{6}$

5. $5 \div \frac{1}{3}$

6. $5 \div \frac{1}{2}$

Find each quotient.

7. $\frac{2}{3} \div \frac{1}{3}$

8. $\frac{4}{3} \div \frac{1}{6}$

9. $2\frac{3}{4} \div \frac{1}{8}$

10. $\frac{4}{5} \div 3$

11. $\frac{5}{4} \div \frac{1}{3}$

12. $4\frac{1}{8} \div \frac{1}{4}$

13. $4\frac{1}{2} \div 3\frac{3}{4}$

14. $2\frac{4}{5} \div \frac{7}{8}$

Extending Concepts

Ms. Marrero is a math teacher who wants to celebrate pi day by buying pies for her class. (Pi day is March 14 or 3–14.) She has 27 students in the class. How much pie will each student get if she buys each of the following number of pies?

15. 3 pies

16. 5 pies

17. 9 pies

18. 30 pies

19. Maria has 7 pounds of mints. She wants to make $\frac{1}{4}$-pound packages for party favors. How many packages can she make?

20. Troy needs to cut a zucchini into slices that measure $\frac{3}{8}$ inch thick. If the zucchini is $6\frac{3}{4}$ inches long, how many slices will he have?

Writing

21. Imagine that your friend has been absent from school with a bad cold. You are taking the homework assignments to your friend's house. To do the math homework, your friend needs to understand how to divide fractions. To help your friend, draw a picture to show $4 \div \frac{1}{2} = 8$ and write a short note to explain the drawing.

Homework 19 Understanding Fraction Division

Applying Skills

For each problem, decide if the answer will be greater or less than the dividend. Then solve the problem to see if you are correct.

1. $\frac{5}{4} \div \frac{1}{2}$

2. $3\frac{2}{5} \div \frac{1}{5}$

3. $\frac{4}{5} \div 3$

4. $\frac{6}{5} \div 2\frac{1}{3}$

5. $\frac{9}{8} \div \frac{1}{4}$

6. $6 \div \frac{2}{3}$

7. $\frac{3}{4} \div 1\frac{1}{8}$

8. $\frac{2}{3} \div \frac{1}{6}$

9. $\frac{4}{3} \div \frac{1}{2}$

10. $3\frac{1}{2} \div 1\frac{1}{4}$

11. $\frac{1}{3} \div \frac{3}{5}$

12. $\frac{2}{5} \div 4$

13. $2 \div \frac{1}{6}$

14. $3 \div 4\frac{1}{2}$

15. $6\frac{1}{4} \div \frac{1}{2}$

16. $\frac{4}{5} \div \frac{6}{5}$

17. $4\frac{3}{4} \div \frac{5}{8}$

18. $5 \div \frac{2}{3}$

19. $5 \div 6\frac{1}{4}$

20. $1\frac{3}{8} \div \frac{3}{4}$

21. Is $\frac{7}{8} \div \frac{1}{2}$ greater or less than 1?

22. Is $\frac{3}{5} \div \frac{5}{6}$ greater or less than 1?

23. Find $\frac{3}{4} \div \frac{5}{6}$.

24. Find $3\frac{1}{4} \div 2\frac{1}{2}$.

Extending Concepts

To make two dozen muffins, you need $1\frac{1}{2}$ cups of flour. How many muffins can you make with each amount of flour?

25. 1 cup of flour

26. $\frac{1}{2}$ cup of flour

27. $\frac{1}{4}$ cup of flour

28. 2 cups of flour

29. $3\frac{1}{2}$ cups of flour

Writing

30. Answer the letter to Dr. Math.

> Dear Dr. Math,
> I have been doing math for a lot of years now, and I know a thing or two. One thing I know for sure is that when you do a division problem, the answer is always less than the first number in the problem. For example, 24 ÷ 4 equals 6. Six is less than 24. That is how division works. So now, they are telling me that 10 ÷ $\frac{1}{2}$ equals 20. I say no way, because 20 is greater than 10! I am right, right?
> Sincerely,
> M. Shure

20 Homework

Multiplication vs. Division

Applying Skills

Find each value.

1. $\frac{1}{4} \times 4\frac{1}{4}$

2. $\frac{1}{5} \times \frac{2}{9}$

3. $\frac{4}{5} \div \frac{1}{10}$

4. $\frac{9}{8} \div \frac{1}{16}$

5. $\frac{4}{3} \times \frac{3}{4}$

6. $\frac{1}{2} \times \frac{1}{4} \div \frac{1}{8}$

For each pair of numbers, decide whether to multiply or divide to get the least number. Then, solve the problem. You must use the numbers in the order they are given.

7. $\frac{1}{4}$ and $\frac{1}{8}$

8. $\frac{2}{3}$ and $3\frac{1}{2}$

9. 5 and $\frac{1}{9}$

10. $\frac{2}{3}$ and 21

11. $\frac{5}{4}$ and $\frac{3}{8}$

12. $2\frac{1}{2}$ and $\frac{5}{2}$

Solve each problem.

13. Frank's bread recipe calls for $3\frac{1}{2}$ cups of flour and yields one loaf of bread.

a. How much flour will Frank need to make 7 loaves of bread?

b. How much flour will he need to make 13 loaves of bread?

c. Frank has $5\frac{1}{4}$ cups of flour. How many loaves of bread can he make?

d. How many loaves of bread can he make with 14 cups of flour?

14. There are 20 stamps on a sheet of stamps. Eli needs to mail 230 invitations. How many sheets of stamps will Eli use to mail the invitations?

15. Hernando has some cantaloupes that he plans to slice into 10 equal servings.

a. How many whole cantaloupes will he need to slice to have enough to give one slice to each of 27 people?

b. How much of a whole cantaloupe will be left?

16. About $\frac{7}{10}$ of the human body is water. If a person weighs 110 pounds, how many pounds are water?

Extending Concepts

For each set of numbers, write four true multiplication and division equations.

17. $\frac{1}{4}$, $1\frac{1}{2}$, and 6

18. $1\frac{1}{3}$, $1\frac{1}{2}$, and 2

Making Connections

19. The woodchuck is a large rodent native to North America. How much wood does a woodchuck chuck? Well, who knows? However, if a woodchuck *could* chuck wood, let's say it would chuck half a log of wood in 5 minutes. How much wood would a woodchuck chuck in $\frac{5}{12}$ of an hour?

DESIGNING SPACES, INC.

Welcome Designing Spaces, Inc. Consultants!

Designing a house involves the mathematics of geometry and spatial visualization. To make sure that the homes appeal to people from different cultures, it is also important to understand how people from around the world use construction shapes to build structures that reflect their own cultural experiences.

How can you describe houses from around the world?

DESIGNING SPACES

PHASE**ONE**

Visualizing and Representing Cube Structures

Your job as a consultant for Designing Spaces, Inc. will begin with exploring different ways to represent three-dimensional structures. In this first phase, you will build houses made from cubes. Then you will create building plans of a house. The true test of your plans will be whether another person can follow them to build the house. The skills you develop in this phase are an important basis for understanding geometry.

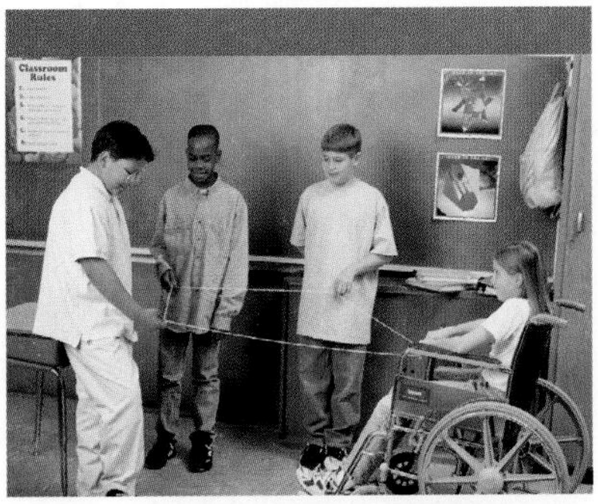

PHASE**TWO**

Functions and Properties of Shapes

What shapes do you see in the houses pictured on these pages? In this phase, you will explore the properties of two-dimensional shapes. You will apply what you learn to decode the formulas for making the construction shapes developed by the Designing Spaces, Inc. technical research group. At the end of the phase, Designing Spaces, Inc. will turn to you for advice about the dimensions of their construction shapes.

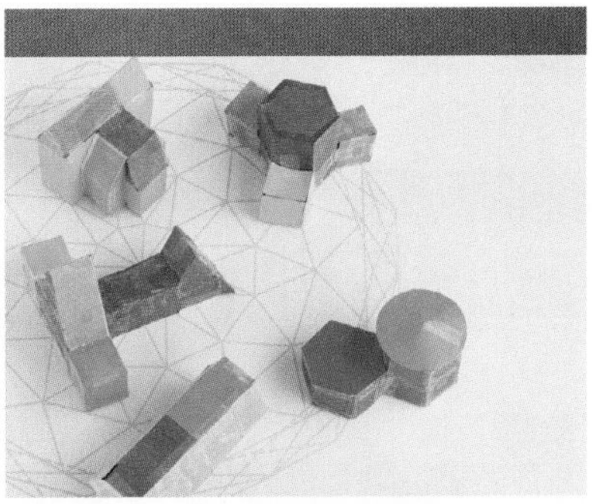

PHASE**THREE**

Visualizing and Representing Polyhedrons

In Phase Three, you will explore how you can use the building shapes you helped design in Phase Two to make a variety of three-dimensional structures. You'll learn names for the structures you create and make drawings of them. The phase ends with a final project. You will use the building shapes to design a home for a cold and snowy climate, or a warm and rainy climate. Then you'll create building plans for your design.

PHASE ONE

To: House Designers
From: General Manager
Designing Spaces, Inc.

Welcome to Designing Spaces, Inc.! In your new job as a house designer, you will be asked to design different kinds of homes for our customers.

One kind of home our company designs is a low-cost modular home. These homes are made from cube-shaped rooms that are all the same size. We can make many different kinds of modular homes because the rooms fit together in so many different ways.

In Phase One, you will take on the special assignment of designing houses made of cube-shaped rooms. You will learn helpful ways to make building plans for your structures. Your plans must be clear. Someone else should be able to build your house from your plans.

Visualizing and Representing Cube Structures

WHAT'S THE MATH?

Investigations in this section focus on:

MULTIPLE REPRESENTATIONS

- Represent three-dimensional structures with isometric and orthogonal drawings.

PROPERTIES and COMPONENTS of SHAPES

- Identify two-dimensional shapes that make up three-dimensional structures.
- Describe properties of structures in writing so that someone else can build the structure.

VISUALIZATION

- Build three-dimensional structures from two-dimensional representations.

MathScape Online
mathscape1.com/self_check_quiz

1 Planning and Building a Modular House

CREATING TWO-DIMENSIONAL REPRESENTATIONS

A modular house is made of parts that can be put together in different ways. You can use cubes to design a modular house model. Then you can record your design in a set of building plans. These plans are a way of communicating about your house design so that someone else could build it.

Use Cubes to Design a House

What kind of structure can you make from cubes that meets the design guidelines?

One kind of home that Designing Spaces, Inc. designs is a low-cost modular home made from cube-shaped rooms. For your first assignment, use eight to ten cubes to design a modular house model. Let each cube represent a room in the house. Follow the Design Guidelines for Modular Houses.

Cubes in the structure you design can be arranged like these.

Cubes cannot be arranged like these.

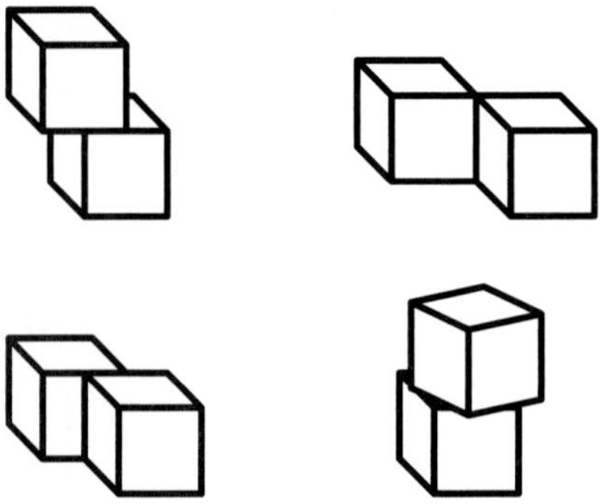

Design Guidelines for Modular Houses

- A face of one cube must line up exactly with a face of at least one other cube so that the rooms can be connected by stairways and doorways.
- No rooms can defy gravity. Each cube must rest on the desktop or directly on top of another cube.

Create Building Plans

Now that you have designed and built your modular house model, make a set of plans that someone else could use to build the same structure. Your plans should include the following:

- Create at least one drawing of the house.
- Write a description of the steps someone would follow to build the house.

How can you make two-dimensional drawings of three-dimensional structures?

Add to the Visual Glossary

In this unit, you will create your own Visual Glossary of geometric terms that describe shapes. These terms will help you to share your building plans with others. One example is the term *face.* The term is used in the Design Guidelines for Modular Houses.

- After your class discusses the meaning of the term *face,* add the class definition to your Visual Glossary.
- Include drawings to illustrate your definition.

hot words | face

Homework
page 194

2 Seeing Around the Corners

REPRESENTING THREE DIMENSIONS IN ISOMETRIC DRAWINGS

How many different houses do you think you can make with three cubes? How many can you make with four cubes? As you explore the possibilities, you will find that a special type of drawing called *isometric drawing* can help you record the different structures you make.

How many different structures is it possible to build with three cubes?

Build and Represent Three-Room Houses

Design as many different modular houses as you can using three cubes.

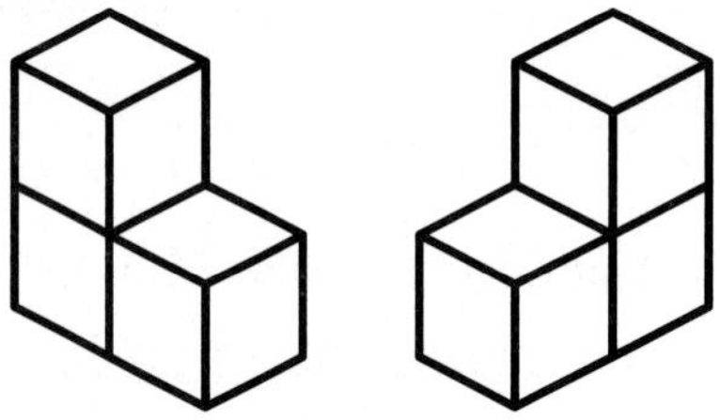

These two houses are the same. You can rotate one to be just like the other, without lifting it.

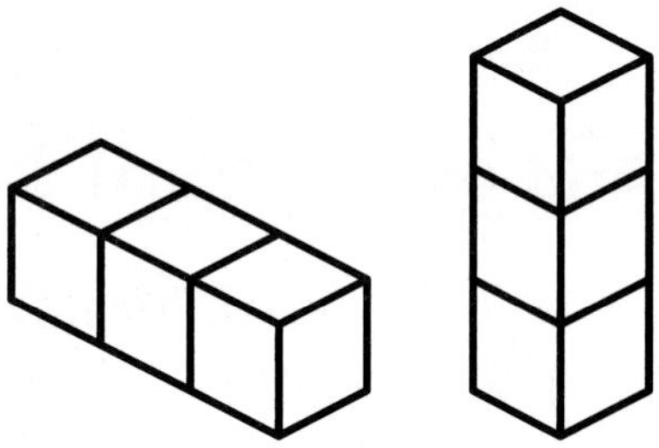

These two houses are different. You have to lift one house to make it just like the other.

- Follow the Design Guidelines for Modular Houses presented on page 166.
- Make an isometric drawing to record each different structure you make.

Sample Isometric Drawings

Isometric drawings show three faces of a structure in one sketch.

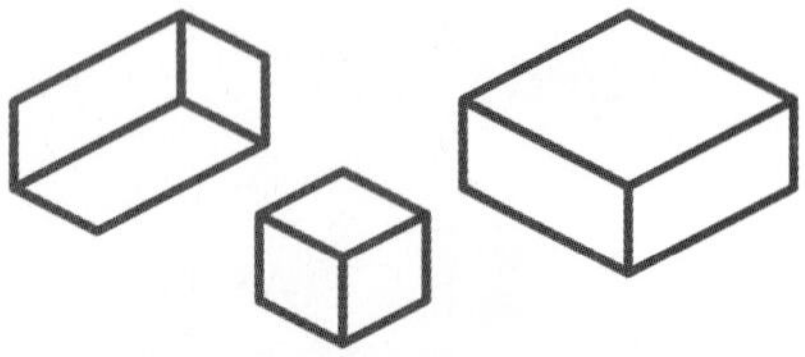

You can use isometric dot paper to help you in making isometric drawings.

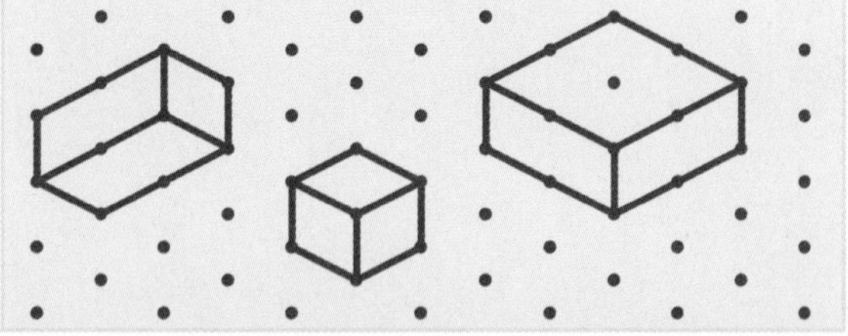

Build and Represent Four-Room Houses

You have explored the number of houses it is possible to build with three cubes. Now try using four cubes to design as many different houses as you can.

- Follow the Design Guidelines for Modular Houses presented on page 166.
- Record each different structure you make as an isometric drawing or another type of drawing.

How many different houses can you make with four cubes?

How many different structures is it possible to build with four cubes?

Add to the Visual Glossary

Think about the isometric drawings you made in this lesson. Look at the illustrations shown on this page. Then compare the lengths of the sides and the sizes of the angles in both kinds of drawings.

- After discussing with the class, add the class definition of the term *isometric drawing* to your Visual Glossary.
- Be sure to include drawings to illustrate the definition.

Isometric Drawings

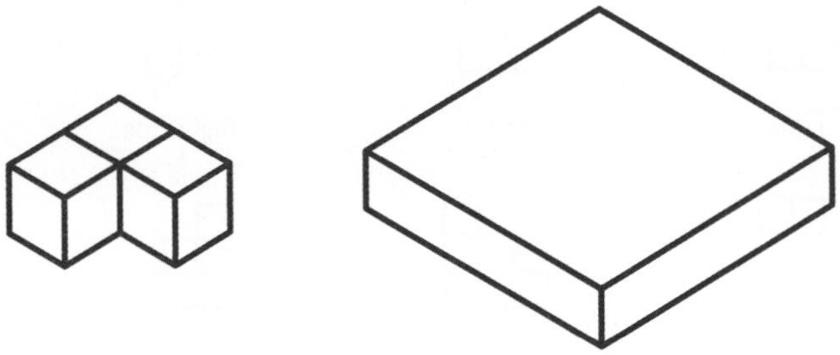

These are isometric drawings.

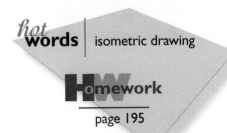

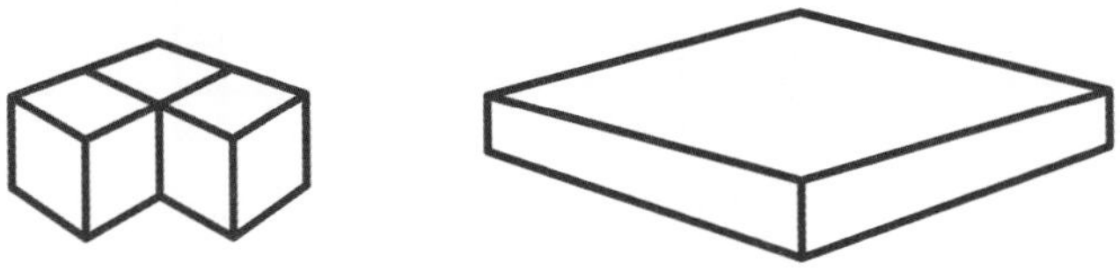

These are **not** isometric drawings.

hot words | isometric drawing

Homework
page 195

3 Seeing All Possibilities

TRANSLATING BETWEEN ORTHOGONAL AND ISOMETRIC DRAWINGS

You have explored using isometric drawings to show three-dimensional structures on paper. Here you will try another drawing method called *orthogonal drawing*. You need to understand both types of drawings, so you can read building plans and create plans of your own.

How can you use orthogonal drawings to build three-dimensional structures?

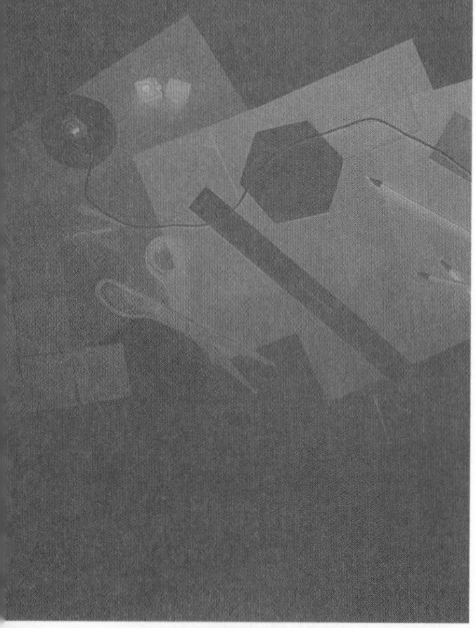

Construct Houses from Orthogonal Drawings

Four sets of building plans for modular houses are shown here. Each plan is represented with orthogonal drawings that show three views of the house. Your job is to build each house with cubes and record the least number of cubes you needed to build the house.

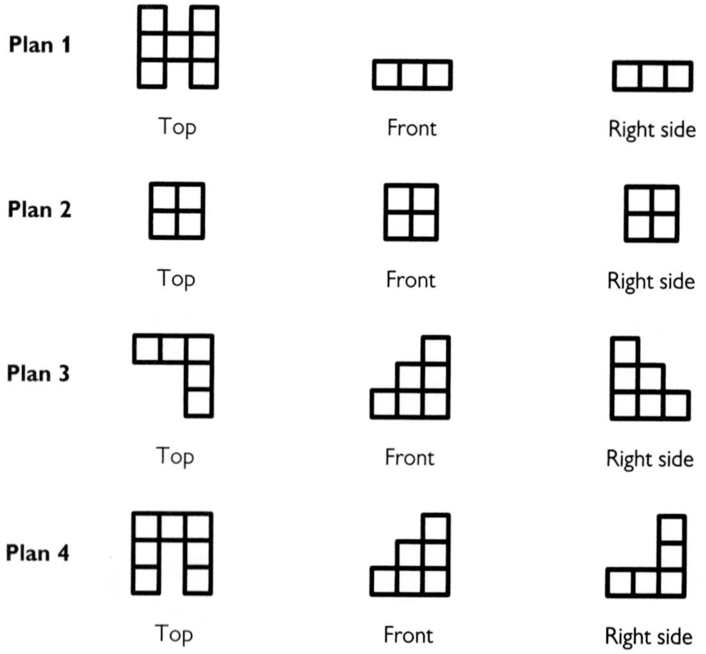

Make Orthogonal Drawings

These plans are isometric drawings of several houses. Your job is to make orthogonal drawings showing the top, front, and right side of each house.

Plan 1

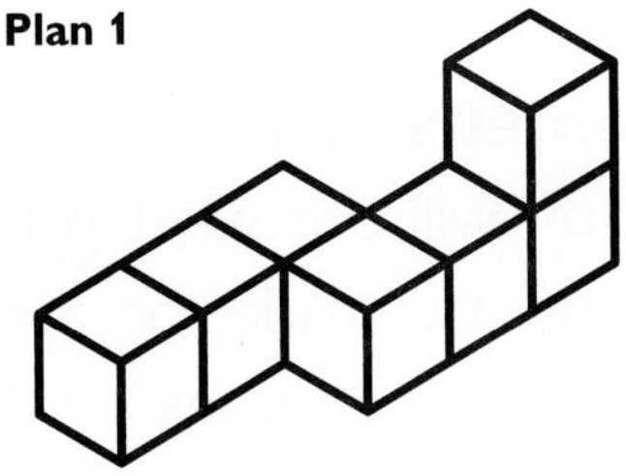

Plan 2

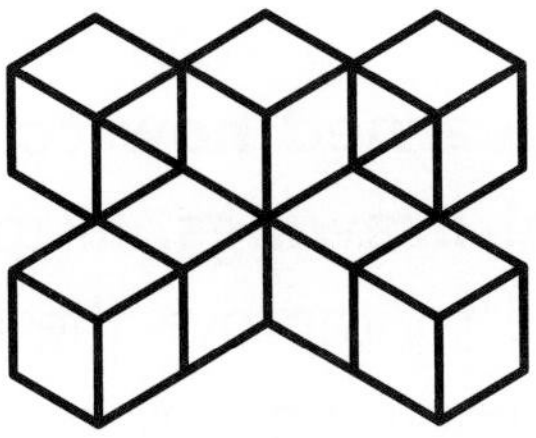

Plan 3

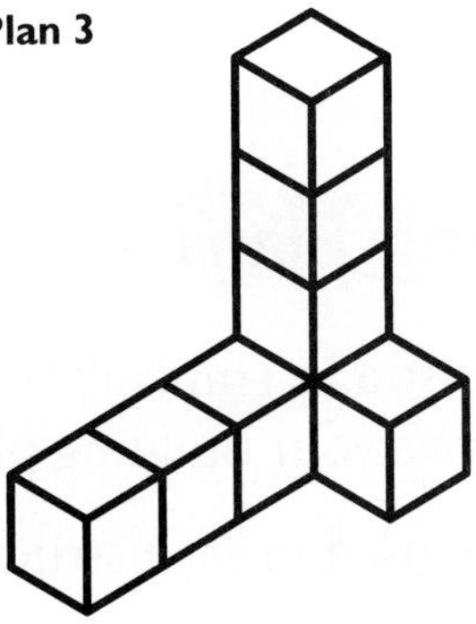

Plan 4

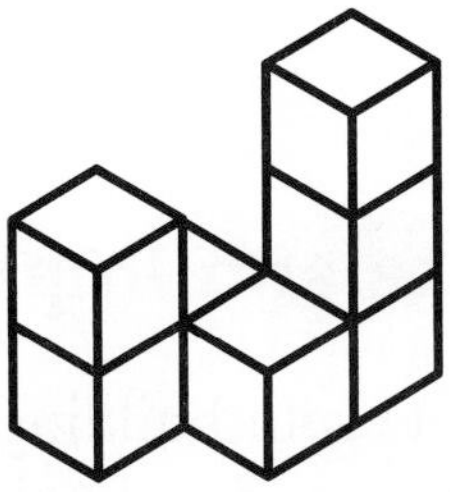

How can you make orthogonal drawings from an isometric drawing of a house?

Add to the Visual Glossary

Think about the orthogonal drawings you made. Look at the illustrations showing drawings that are orthogonal and drawings that are not orthogonal.

- After class discussion, add the class definition of the term *orthogonal drawing* to your Visual Glossary.
- Be sure to include drawings to illustrate the definition.

Orthogonal Drawing

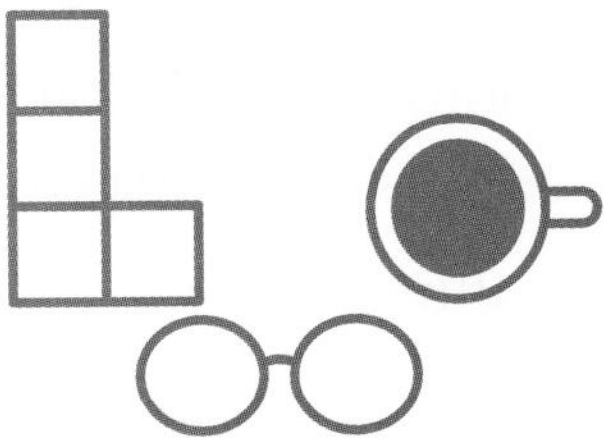

These are orthogonal drawings.

These are **not** orthogonal drawings.

hot words | orthogonal drawing

Homework

page 196

4 Picture This

COMMUNICATING WITH DRAWINGS

You have learned how to make isometric and orthogonal drawings. In this lesson, you will use what you have learned to improve the building plans you made for your first assignment. One way to check how well your plans communicate is to get someone else's comments on your plans.

How can you apply what you know to improve how well your plans communicate?

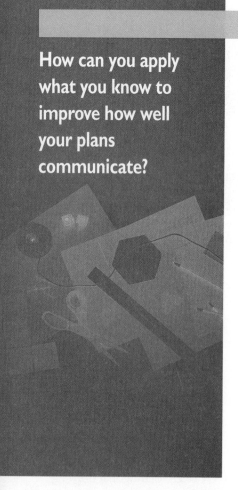

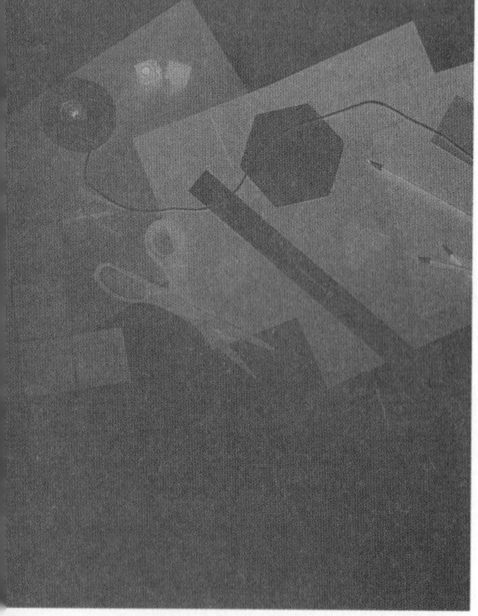

Evaluate and Improve Building Plans

1 Evaluate your building plans from Lesson 1. Think about these questions and make changes to improve your building plans.

a. Can you tell how many cubes were used in each structure? How do you know?

b. Can you tell how the cubes should be arranged? How would you make the drawings clearer?

c. Do the written plans include a step-by-step description of the building process? Are any steps missing?

d. What are one or two things that you think are done well in the plans? What are one or two things that you think need to be changed in the plans?

2 Exchange your plans with a partner. Carefully follow your partner's plans and build the structure. Write down any suggestions that you think would improve your partner's building plans.

a. Are the drawings clear? How could you make them clearer?

b. Are the step-by-step written instructions easy to follow? What suggestions can you give to improve the instructions?

c. What are one or two things that you think are done well in the plans?

Use Feedback to Revise Building Plans

Carefully review the feedback you received from your partner. Use it to make final changes to your plans.

- Keep in mind that feedback is suggestions and opinions that can help you improve your work.
- It is up to you to decide whether to use the feedback and how to use it.

How can you improve your plans so that they are easier for someone else to use?

Write About the Changes

When you finish changing your building plans, write a memo to the General Manager of Designing Spaces, Inc. The memo should summarize the revisions you made.

- Describe the changes you made in your drawings.
- Describe the steps it took to build your structure.

A home in the United States

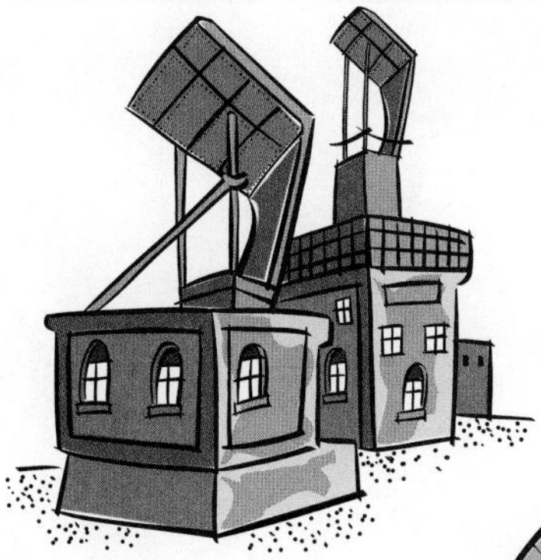

A house in Pakistan

An igloo of the Inuit

hot words | three-dimensional
two-dimensional

Homework
page 197

PHASE TWO

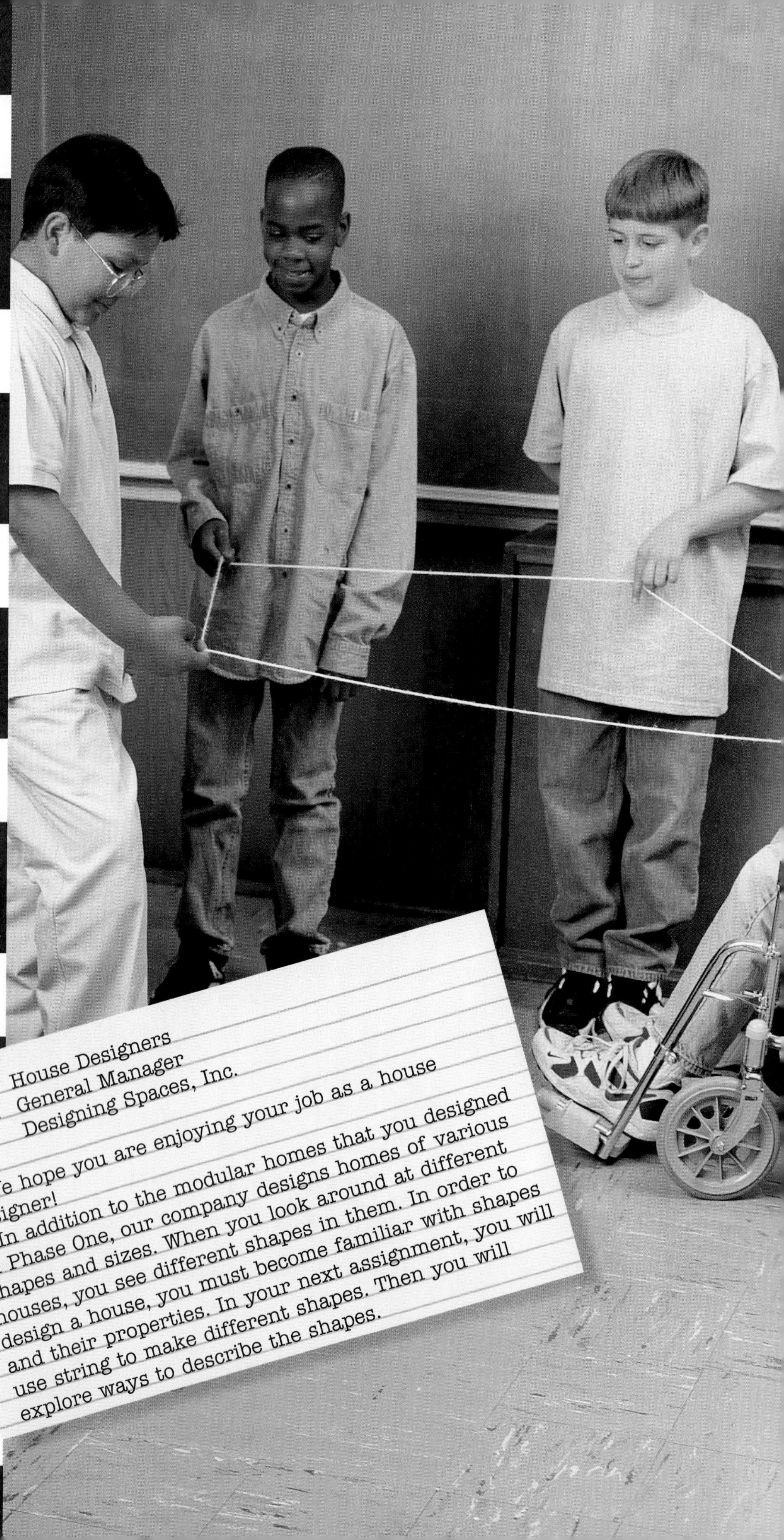

To: House Designers
From: General Manager
Designing Spaces, Inc.

We hope you are enjoying your job as a house designer!

In addition to the modular homes that you designed in Phase One, our company designs homes of various shapes and sizes. When you look around at different houses, you see different shapes in them. In order to design a house, you must become familiar with shapes and their properties. In your next assignment, you will use string to make different shapes. Then you will explore ways to describe the shapes.

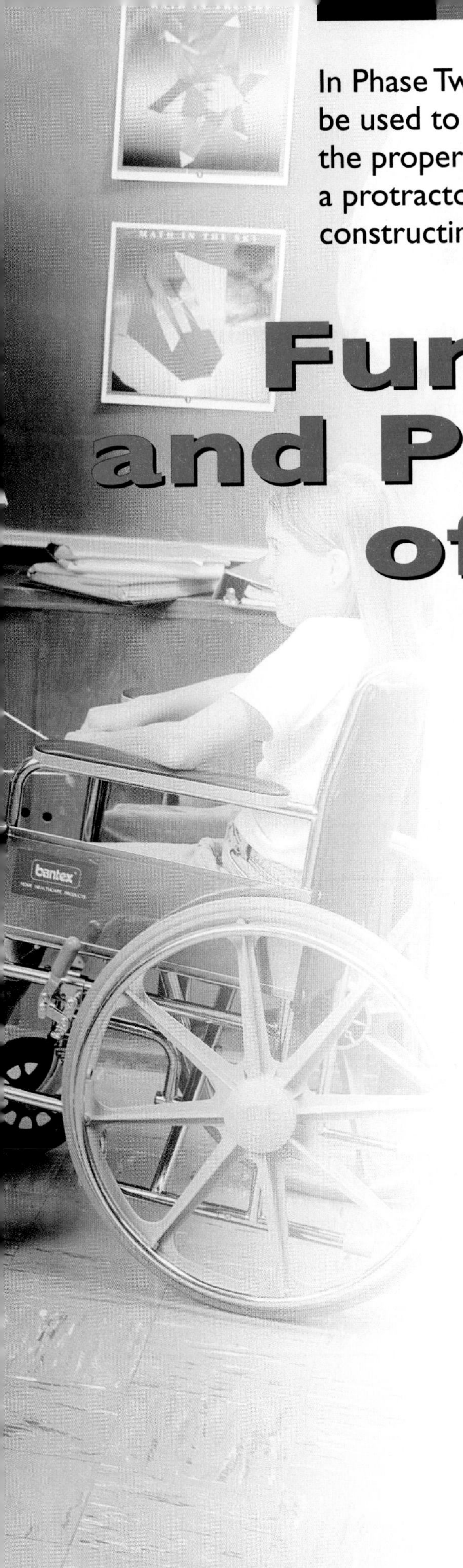

In Phase Two, you will investigate various shapes that can be used to design and build houses. You will learn about the properties of shapes and how to measure angles with a protractor—ideas that are good to know when constructing a house.

Functions and Properties of Shapes

WHAT'S THE MATH?

Investigations in this section focus on:

PROPERTIES and COMPONENTS of SHAPES

- Identify two-dimensional shapes and their properties.
- Measure sides and angles of shapes.
- Estimate area and perimeter of shapes.
- Use geometric notation to indicate relationships between angles and sides in shapes.
- Describe properties of shapes in writing.
- Expand vocabulary for describing shapes.

VISUALIZATION

- Visualize shapes from clues about their sides and angles.
- Perform visual and mental experiments with shapes.

MathScape Online
mathscape1.com/self_check_quiz

5 String Shapes

INVESTIGATING PROPERTIES OF SIDES

What shapes can you find in the houses in your neighborhood? Shapes that have three or more sides are called **polygons.** As you make shapes from string, you will learn some special properties of sides of polygons. You'll also learn mathematical names for the shapes you create.

Make Shapes from Clues

Can you find a shape that satisfies clues about parallel and equilateral sides?

Do the following for each shape clue given:

1. Try making a shape that fits the description in the clue.
2. If you can make the shape, record it. Then label the equal sides and the parallel sides. (The top of the handout Naming and Labeling Polygons will help you with this.) If you can't make the shape, write "Impossible."
3. Label the shape you draw with a mathematical name. If you don't know a name for the shape, make up a name that you think describes the properties of the shape. (The bottom of the handout Naming and Labeling Polygons will help you with this.)

Shape Clue 1: An equilateral shape with more than 3 sides and no parallel sides.	**Shape Clue 2:** A shape with 2 sides equal and 2 different sides parallel but not equal. (Hint: Your shape can have more than 4 sides.)
Shape Clue 3: A quadrilateral with 2 pairs of parallel sides and only 2 sides equal.	**Shape Clue 4:** A shape with at least 2 pairs of parallel sides that is not an equilateral shape. (Hint: Your shape can have more than 4 sides.)
Shape Clue 5: A quadrilateral that is equilateral and has no parallel sides.	**Shape Clue 6:** Make up a clue of your own and write it down. Test it to see if you can make the shape, and write a sentence telling why you can or cannot make it.

Make Animated Shapes

Animated shapes are shapes that a group makes with string. The group starts with one shape, and then one member of the group changes positions to change the shape as the rest stand still. Make each of the animated shapes on the handout Animated Shapes.

1. Try to make the shape by moving the fewest people.
2. Record how you made the shape with drawings or words. Include your starting positions, who moved, and where.

Using string, how can you change a square into a triangle? a trapezoid into a square?

Add to the Visual Glossary

Think about how you used the terms *parallel* and *equilateral* to describe the sides of shapes. Then look at the illustrations showing lines and walls that are parallel and lines and walls that are not parallel.

- After some class discussion, write the class's description of what is meant by the terms *parallel* and *equilateral* in your Visual Glossary.
- Be sure to include drawings to illustrate your description.

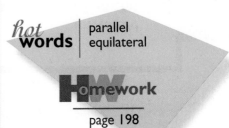

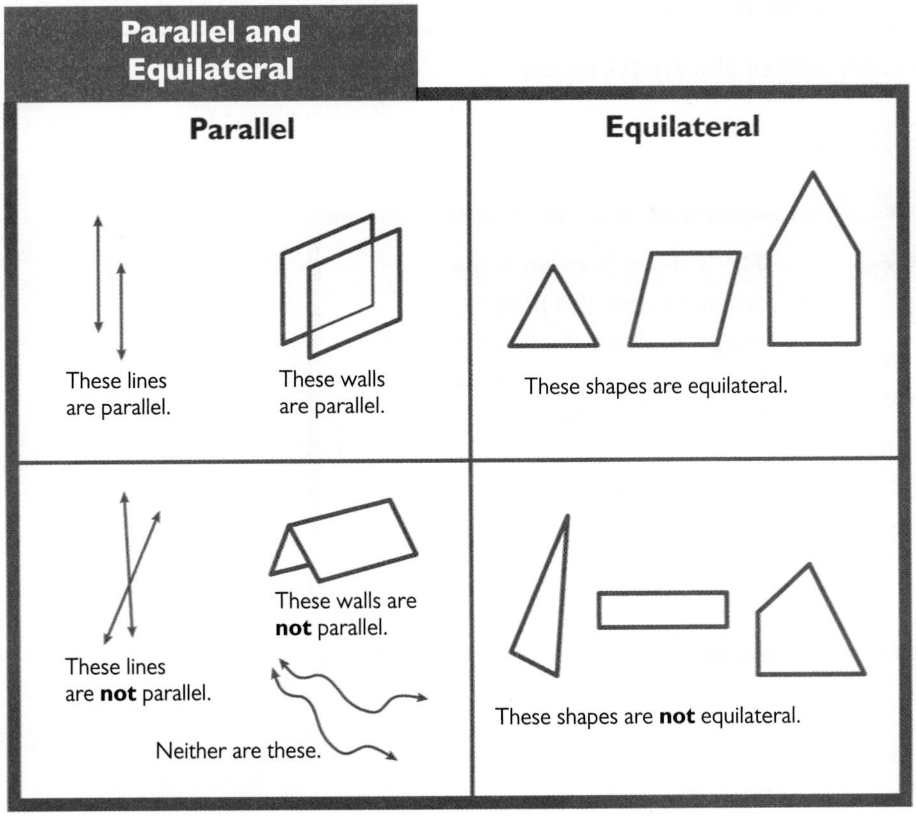
Parallel and Equilateral
Parallel
Equilateral
These lines are parallel.
These walls are parallel.
These shapes are equilateral.
These walls are not parallel.
These lines are not parallel.
Neither are these.
These shapes are not equilateral.

hot words parallel, equilateral

Homework page 198

6 Polygon Paths

USING DISTANCE AND ANGLE MEASURES TO DESCRIBE POLYGONS

Where the sides of a shape meet, they form angles. Building plans for houses have to show the measures of angles and the lengths of sides. Here you will explore how measuring angles is different from measuring lengths. Then you will use what you have learned to describe polygons.

Measure Angles

How can you tell if two angles are equal?

Use a protractor to measure the angles in this polygon. Be sure to refer to the guidelines on How to Use a Protractor.

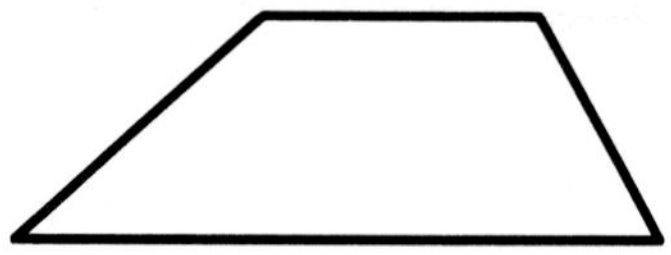

- Which is the smallest angle? How did you measure it?
- Which is the largest angle? How did you measure it?
- Are any of the angles equal?

How to Use a Protractor

- Place the 0° line along one side of the angle.
- Center the protractor where the sides meet at the vertex.
- Read the degree mark that is closest to where the other side of the angle crosses the protractor.

TIP: If the side of an angle doesn't cross the protractor, imagine where it would cross if the line was longer. Or copy the figure and extend the line.

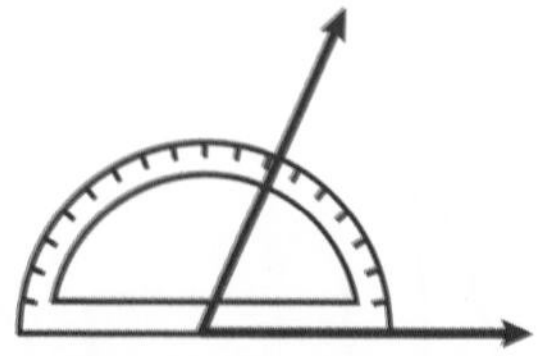

Write Polygon Path Instructions

Polygon path instructions describe the steps you take when drawing a polygon. Study the example shown. Then write polygon path instructions for the shapes on the handout.

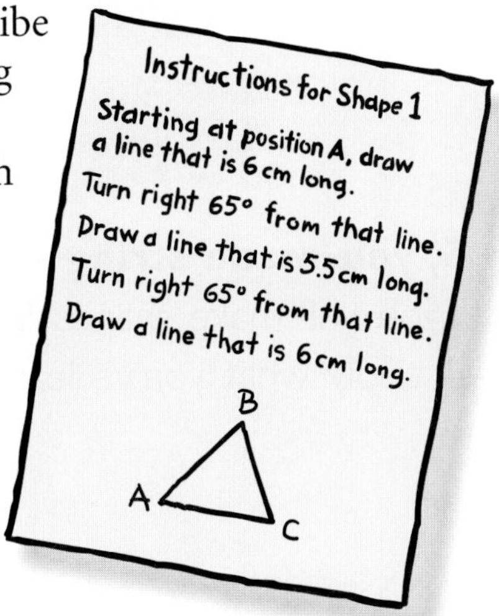

Instructions for Shape 1
Starting at position A, draw a line that is 6 cm long.
Turn right 65° from that line.
Draw a line that is 5.5 cm long.
Turn right 65° from that line.
Draw a line that is 6 cm long.
B
A
C

How can you use measures of sides and angles to describe a polygon?

Add to the Visual Glossary

Think about how you have used the terms *equal angles* and *right angles* to describe angles in this lesson. Study the pictures of equal angles and right angles.

- After the class discussion, write in your Visual Glossary the meanings of *equal angles* and *right angles* that your class developed.
- Include drawings to illustrate the definitions.

Types of Angles

Equal Angles		Right Angles
 These are equal angles.	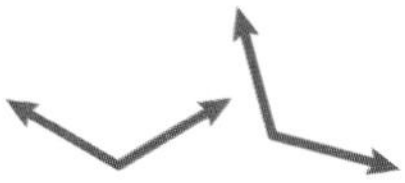 These are equal angles.	 These angles are right angles.
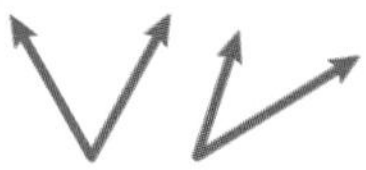 These are **not** equal angles.	 These are **not** equal angles.	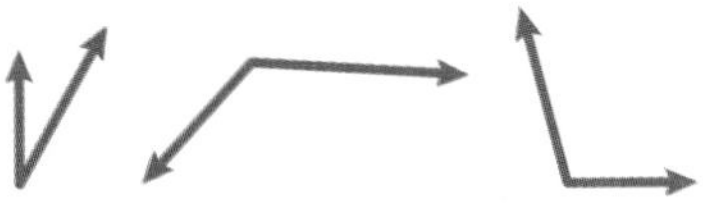 These angles are **not** right angles.

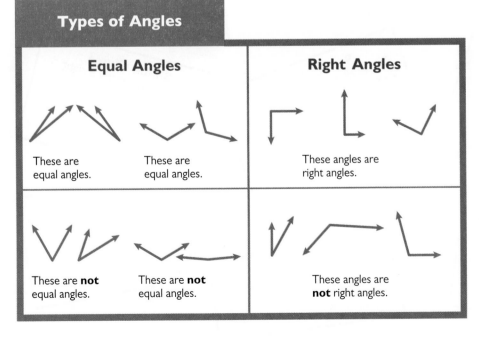

hot words equal angles, right angles

Homework

page 199

7 Shaping Up

INVESTIGATING PROPERTIES OF SIDES AND ANGLES

In Lesson 5, you made string shapes from clues about sides. Here, you will make shapes from clues about angles. Then you will apply what you've learned so far in the Sides and Angles Game.

Make Shapes from Clues

How can you use what you know about the properties of sides and angles to make shapes?

Do the following for each shape clue given:

1. Try making a shape that fits the description in the clue.
2. If you can make the shape, record it. Then label the equal angles and the right angles. (The top of the handout Naming and Labeling Polygons will help you with this.) If you can't make the shape, write "Impossible."
3. Label the shape you draw with a mathematical name. If you don't know a name for the shape, make up a name that you think describes the properties of the shape. (The bottom of the handout Naming and Labeling Polygons will help you with this.)

Shape Clue 1
A quadrilateral with exactly 2 right angles.

Shape Clue 2
A quadrilateral with exactly 2 right angles that are also opposite angles.

Shape Clue 3
A shape with exactly 3 right angles.

Shape Clue 4
A parallelogram with only 1 pair of opposite angles that are equal.

Shape Clue 5
A shape with 5 equal angles.

Shape Clue 6
A quadrilateral in which each pair of opposite angles are the same size and at least 1 pair are right angles.

Shape Clue 7
Make up a clue of your own and write it down. Test it to see if you can make the shape, and write a sentence telling why you can or cannot make it.

Play the Sides and Angles Game

In the Sides and Angles Game, you will try to make shapes that fit two different descriptions. One description is of the sides of the shape. The other is of the angles. Be sure to read the rules carefully before beginning.

What shapes can you make from clues about sides and angles?

The Sides and Angles Game Rules

Before beginning the game, place the Sides cards face-down in one pile. Place the Angles cards face-down in another pile.

1. The first player takes one card from each pile.
2. The player tries to draw a shape that fits the descriptions on the two cards. Then the player marks the drawing to show any parallel sides, equal sides, equal angles, or right angles. The player should label the shape with a mathematical name or create a name that fits. If the player cannot make a shape, the player should describe why.
3. Players take turns trying to find another way to draw the shape, following the directions in Step 2. When the group draws four shapes or runs out of possible drawings, the next player picks two new cards. Play begins again.

Add to the Visual Glossary

Think how you have used the terms *regular shape* and *opposite angles* to describe the shapes you made in this lesson.

- After class discussion, write the class definition for each term in your Visual Glossary.
- Include drawings to illustrate the definitions.

Shapes and Angles

These are regular shapes.

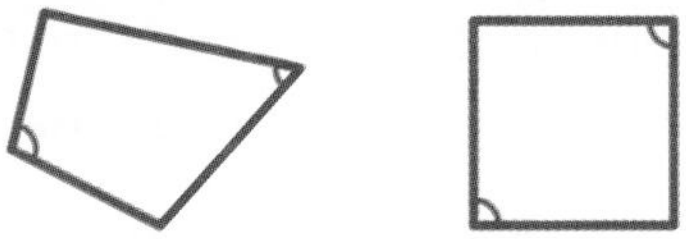
These are opposite angles.

These are **not** regular shapes.

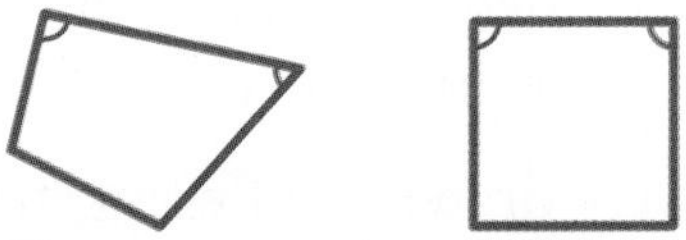
These are **not** opposite angles.

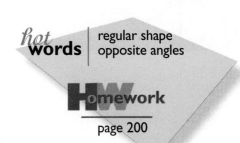

hot words | regular shape, opposite angles

Homework page 200

8 Assembling the Pieces

DESCRIBING POLYGONS

You've explored sides, angles, and shape names when describing two-dimensional shapes. In this lesson, you will investigate two more properties of shapes—perimeter and area. Then you will apply everything you've learned to describe six shapes. You will use the shapes in the next phase to build a model home.

Investigate Perimeter and Area

How many shapes can you draw with a perimeter of 10 cm? 15 cm?

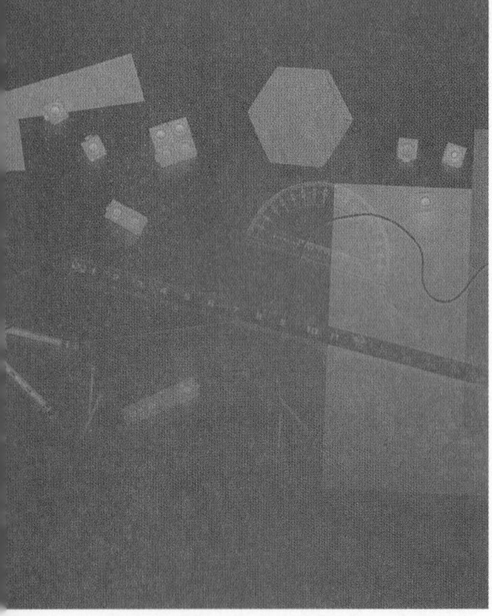

Perimeter and *area* are properties of shapes. **Perimeter** is the distance around a shape. **Area** is the number of square units a shape contains. Follow the directions given to investigate these two properties. You'll need a ruler for this activity.

The area of the small rectangle is 12 cm^2. The perimeter is 14 cm. Can you estimate the area and perimeter of the shaded triangle?

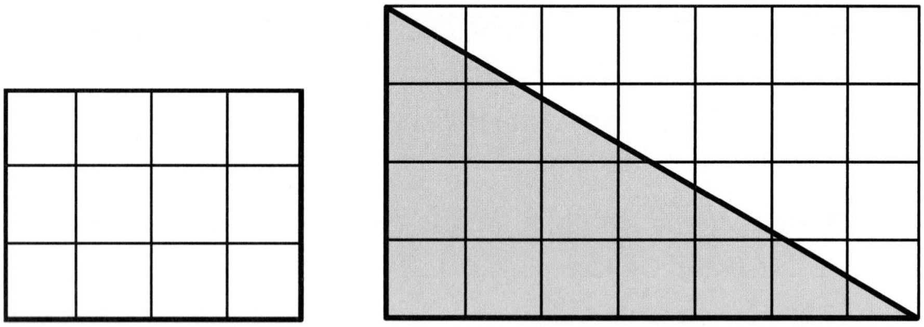

1. Choose a perimeter between 8 cm and 16 cm. Make at least four different shapes with that perimeter. Then record each shape you made and label the lengths of the sides.

2. Estimate the area in square centimeters of some of the shapes you recorded. You may find it helpful to trace the shapes on centimeter grid paper.

As you were drawing different shapes with the same perimeter, what did you notice about the area of the shapes? Why do you think this is true?

Describe Construction Shapes

Later in the unit, you will use six shapes to design and build a house. Your teacher will give you samples of these six shapes.

- Pretend that you are explaining to a manufacturer the shapes and the sizes you will be using.
- Describe each shape in as many ways as you can.
- Be sure to refer to the assessment criteria when writing your descriptions.

How can you use what you've learned to describe a shape so that someone else could draw it?

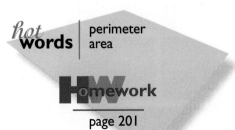

hot words | perimeter, area

Homework

page 201

PHASE THREE

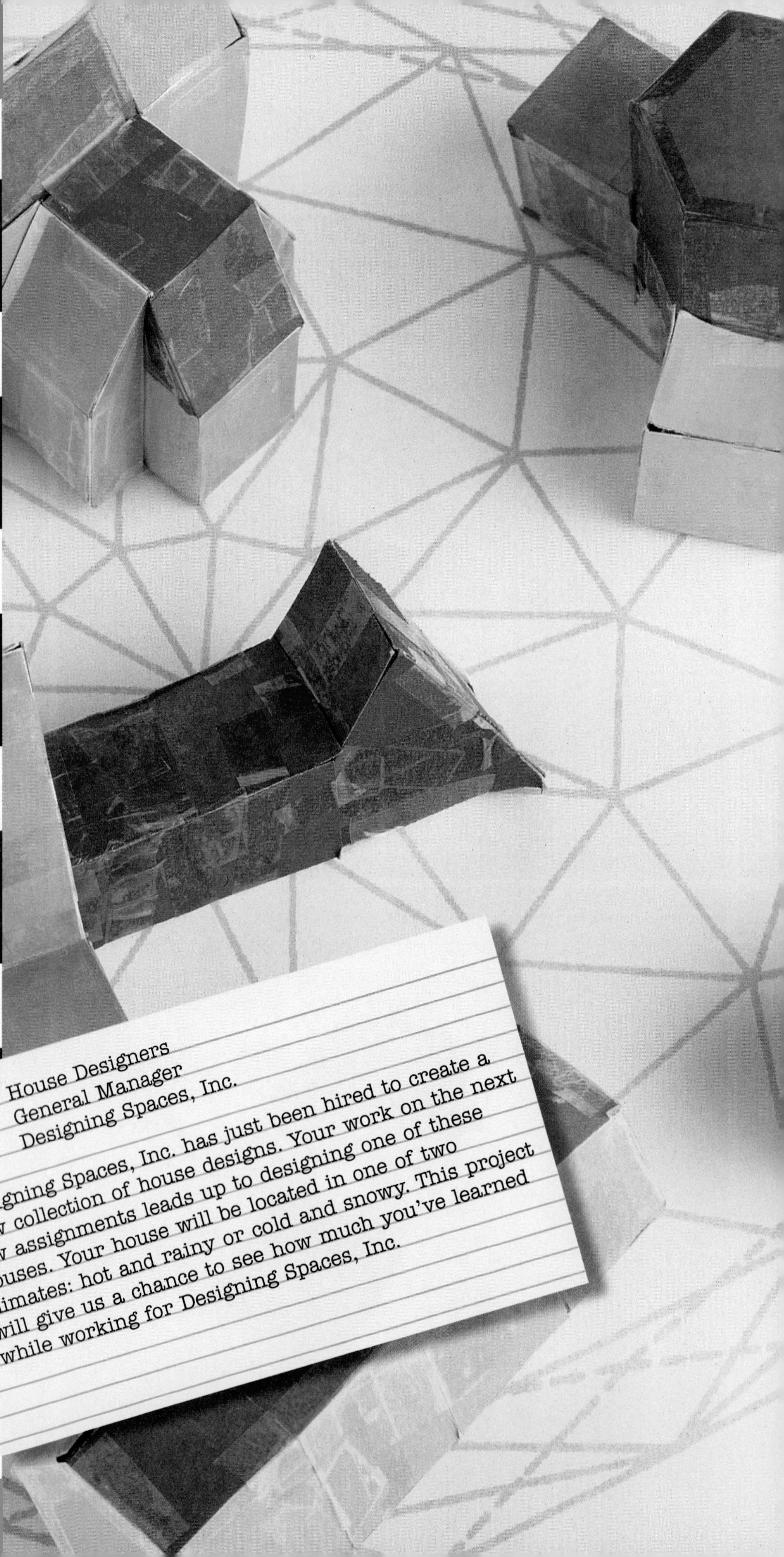

To: House Designers
From: General Manager
Designing Spaces, Inc.

Designing Spaces, Inc. has just been hired to create a new collection of house designs. Your work on the next few assignments leads up to designing one of these houses. Your house will be located in one of two climates: hot and rainy or cold and snowy. This project will give us a chance to see how much you've learned while working for Designing Spaces, Inc.

In Phase Two, you explored the properties of two-dimensional shapes that make up three-dimensional structures. In this phase, you will investigate the properties of the three-dimensional structures themselves. Using this information in the final project, you will design, build, and make plans for a model home!

Visualizing and Representing Polyhedrons

WHAT'S THE MATH?

Investigations in this section focus on:

MULTIPLE REPRESENTATIONS

- Use perspective drawing techniques to represent prisms and pyramids.
- Identify orthogonal views of a structure.

PROPERTIES and COMPONENTS of SHAPES

- Identify prisms and pyramids.
- Investigate the properties of edge, vertex, and face.
- Explore the relationships among the properties of two- and three-dimensional shapes.

VISUALIZATION

- Form visual images of structures in your mind from clues given about the structure.

MathScape Online
mathscape1.com/self_check_quiz

9 Beyond Boxes

MOVING FROM POLYGONS TO POLYHEDRONS

In Phase One, you designed houses with cubes. If you look at houses around the world, however, you will see many different shapes. Here you will use the shapes you made in Lesson 8 to build three-dimensional structures.

Construct Closed Three-Dimensional Structures

What three-dimensional structures can you build by choosing from six different shapes?

A **polyhedron** is any solid shape whose surface is made up of polygons. Using the shapes from Lesson 8, you will build three-dimensional models of polyhedrons.

1 Use the Shape Tracers to trace four of each of the following shapes on heavy paper: triangles, rectangles, squares, rhombi, trapezoids, and hexagons. Then carefully cut out the shapes.

2 Build at least two different structures from the shapes you cut out. Be sure to follow the Building Guidelines.

A thatched roof house in Central America

Building Guidelines

- Use 3–15 pieces for each structure that you build. If you need more pieces, cut them out.
- The base, or bottom, of the structure can be made of only **one** piece.
- The pieces must **not** overlap.
- The structure must be **closed.** No gaps are allowed. (Use tape to hold the shapes together.)
- No hidden pieces are allowed; you must be able to see them all.

Record and Describe Properties of Structures

In Phase Two, you used properties of sides and angles to describe polygons. You can use properties of faces, vertices, and edges to describe polyhedrons.

What are the properties of the three-dimensional structures you built?

1 For each structure you made, create a table of properties. Record the following numbers:

a. faces **b.** vertices **c.** edges

d. sets of parallel faces **e.** sets of parallel edges

2 Write a description of one of the structures you built. Include enough information so that your classmates would be able to pick out your structure from all the others. You can include drawings. Think about the following when you write your description:

a. What shapes did you use in the structure? What shape is the base?

b. How many faces, vertices, and edges does the structure have?

c. Are any edges parallel to each other?

d. Are any faces parallel to each other?

Table of Properties

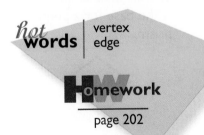

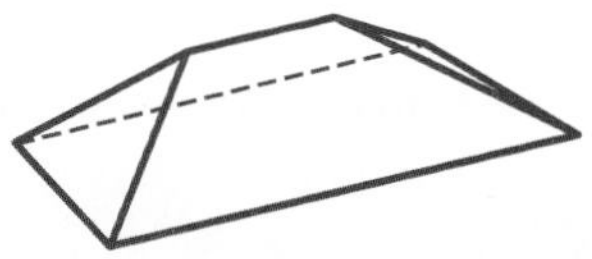

Shape of Base	Shapes Used	No. of Faces	No. of Vertices (Corners)	No. of Edges	Sets of Parallel Faces?	Sets of Parallel Edges?
Rectangle	1 rectangle 2 triangles 2 trapezoids	5	6	9	No (0)	Yes (2)

Add to the Visual Glossary

In this lesson, you used the terms *vertex* and *edge* to describe properties of polyhedrons. Write definitions for vertex and edge.

- After class discussion, add the class definitions of the terms *vertex* and *edge* to your Visual Glossary.
- Be sure to add drawings to illustrate your definitions.

hot words | vertex, edge

Homework page 202

10 Drawing Tricks

VISUALIZING AND DRAWING PRISMS AND PYRAMIDS

Architects draw many different views of their house designs. In Phase One, you learned to create isometric drawings of houses made of cubes. The methods you learn here for drawing prisms and pyramids will prepare you for drawing house plans that include other shapes.

Draw Prisms

How can you draw a prism in two dimensions?

A **prism** has two parallel faces that can be any shape. These are its **bases.** A prism gets its name from the shape of its bases. For example, if the base is a square, it is called a square prism. Read the tips on How to Draw a Prism.

- Draw some prisms on your own.
- Label each prism you draw with its name.

How to Draw a Prism

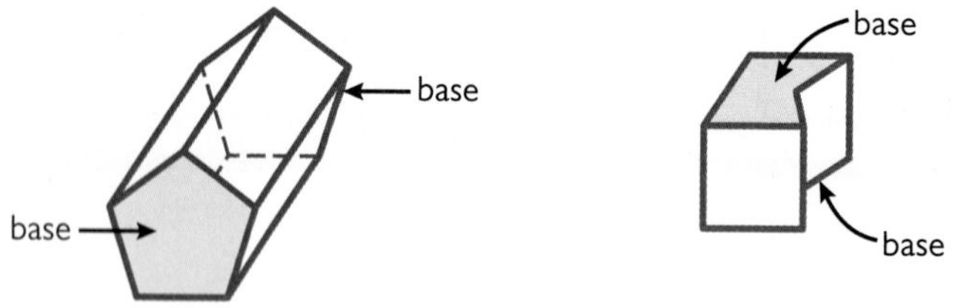

1. Draw the base of the prism. The example shown is the base for a pentagonal prism.

2. Draw the second base by making a copy of the first base. Keep corresponding sides parallel.

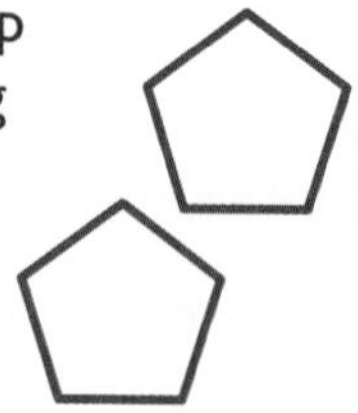

3. Connect the corresponding vertices of the two bases. This will produce a collection of parallel edges with the same length.

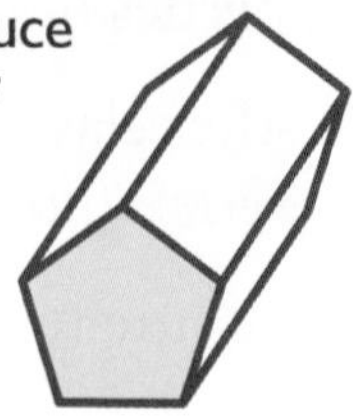

Draw Pyramids

In a **pyramid,** the base can be any shape. All the other faces are triangles. A pyramid gets its name from the shape of its base. For example, if the base is a triangle, it is called a triangular pyramid. Read the tips for How to Draw a Pyramid.

- Draw some pyramids on your own. Try some with different shapes as their bases.
- Label each pyramid you draw with its name.

How can you draw a pyramid in two dimensions?

How to Draw a Pyramid

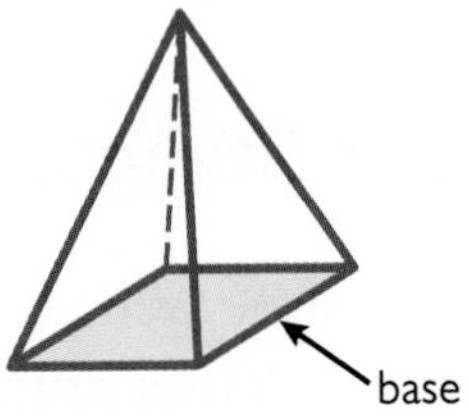

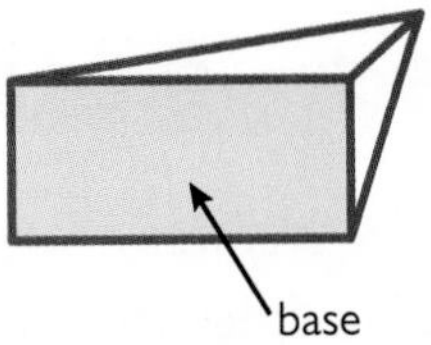

1. Draw the base. The example shown is the base for a pentagonal pyramid.

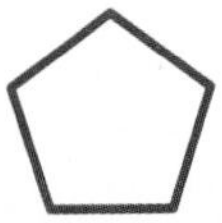

2. Mark any point outside the base.

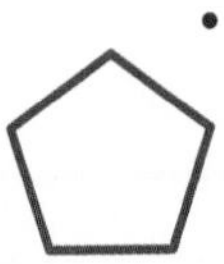

3. Draw line segments from each vertex of the base to the point.

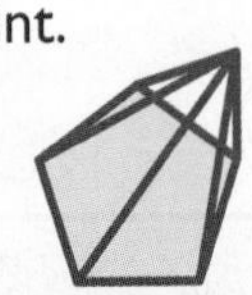

Add to the Visual Glossary

Think about the drawings you made of pyramids and prisms. What do all pyramids have in common? What do all prisms have in common?

- After class discussion, add the class definitions of the terms *prism* and *pyramid* to your Visual Glossary.
- Be sure to add drawings to illustrate your definitions.

hot words prism, pyramid

Homework
page 203

11 Mystery Structures

VISUALIZING THREE-DIMENSIONAL STRUCTURES

Have you ever solved a puzzle from a set of clues? That's what you will do to build three-dimensional Mystery Structures. Then you will build your own Mystery Structure and write clues to go with it. To create an answer key for your clues, you will use the drawing skills you have learned.

Solve the Mystery Structures Game

How can you apply what you know about two-dimensional shapes to solve clues about three-dimensional structures?

The Mystery Structures Game pulls together all of the geometric concepts you have learned in this unit. Play the Mystery Structures Game with your group. You will need the following: Round 1 clues, a set of shapes (4 each of the triangle, rhombus, trapezoid, and hexagon; 6 each of the rectangle and square), and tape.

The Mystery Structures Game

How to play:

1. Each player reads, but does not show, one of the clues to the group. If your clue has a picture on it, describe the picture.
2. Discuss what the structure might look like.
3. Build a structure that matches all the clues. Recheck each clue to make sure the structure satisfies each one. You may need to revise the structure several times.
4. Make a drawing of the structure that shows depth.

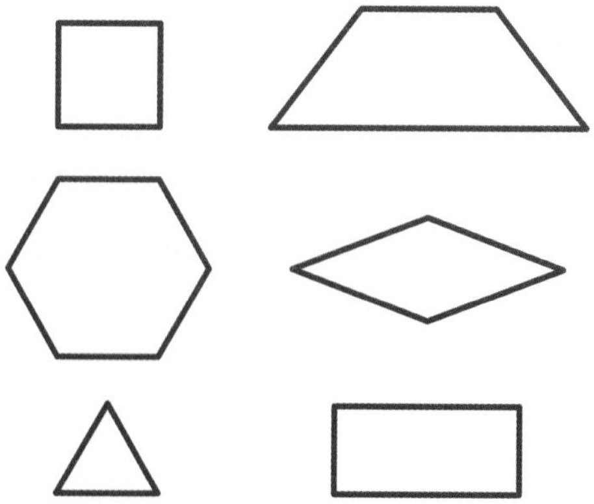

Create Clues

How can you use what you have learned about shapes to describe three-dimensional structures?

Now that you know how the Mystery Structures Game works, follow these steps to write your own set of clues for the game:

1. Build a closed, three-dimensional shape with up to 12 shape pieces. Make a structure that will be an interesting project for someone else to build.
2. Write a set of four clues about your structure. Write each clue on a separate sheet of paper. Use at least one orthogonal drawing. From your four clues, someone else should be able to build your structure.
3. Make a drawing of the structure showing depth to serve as an answer key for other students who try to follow your clues.

Give Feedback on Clues

Exchange clues with a partner. Then try to build your partner's structure. When you are done, compare your structure with the answer key. Write feedback on how your partner's clues might be improved.

- Are the drawings clear? How could you make them clearer?
- Are the clues easy to follow? What suggestions can you give to improve the clues?
- What are one or two things that you think are done well?

Houses in the Syrian Desert

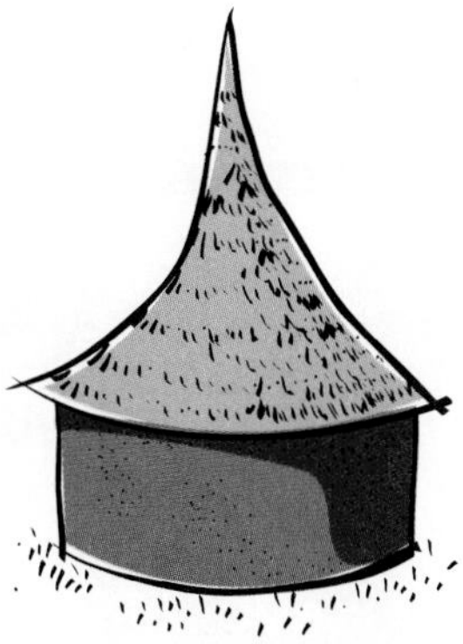

A house of the Kirdi people in Cameroon

A yurt of the Kazakh

hot words polygon, polyhedron

Homework page 204

12 Putting the Pieces Together

FINAL PROJECT

The climate in which a house is built can influence the shape of the house and the materials used to build the house. For your final project, you will research houses in different climates around the world. Then you will apply what you have learned to design a house model for a specific climate.

What kind of three-dimensional structure can you make that meets the building guidelines?

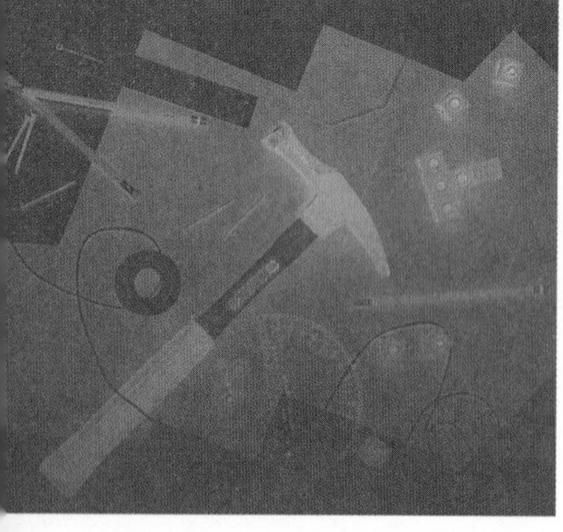

Build a House

Before you start building your house, read about home designs in different climates. Choose the climate for your house model: a hot and rainy climate or a cold and snowy climate. Think about the design features that your house should have for that climate. Then build the house model.

1. Use the Shape Tracers to trace and cut out a set of building shapes.
2. Build a house using the shape pieces. Be sure to follow the Building Guidelines when constructing your house.

Building Guidelines

- The home must have a roof. The roof should be constructed carefully so that it will not leak.
- The home must be able to stand on its own.
- You must use 20–24 shape pieces. You do not need to use each type of shape.
- You may add one new type of shape.

Create a Set of Plans

After building your house, make a set of plans. Another person should be able to use the plans to build your house. Use the following guidelines to help you create your plans:

- Include both orthogonal and isometric drawings of your house. Be sure to label the drawings. The labels will make your plans easier to understand.
- Include a written, step-by-step description of the building process so that someone else could build your house. Use the names and properties of shapes to make your description clear and precise.

How can you represent your house in two dimensions?

Write Design Specifications

Write a memo to the Directors of Designing Spaces, Inc. describing your house. The specifications should explain the design clearly. The Marketing Department should be able to understand and sell the design. The designers should be able to make the shapes and calculate the approximate costs. The memo should answer the following questions:

- What does your home look like?
- How would you describe the shape of the entire house?
- How would you describe the shape of its base, its roof, and any special features?
- What climate is your house designed for?
- What features does your home have to make it well suited to the climate?
- Where in the world could your house be located?
- What shapes are used in your house?
- How many of each shape are there?
- How many edges and vertices does your house have?
- Does your house have any parallel faces? If so, how many?

hot words
isometric drawing
orthogonal drawing

Homework
page 205

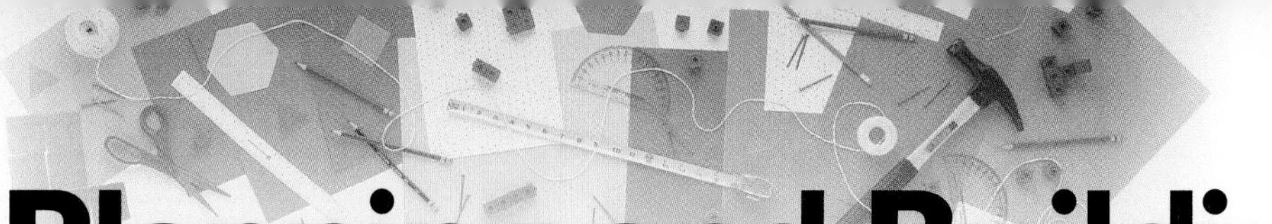

Homework 1: Planning and Building a Modular House

Applying Skills

How many faces does each structure have?

1. 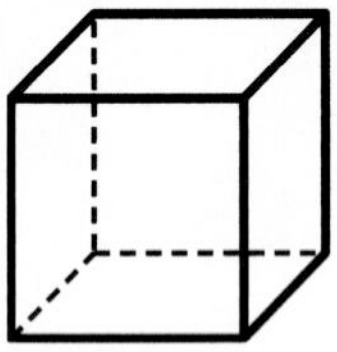

2. 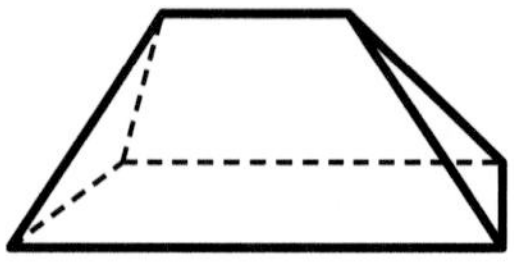

3. 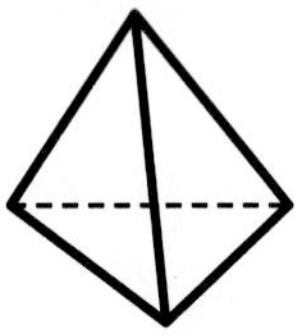

4. 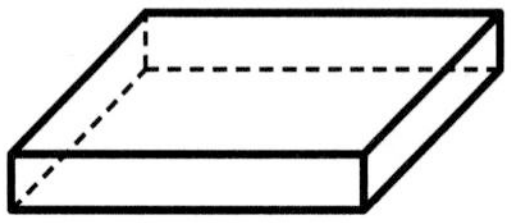

How many cubes are in each model? Remember that each cube on an upper level must have a cube below to support it.

5. 

6. 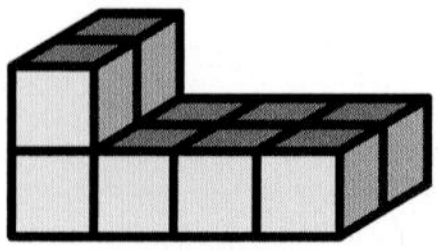

7. 

8. 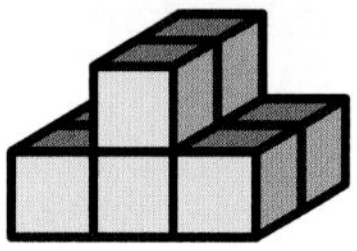

9. Of models 5–8, which ones have the same bottom layer?

Extending Concepts

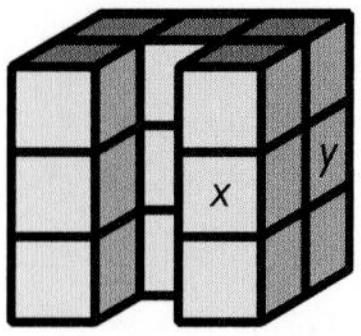

Look at this model house. Suppose you painted all the outside surfaces of the model, including the underside.

10. How many squares would you paint?
11. How many faces would be painted on the cube marked *x*? on the cube marked *y*?
12. Would any cube have 4 faces painted? 5 faces painted? Why or why not?
13. Draw a 6-face figure that is not a cube.

Writing

14. Look in the real estate section of a newspaper or in magazines about housing to find pictures that show houses or buildings in different ways. Find two or three examples of different ways to show structures. Tell why you think the artist chose each one.

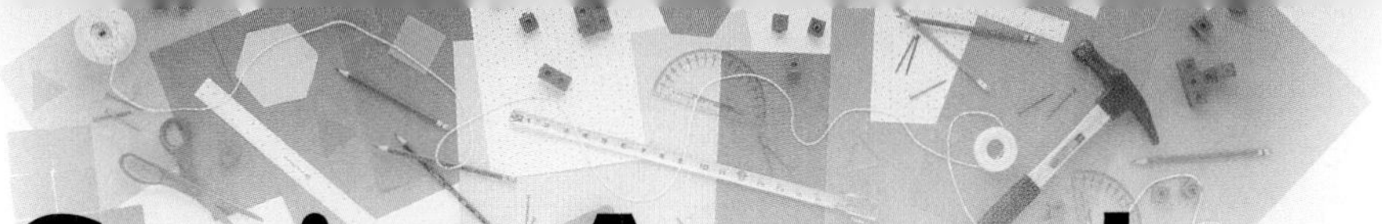

Homework 2: Seeing Around the Corners

Applying Skills

Look at the first structure in each box. Tell which of the other structures are rotations of the sample. Answer *yes* or *no* for each structure.

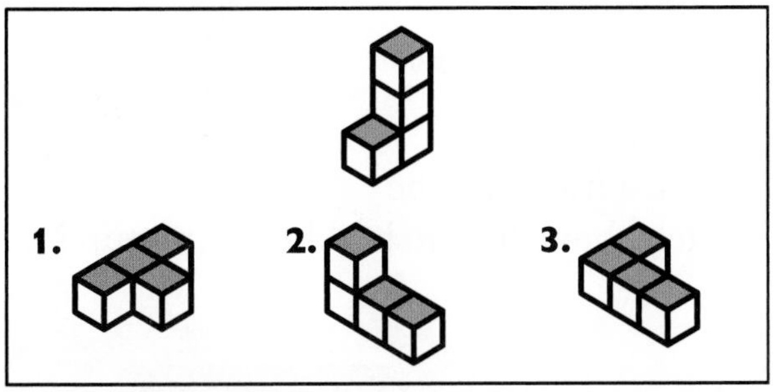

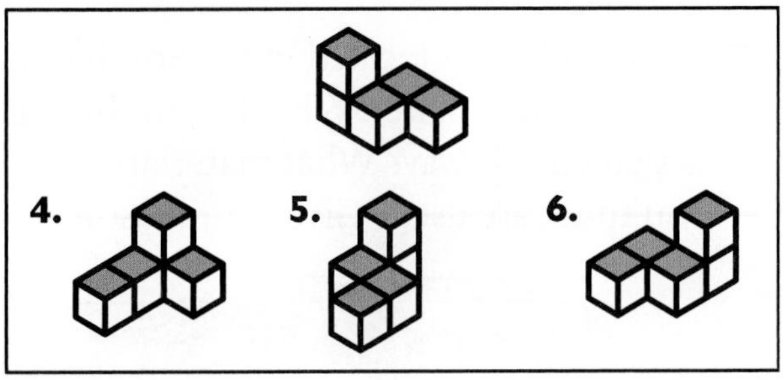

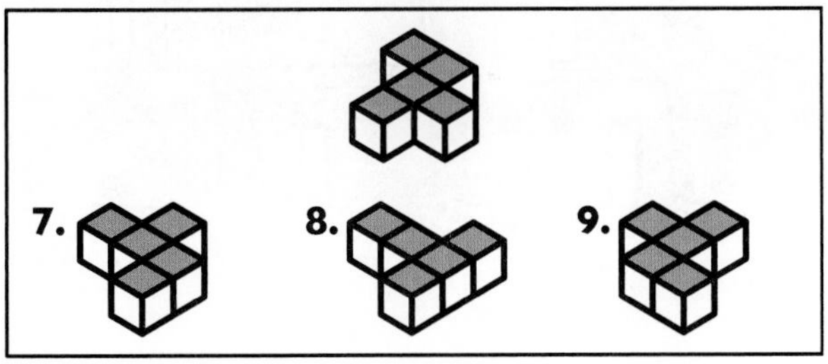

Extending Concepts

10. Draw this structure in a different position. Show how it would look:

a. rotated halfway around

b. lifted up, standing on its side

Making Connections

11. Draw the next structure in this pattern.

12. Make an isometric drawing of your house, school, or other building as if you were looking at it from the front. Then, make an isometric drawing of the structure as if you were looking at it from the left side.

Homework 3

Seeing All Possibilities

Applying Skills

Look carefully at each isometric drawing. The top side is shaded. Write *front, back, top, side,* or *not possible* for each orthogonal view.

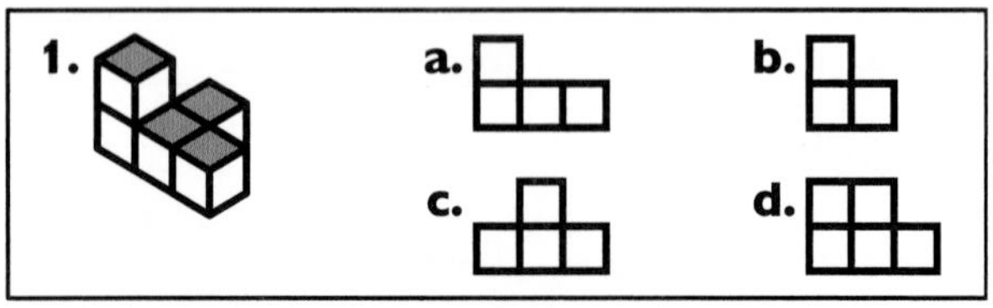

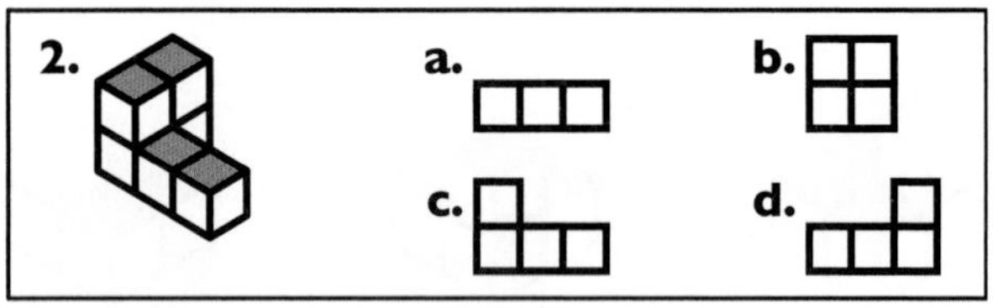

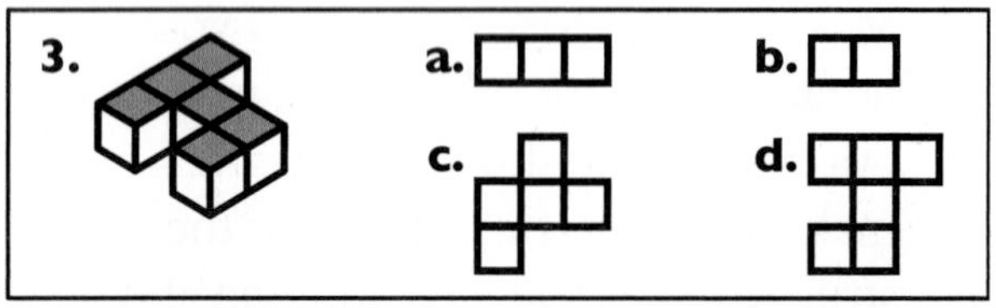

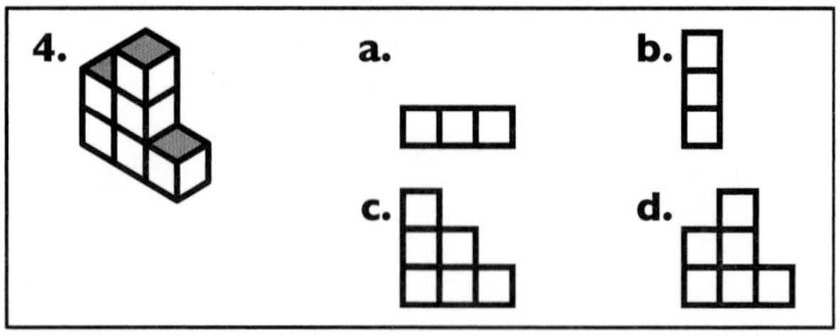

Extending Concepts

5. Make orthogonal drawings of this model that show these views:

a. top view

b. back view

c. right side view

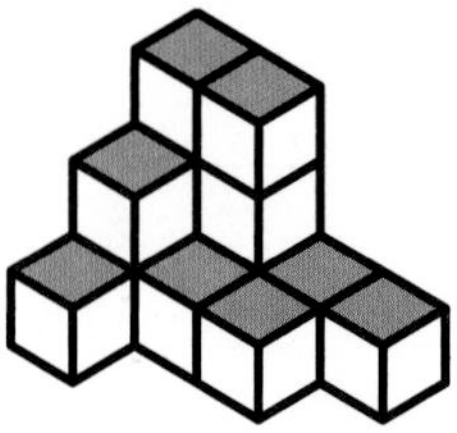

6. How would you tell a friend on the telephone how to build this structure?

Writing

7. This dwelling is located in the Southwest of the United States. Why do you think it is shaped this way? What materials do you think are used? Give your reasons.

Homework 4: Picture This

Applying Skills

Look carefully at each set of orthogonal drawings. Choose which isometric drawings could show the same structure. Answer *possible* or *not possible.* Some structures may be rotations.

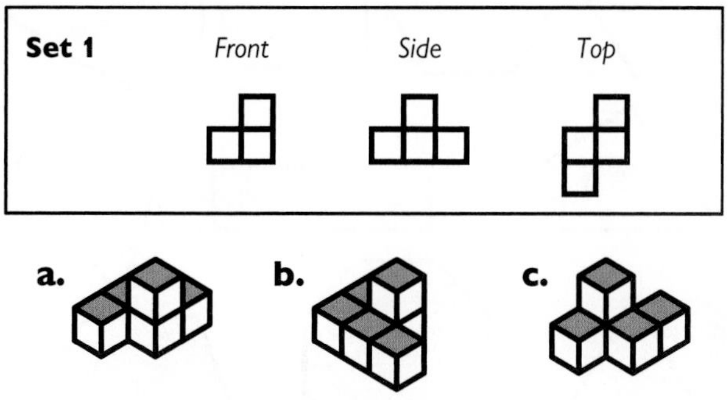

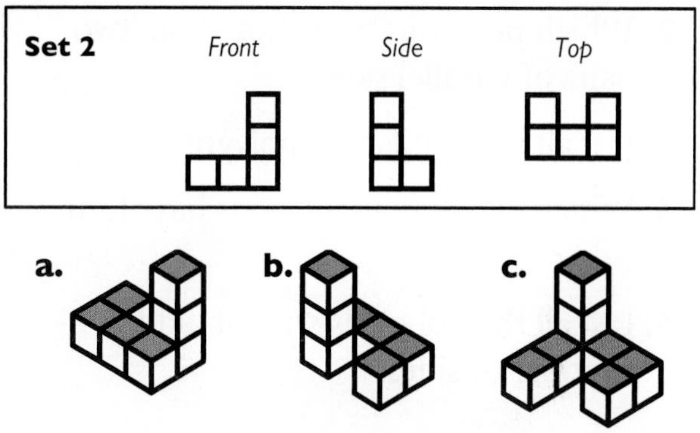

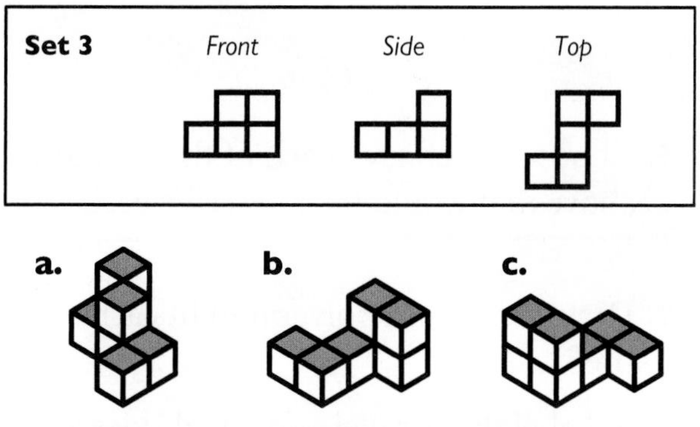

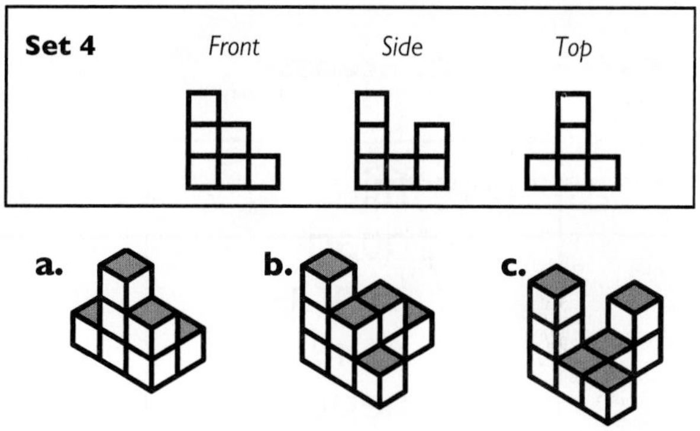

Extending Concepts

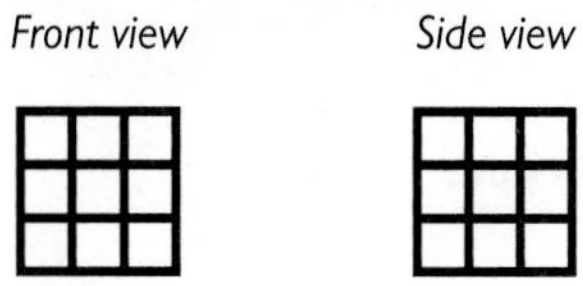

5. Draw at least three different top views of structures that all could have this front view and right side view.

Making Connections

6. Make orthogonal sketches of your school building. Show how you think it looks from the top, front, and right side.

Homework 5

String Shapes

Applying Skills

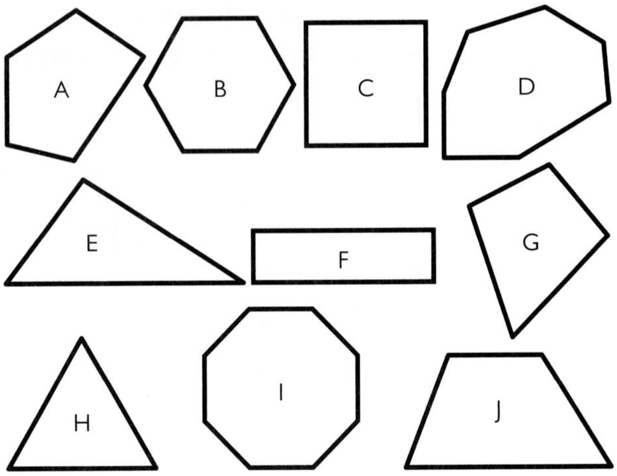

1. Which polygons have at least two parallel sides?
2. Which polygons have more than two pairs of parallel sides?
3. List all the equilateral polygons.
4. Which equilateral polygons have four sides?
5. List all the polygons with an odd number of sides.

Extending Concepts

tri- quad- penta- hexa- hepta- octo-

6. Choose a prefix for each of the polygons above that would help describe each figure.
7. Draw a different polygon to fit each prefix.
   a. Label the parallel and equal sides.
   b. Write clues to help someone identify each of your polygons.

Making Connections

8. Match the titles to the pictures. Tell why you made each choice.

The Pentagon
Tripod
Hexagram
Octopus
Quartet

Homework 6

Polygon Paths

Applying Skills

For items 1 through 6, use a protractor and ruler.

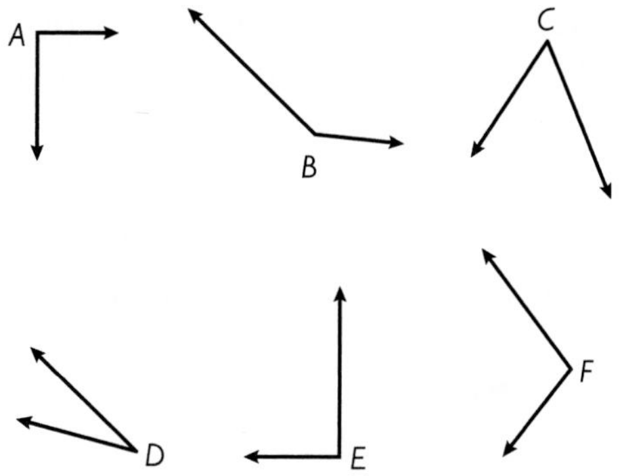

1. Which angles measure less than 90°?

2. Which angles measure greater than 90°?

3. Which angles measure exactly 90°?

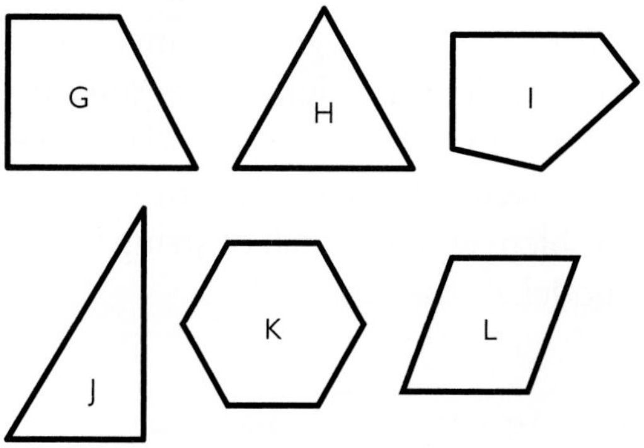

4. Which polygons have at least one right angle?

5. Which polygons have at least two equal angles?

6. Which polygons have all equal angles?

Extending Concepts

7. Use a protractor and ruler. Design and draw a path to create a polygon for each of the following specifications. What is the name of each polygon?

a. four sides

at least two sides parallel

no 90° angles

b. three sides

one 90° angle

two sides of equal length

Writing

8. Draw an imaginary neighborhood that includes a path made up of straight line segments (going from one end of the neighborhood to the other). Describe the path by giving the length of each line segment and the angle and direction of each turn that makes a path through the neighborhood.

Example:

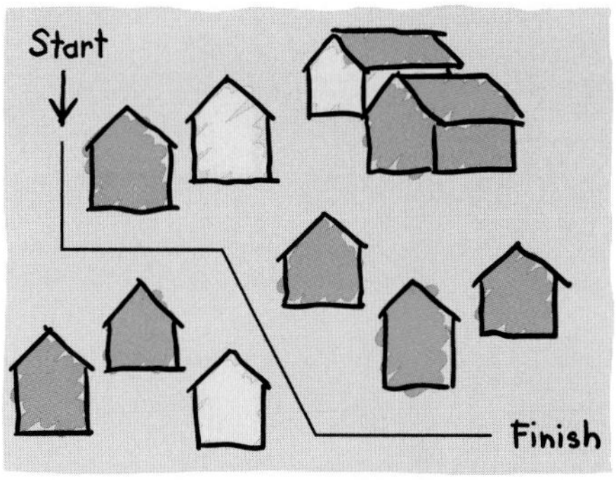

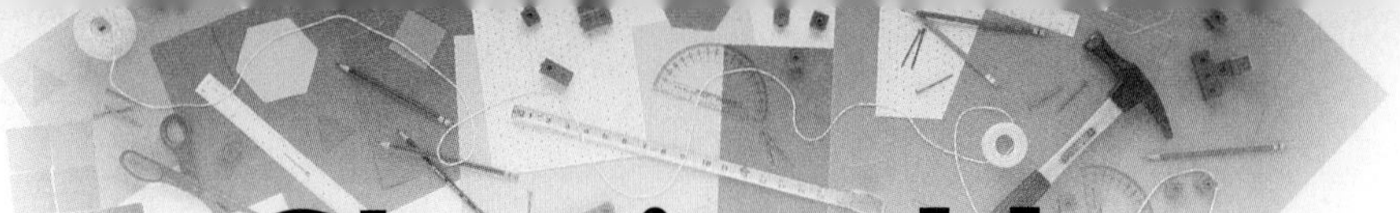

7 Homework

Shaping Up

Applying Skills

Tell whether each polygon is *regular* or *not regular.*

1. 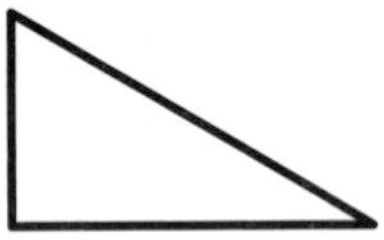

2. 

3. 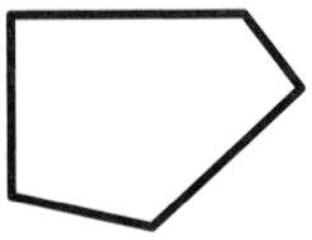

4. 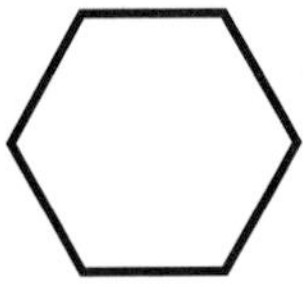

5. 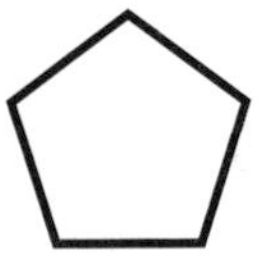

6. 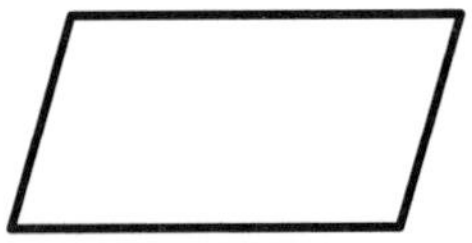

7. 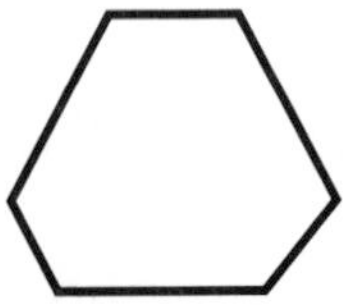

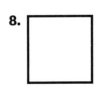

8. 

Copy each quadrilateral and mark one pair of opposite angles.

9. 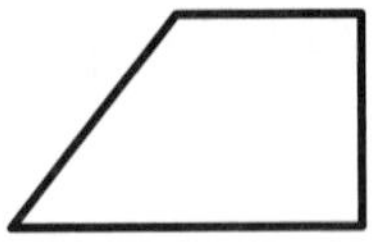

10. 

11. 

12. 

Extending Concepts

String or rope can be used as a tool for making right angles.

Method: A twelve-foot length of rope is marked at every foot and staked into a triangle. The sides are adjusted until one side is 3 feet, one side is 4 feet, and the third side is 5 feet. When these sides are exact, the triangle will contain a right angle.

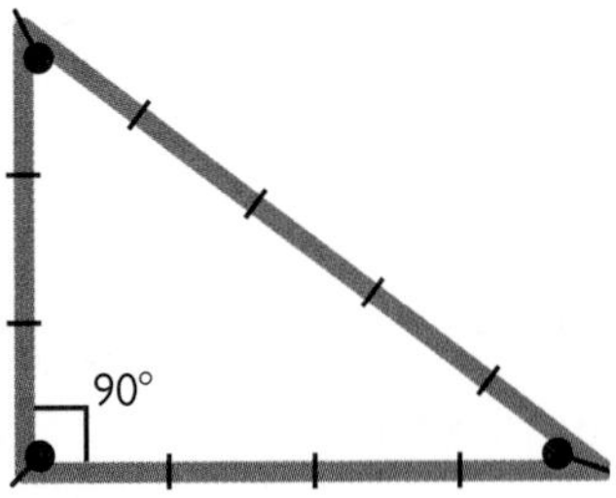

13. Find another trio of side lengths that will create a right triangle. Explain how you know you have a right angle. You can use string or rope to make a model. (An easy scale factor to use is 1 foot of rope in the problem equals 1 inch of string in your model.)

Making Connections

14. Find two other trios of side lengths that will create a right angle in a triangle. There are patterns to some of these trios of side lengths. If you see a pattern, describe it and make a prediction for how you would find other side lengths.

Assembling the Pieces

Applying Skills

1. Which of rectangles A–H have the same perimeter?

2. Which of rectangles A–H have the same area?

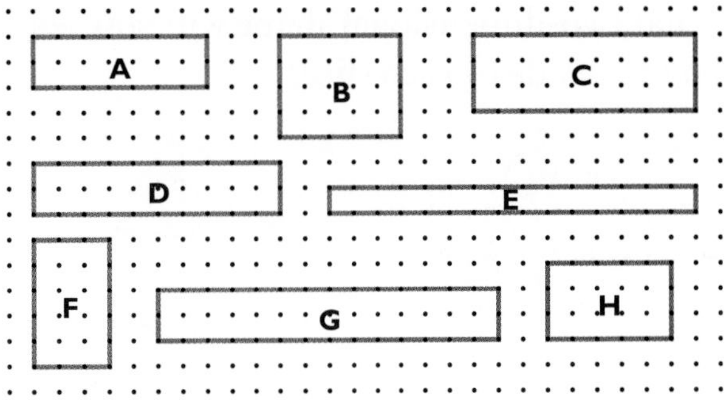

Use centimeter grid or dot paper.

3. Draw a polygon that is not regular and has the same perimeter as rectangle B.

4. Draw a regular polygon that has the same perimeter as rectangle F.

Extending Concepts

5. Use a piece of string about 20 cm long. Knot it into a loop. Use the loop as the perimeter of at least three different triangles. Sketch your triangles on grid paper or dot paper to compare them. Estimate the area of each triangle.

6. Look over these definitions. Do they exactly describe each polygon? Can you improve the lists of clues?

a.

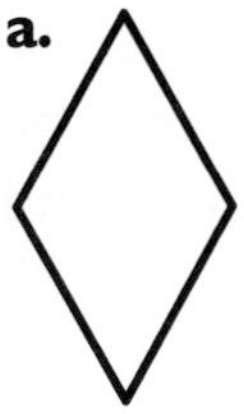

It has equal sides.
It has parallel sides.
It has four sides.

b.

It is not regular.
Its sides are not equal.
It has no parallel sides.

Writing

7. Answer the letter to Dr. Math.

> Dear Dr. Math:
>
> I don't understand how two rectangles with exactly the same perimeter can enclose different areas. Can you explain that to me?
>
> Perry Mitter

Homework 9

Beyond Boxes

Applying Skills

What two-dimensional shape will you see on the inside of each slice?

1. 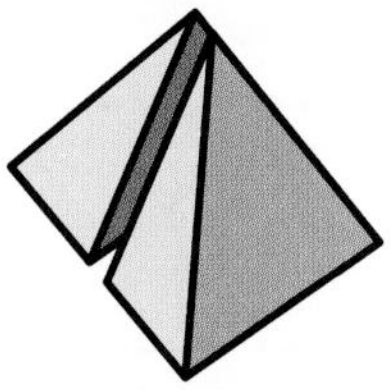

2. 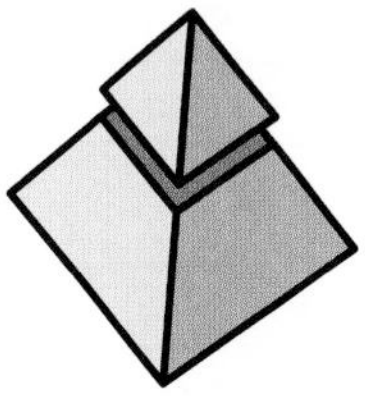

3. 

4. 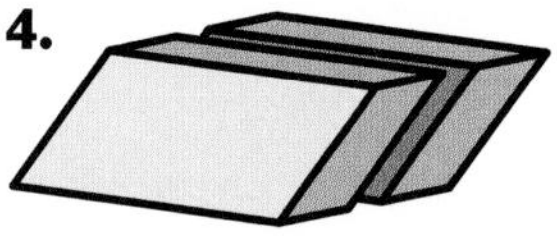

Copy the chart and fill it in for each structure.

5.  6.  7. 

	5.	6.	7.
Shape of base			
Shapes used			
Number of faces			
Number of vertices			
Number of edges			
Sets of parallel faces			
Sets of parallel edges			

Extending Concepts

8. Draw three more three-dimensional figures. Add them to the chart you made for items **5–7.** Can you find a pattern in the relationship between the numbers of edges, vertices, and faces in the shapes? Tell all the steps in your thinking.

9. Can you draw a figure that does not follow the pattern you described in item **8**?

Writing

10. Imagine a penny. Now imagine ten pennies stacked carefully so that all the edges line up evenly.

a. Describe the shape of the stack. What properties does it have?

b. Imagine slicing that stack of pennies down the middle and opening the two halves. What is the two-dimensional shape of the new faces that are formed by the slice?

c. Think of another example of stacking a group of flat objects to form a new shape. Describe the new shape and how you would make it.

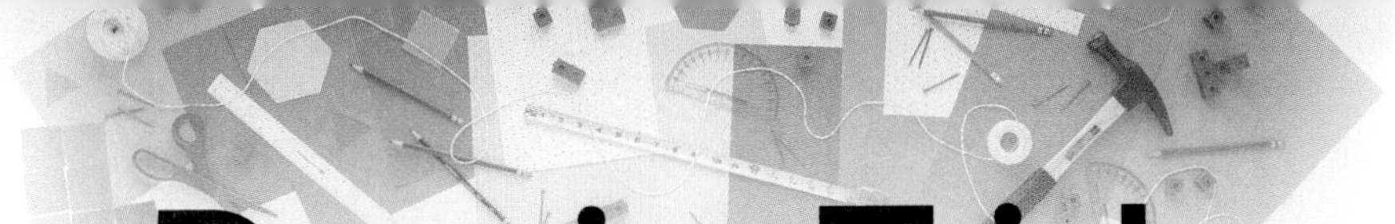

Drawing Tricks

Homework 10

Applying Skills

Is the figure a prism or a pyramid? Name the polygon forming the base.

1. 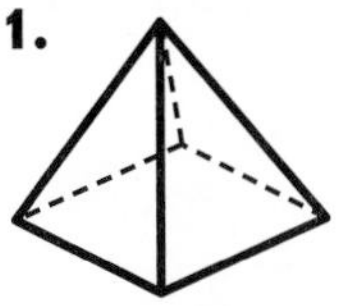 **2.** 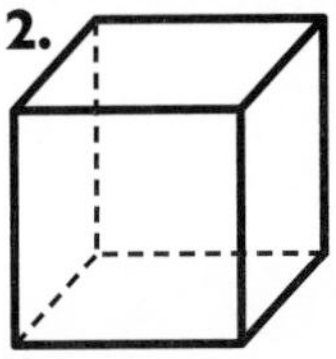 **3.** 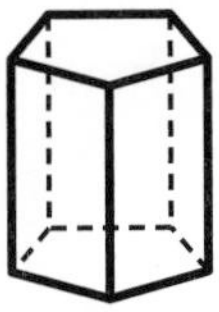

4. 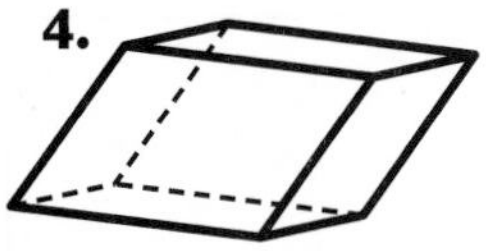 **5.** 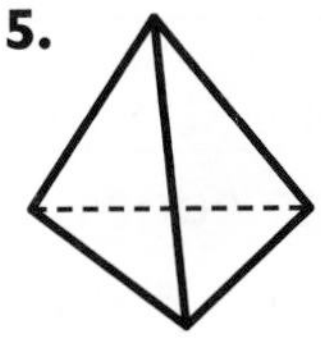 **6.** 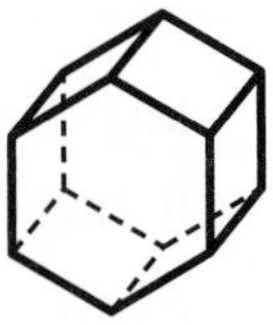

7.  **8.** 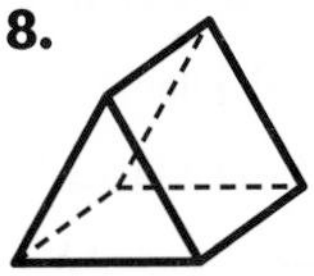 **9.** 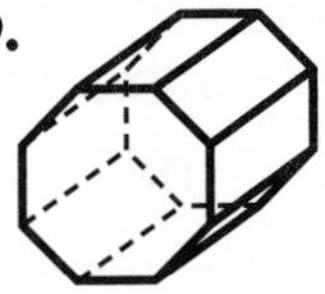

10. 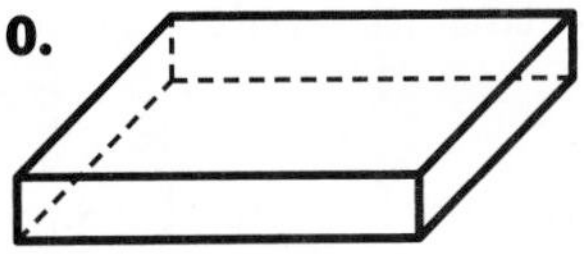

Draw each polyhedron. Tell how many faces it has.

11. Pentagonal pyramid

12. Pentagonal prism

13. Triangular prism

14. Triangular pyramid

Extending Concepts

15. Imagine that you are walking around the group of buildings in the pictures. Choose one picture as a starting point. List the pictures in the order that shows what you would see as you circle the buildings.

Writing

16. Tell how you visualized moving around the group of buildings.

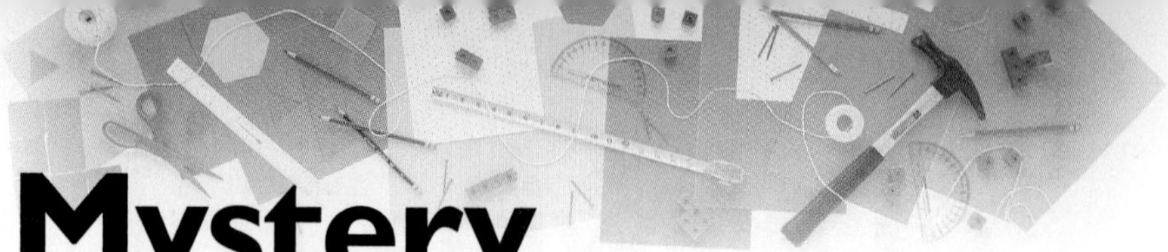

Homework 11 Mystery Structures

Applying Skills

For items 1–8, give the letters of the polyhedrons that fit the clues.

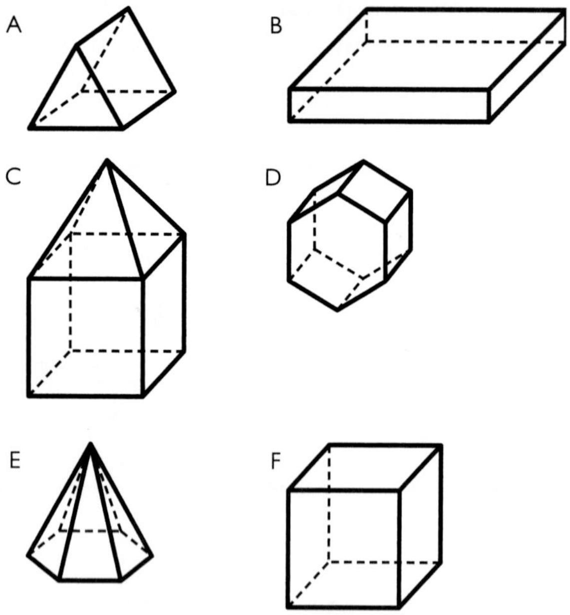

1. It has 7 vertices.

2. Its base is a square.

3. The structure is a prism.

4. It has exactly 3 sets of parallel faces.

5. All sides are equilateral.

6. It has 16 edges.

7. It has 4 sets of parallel faces.

8. It has no sets of parallel faces.

Extending Concepts

9. Draw two different polyhedrons with quadrilateral bases. Make a list of clues for each structure to help someone tell them apart.

Writing

> <u>My Mystery Structure</u>
> It has some edges.
> All the edges are the same length.
> It comes to a point on top.
> Some of the pieces are triangles.
> The base is a square.

10. Do you think this student wrote a good list of clues?

a. Draw a structure that fits the clues.

b. Could you draw a different structure that fits the same clues?

c. Rewrite this list of clues if you think it can be improved. Explain your changes.

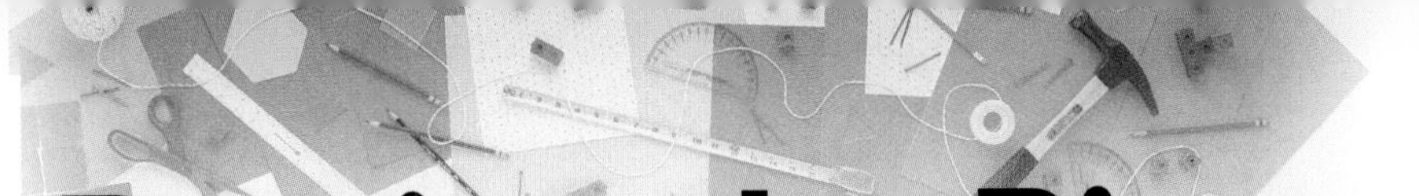

Putting the Pieces Together

Applying Skills

Make a chart like the one shown. Fill in the chart to describe each shape or structure.

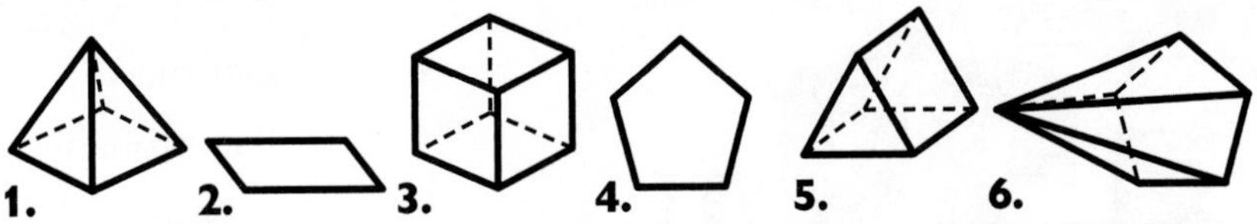

1. 2. 3. 4. 5. 6.

	1.	2.	3.	4.	5.	6.
Name of shape or structure						
Polygon or polyhedron						
Number of sides or faces						
Number of vertices						
Number of right angles						
Sets of parallel sides or edges						

Extending Concepts

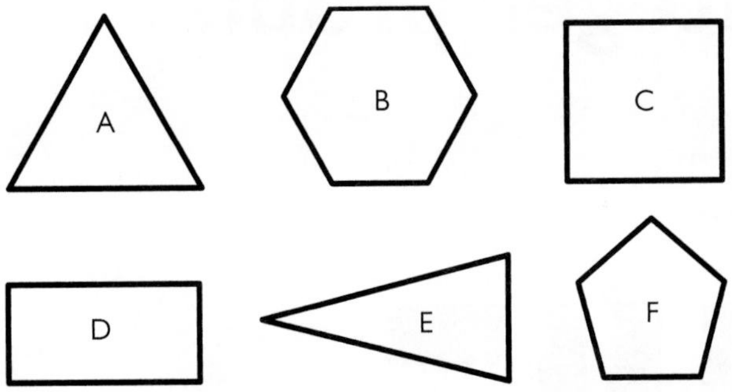

Using the pieces shown, tell how many of each piece you would need to make each polyhedron. (Assume that you can make matching edges fit.)

7. Hexagonal pyramid

8. Triangular prism

9. Cube

10. Pentagonal prism

Writing

11. Write a letter giving some good advice for next year's designers. Tell about some things that helped you with the investigations.

PHASE**ONE**

Decimals

You probably have some change in your pocket right now. In this phase, you will investigate how money relates to our decimal system. You will be building on what you know about equivalent fractions to help you make "cents" of it all.

How can you use your own number sense to solve fraction, decimal, percent, and integer problems?

BESIDE THE POINT

PHASE**TWO**
Computing with Decimals

Decimals are all around us in the real world. In this phase, you will work on computation strategies for addition, subtraction, multiplication, and division of decimals. You will use what you know about the relationship between fractions and decimals to help you find efficient computation methods.

PHASE**THREE**
Percents

This phase explores the relationship among fractions, decimals, and percents. You will calculate percents to help interpret data. You will investigate percents that are less than 1% and percents that are greater than 100%. Finally, you will investigate how percents are commonly used and how they are sometimes misused.

PHASE**FOUR**
The Integers

You have probably heard negative numbers being used when discussing topics such as games and temperatures. In this phase, you will learn how negative numbers complete the integers. You will use number lines and cubes to learn how to add and subtract with negative numbers.

PHASE ONE

How is all your loose change related to decimals? How can you take what you know about money to help to make sense of decimals? This phase will help you make connections between fractions and decimals.

Decimals

WHAT'S THE MATH?

Investigations in this section focus on:

NUMBER AND COMPUTATION

- Expressing money in decimal notation
- Using area models to interpret and represent decimals
- Writing fractions as decimals and decimals as fractions
- Comparing fractions and decimals

ESTIMATION

- Rounding decimals

MathScape Online
mathscape1.com/self_check_quiz

1 The Fraction-Decimal Connection

USING DECIMALS IN MONEY NOTATION

You are already familiar with decimals, because you often see them in daily life. Every time you use money or compare prices, you use decimals. To begin this in-depth study of decimals, you will relate familiar coins to decimal numbers.

How are fractions and decimals related?

Make Penny Rectangles

Using the Paper Pennies handout, work with a partner to determine how many different ways you can arrange 100 pennies in the shape of a rectangle. Each row must have the same number of pennies.

Make a table like the one below. Use it to organize your answers. Some answers may be the same rectangle, oriented a different way.

Describe the rectangle made with 100 pennies.	Write one row as a fraction of the whole.	Write the simplified fraction.	Write the value of one row of money.
4 rows of 25 pennies 4×25	$\frac{25}{100}$	$\frac{1}{4}$	\$0.25

Use Area Models

How can decimals be represented using area models?

You can use area models to represent fractions and decimals.

1 Write the fraction represented by the shaded area of each square or squares. Then, write the corresponding decimal.

a. 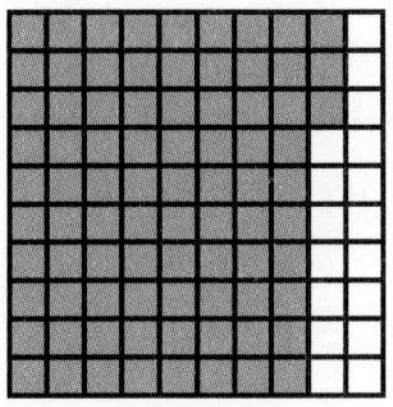

b. 

c. 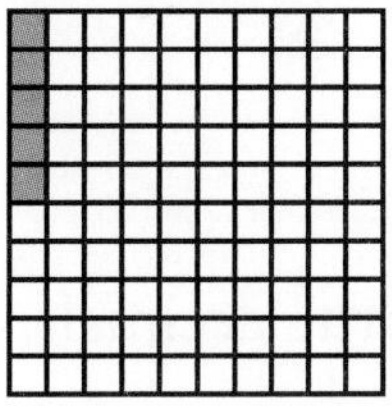

d. 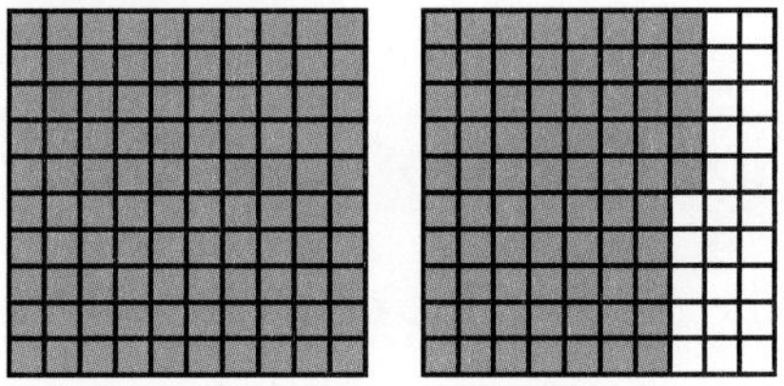

2 Draw an area model for each decimal.

a. 0.41 **b.** 0.95 **c.** 1.10 **d.** 1.38

Solve a Decimal Puzzle

Now, you and a partner will make and solve decimal puzzles. Cut out squares that are 10 centimeters by 10 centimeters.

- Divide your square into at least four parts. Shade each part a different color. Try to make each part a different shape and size.
- Carefully cut out the parts along the lines to create puzzle pieces. Trade puzzles with your partner.
- Describe each piece of your partner's puzzle as a fraction of the whole and as a decimal. Then, try to put the puzzle together.
- Write two number sentences that describe the puzzle. Use fractions in one number sentence and decimals in the other.

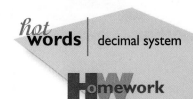

hot **words** | decimal system

Homework page 256

2 What's the Point?

WRITING FRACTIONS AS DECIMALS

To write fractions as decimals in Lesson 1, you had to use place value. Now, you will write some more fractions as decimals. But, some of these decimals will require more decimal places.

Write Fractions as Decimals

How can you write a fraction as a decimal?

Use the place-value chart below to help you write decimals.

1 Write two different numbers that both have a 7 in the tens place, a 3 in the hundreds place, a 0 in the tenths place, a 4 in the hundredths place, and a 2 in the ones place. The two numbers will need to differ in other places.

2 Write each number as a decimal.

a. $13\frac{7}{10}$ **b.** $23\frac{4}{10}$ **c.** $2\frac{13}{100}$

3 Use what you know about equivalent fractions to write each number as a decimal.

a. $25\frac{1}{2}$ **b.** $12\frac{3}{5}$ **c.** $3\frac{8}{25}$

4 How could you write the number 17 so that it has a tenths digit, but still represents the same number?

Decimal Place Value

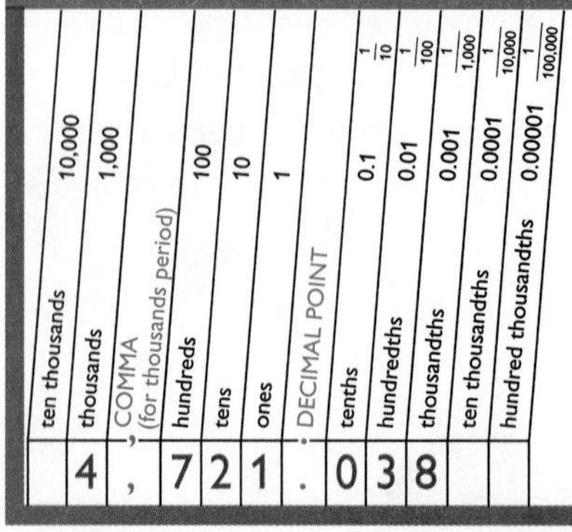

What number is shown in the diagram?

What digit is in the tenths place?

Find a Conversion Method

Study the patterns below.

	Pattern I		Pattern II	
	$4{,}000 \div 5 = 800$	$\frac{4{,}000}{5} = 800$	$1{,}000 \div 8 = 125$	$\frac{1{,}000}{8} = 125$
	$400 \div 5 = 80$	$\frac{400}{5} = 80$	$100 \div 8 = 12.5$	$\frac{100}{8} = 12.5$
	$40 \div 5 = 8$	$\frac{40}{5} = 8$	$10 \div 8 = 1.25$	$\frac{10}{8} = 1.25$
	$4 \div 5 = ?$	$\frac{4}{5} = ?$	$1 \div 8 = ?$	$\frac{1}{8} = ?$

Can you devise a method for changing any fraction to a decimal?

1 How could you change $\frac{4}{5}$ and $\frac{1}{8}$ to a decimal without using equivalent fractions?

2 Use your method to write each fraction as a decimal. Use a calculator to do the calculations.

a. $\frac{5}{8}$ **b.** $3\frac{3}{8}$ **c.** $\frac{9}{16}$

d. $\frac{7}{32}$ **e.** $\frac{13}{40}$ **f.** $\frac{5}{64}$

hot words place value, equivalent fractions

Homework page 257

3 Put Them in Order

COMPARING AND ORDERING DECIMALS

You see decimals used every day, and you frequently need to compare them. In this lesson, you will first complete a handout to learn how to compare decimals using a number line. Then, you will learn to compare decimals without using a number line. Finally, you will play a game in which strategy helps you to create winning numbers.

Compare Decimals

How can you compare decimals?

Chris and Pat's teacher asked them to determine whether 23.7 or 3.17 is greater.

Pat said, "I lined them up and compared the digits. It is just like comparing 237 and 317. Since 3 is greater than 2, I know 3.17 is greater than 23.7."

Chris listened to Pat's explanation and said, "I agree that 317 is greater than 237. But 3.17 is not greater than 23.7, because 3.17 is a little more than 3 and 23.7 is more than 23. I do not think you lined up the numbers correctly."

1. Do you agree with Pat or with Chris? How would you explain which of the two numbers is greater? How would you line up the numbers? Write a rule for comparing decimal numbers. Explain why your rule works.

2. Use your rule to place the following numbers in order from least to greatest. If two numbers are equal, place an equals sign between them.

   18.1 18.01 18.10 1.80494 2.785 0.2785 18.110

Play the "Place Value" Game

How can you create the least and greatest decimals?

The "Place Value" game is a game for two players.

1 Play a sample round of the "Place Value" game. How many points did each player get? Analyze the possibilities.

  a. Suppose your partner's digit placements remained the same. Could you have placed your digits differently and earned two points?

  b. Suppose your digit placements remained the same. Could your partner have placed his or her digits differently and earned two points?

  c. Choosing from the six digits you drew, what is the greatest number you could make? What is the least number?

  d. Choosing from the six digits your partner drew, what is the greatest number he or she could make? What is the least number?

2 Play the game until someone gets 10 points. Discuss strategies for winning the game.

Place Value Game

Each player will need:

- ten pieces of paper each with a digit (0, 1, 2, 3, 4, 5, 6, 7, 8, 9),
- a bag to hold the pieces of paper, and
- the following game layout for each round.

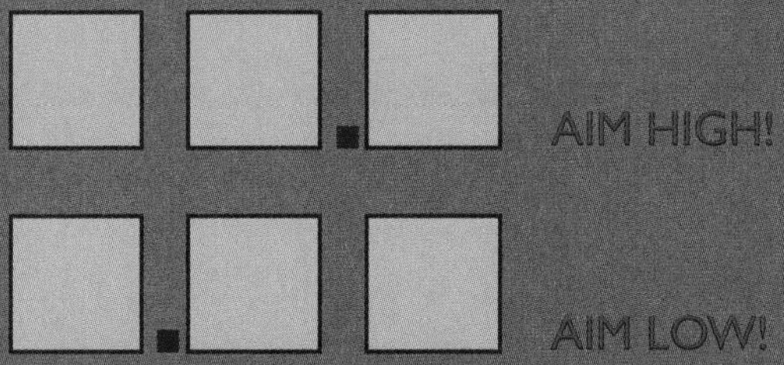

For each round of the game, players take turns drawing single digits from their own bags. After drawing a digit, the player places it in a blank square. The players take turns drawing and placing digits until they have each filled all six blanks in the layout. Once played, the digit's placement cannot be changed. Digits are not returned to the bag. The player with the greatest "Aim High" number and the player with the least "Aim Low" number each earn a point. The first player to get 10 points wins.

hot words | place value

Homework page 258

4 Get It Close Enough

ROUNDING DECIMALS

In everyday life, estimates can sometimes be more useful than exact numbers. To make good estimates, it is important to know how to round decimals. Fortunately, rounding decimals is similar to rounding whole numbers.

Round Decimals

How can you round decimals?

You can use a number line to help you round decimals.

1 Copy the number line below.

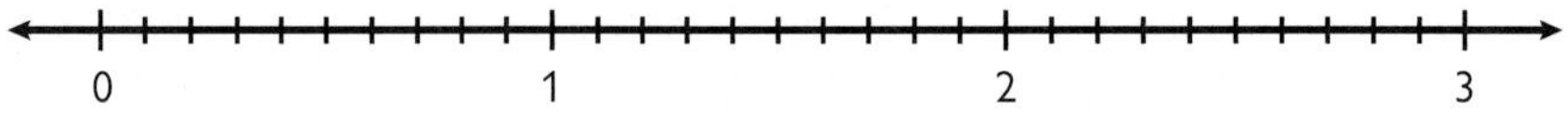

a. Graph 2.7 on the number line.

b. Is the point you graphed closer to 2 or 3?

c. Round 2.7 to the nearest whole number.

d. Use the number line to round 1.3 to the nearest whole number.

Rounding to Tenths

If the hundredths digit of the number is 4 or less, "round down" by eliminating the hundredths digit and all the digits to its right.

If the hundredths digit is 5 or more, "round up" by adding one to the tenths digit. Then, eliminate the hundredths digit and every digit to its right.

2 Round each number to the nearest tenth.

a. 9.3721 **b.** 6.5501 **c.** 19.8397

3 Round 3.1415928 to the nearest hundredth, thousandth, and ten-thousandth.

Order Fractions and Decimals

How can you order fractions and decimals?

Consider the ten numbers below.

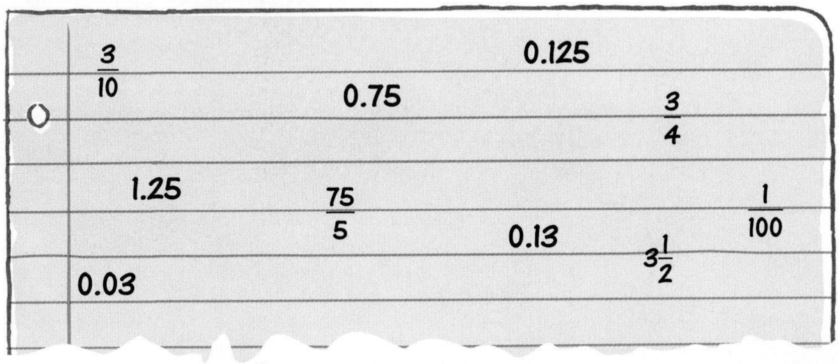

1 Write each of the ten numbers on small pieces of paper so you can move them around. Use any strategy that you find helpful to arrange these numbers in order from least to greatest. If two numbers are equal, place an equals sign between them. As you work, write notes that you can use to convince someone that your order is correct.

2 Once you are satisfied with your order, write the numbers from least to greatest. Place either $<$ or $=$ between the numbers. Make sure that your list contains all ten numbers and that each number is expressed as it was originally shown.

Write about Ordering Numbers

The numbers $\frac{19}{8}$, 2.3, and $2\frac{1}{8}$ are written in different forms. Explain how to determine which is least and which is greatest.

hot **words** | round

HW Homework
page 259

PHASE TWO

In this phase, you will be making sense of decimal operations. Deciding where the decimal point goes when you add, subtract, multiply, or divide decimals is important. Think about where you see decimals in your everyday experience. You will investigate ways to use mental math, estimations, and predictions to determine answers quickly.

Computing with Decimals

WHAT'S THE MATH?

Investigations in this section focus on:

COMPUTATION

- Understanding decimals
- Adding, subtracting, multiplying, and dividing decimals

ESTIMATION

- Using estimation to determine where to place the decimal point
- Using estimation to check whether the result makes sense

MathScape Online
mathscape1.com/self_check_quiz

5 Place the Point

ADDING AND SUBTRACTING DECIMALS

You can use what you have learned about fractions and decimals to add decimals. One way to add decimals is to change the addends to mixed numbers and add them. Then, you can change the sum back to a decimal. In this lesson, you will use this method to develop another rule for adding decimals.

Add Decimals

How can you add decimal numbers?

Use what you know about adding mixed numbers.

1 Carl is trying to add 14.5 and 1.25, but he is not sure about the answer. He thinks his answer could be 0.270, 2.70, 27.0 or 270.

$$\begin{array}{r} 14.5 \\ +\ 1.25 \\ \hline 270 \end{array}$$

a. Use estimation to show that none of Carl's possible answers is correct.

b. Write 14.5 and 1.25 as mixed numbers. Then, add the mixed numbers.

c. Write the sum in part **b** as a decimal.

d. Explain to Carl how to add decimals.

2 Carl's sister Vonda told Carl that he should line up the decimal points. Is she correct? To find out, convert the decimals to mixed numbers, add the mixed numbers, and write the sum as a decimal. Then, use Vonda's method of lining up the decimal points and compare the sums.

a. 12.5 + 4.3 b. 22.25 + 6.4 c. 5.65 + 18.7

Why Do We Line Up the Decimal Points?

When you add decimals, you want to be sure that tenths are added to tenths, hundredths are added to hundredths, and so on.

$$\begin{array}{r} 2.35 \\ +\ 12.4 \\ \hline 14.75 \end{array}$$

	2 ones 3 tenths 5 hundredths
+	1 ten 2 ones 4 tenths
	1 ten 4 ones 7 tenths 5 hundredths

Play the "Place Value" Game

How can you create the greatest sum or difference?

Put ten pieces of paper, numbered 0, 1, 2, 3, 4, 5, 6, 7, 8, and 9, in a bag. You and your partner are going to take turns removing 6 pieces of paper, one at a time, from the bag. Each time you or your partner remove a piece of paper, write the number in one of the squares in a layout like the one below. Then, put the piece of paper back into the bag. The player with the greater sum earns a point. Play nine rounds to determine the winner. Then, answer the following questions.

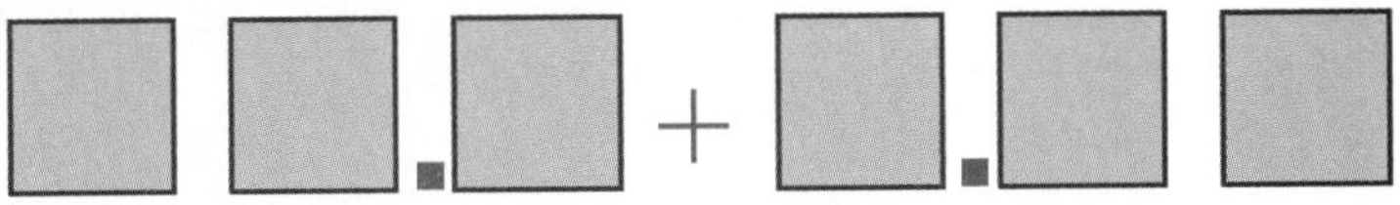

1 If the six numbers you drew were 1, 3, 4, 6, 7, and 9, how could you place the numbers to get the greatest sum?

2 How could you place the digits if you want to get the greatest difference of the two numbers?

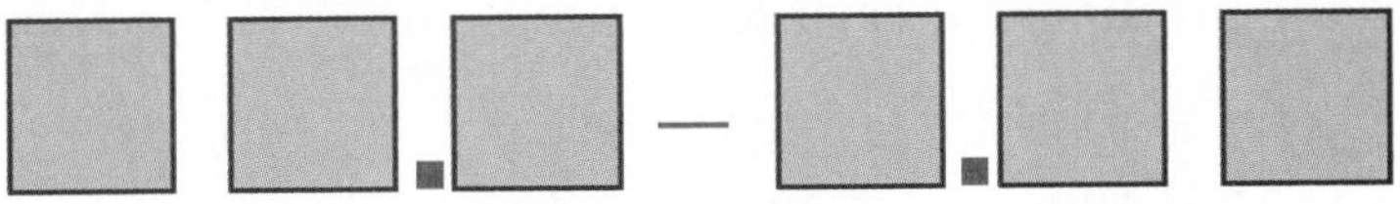

3 Explain your strategy for each "Place Value" game.

What if You Do Not Have Enough Tenths?

When trying to compute 23.25 − 8.64, you line up the decimals as you did in addition. What do you do since there are more tenths in 8.64 than in 23.25? Here is how Vonda did it. Explain what she did. Is her answer correct?

```
  1 12 12
  2̸3̸.2̸5
 − 8.64
 ------
  14.61
```

hot **words**
sum
difference
operation

Homework
page 260

6 More to the Point

MULTIPLYING WHOLE NUMERS BY DECIMALS

Can you multiply decimals like you multiply whole numbers? In the last lesson, you found a way to add and subtract decimals that is very similar to adding and subtracting whole numbers. In this lesson, you will learn to multiply a whole number by a decimal.

Look for a Pattern

Where do you place the decimal point?

You can use a calculator to find patterns when multiplying by special numbers.

1 Find the products for each set of problems.

a. 124.37×10
124.37×100
$124.37 \times 1{,}000$

b. 14.352×10
14.352×100
$14.352 \times 1{,}000$

c. 0.568×10
0.568×100
$0.568 \times 1{,}000$

2 Describe what happens to the decimal point when you multiply by 10, 100, and 1,000.

3 Find the products for each set of problems.

a. 532×0.1
532×0.01
532×0.001

b. $3{,}467 \times 0.1$
$3{,}467 \times 0.01$
$3{,}467 \times 0.001$

c. 72×0.1
72×0.01
72×0.001

4 Describe what happens to the decimal point when you multiply by 0.1, 0.01, and 0.001.

5 Use the patterns that you observed to find each product. Then, use a calculator to check your answers.

a. $4.83 \times 1{,}000$

b. 3.6×100

c. 477×0.01

d. 572×0.001

e. 91×0.1

f. $124.37 \times 10{,}000$

Multiply Whole Numbers by Decimals

You can continue studying patterns to discover how to multiply by decimals.

1 Use a calculator to find the products for each set of problems.

a. 300 × 52
300 × 5.2
300 × 0.52
300 × 0.052

b. 230 × 25
230 × 2.5
230 × 0.25
230 × 0.025

c. 14 × 125
14 × 12.5
14 × 1.25
14 × 0.125

d. 18 × 24
18 × 2.4
18 × 0.24
18 × 0.024

e. 261 × 32
261 × 3.2
261 × 0.32
261 × 0.032

f. 4,300 × 131
4,300 × 13.1
4,300 × 1.31
4,300 × 0.131

2 Study each set of problems. How are the products in each set similar? How are they different?

3 Predict the value of each product. Then, use a calculator to check your answers.

a. 700 × 2.1

b. 700 × 0.21

c. 14 × 1.5

d. 34 × 0.25

e. 710 × 0.021

f. 57 × 0.123

Write about Multiplication by Decimals

Reflect on the problems on this page. How did you determine where to place the decimal point after computing the whole number product? Why does this make sense?

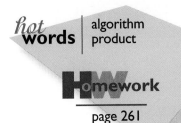

How can you multiply a whole number by a decimal?

hot words | algorithm
product

Homework
page 261

7 Decimal Pinpoint

MULTIPLYING DECIMALS

When you multiply decimals, you can multiply whole numbers and then place the decimal in the correct place. In this lesson, you will learn how to find the product of two decimals.

How can you multiply a decimal by a decimal?

Multiply Decimals

The placement of the decimal point is important in decimal multiplication.

1 The computations for the multiplication problems are given. However, the decimal point has not been placed in the product. Use estimation to determine the location of each decimal point. Use a calculator to check your answer.

a.
```
    25.3
  × 1.25
  ------
    1265
    506
   253
  ------
   31625
```

b.
```
     64.1
  × 0.252
  -------
     1282
    3205
   1282
  -------
   161532
```

c.
```
    1.055
   × 4.52
  -------
     2110
    5275
   4220
  -------
   476860
```

2 How is the number of decimal places in the two factors related to the number of decimal places in the corresponding product?

3 Explain your method for placing the decimal point in the product of two decimal factors.

4 Use your method to find each product. Then, use a calculator to check your answer.

a. 12.5×2.4 **b.** 24.3×1.15 **c.** 8.9×0.003

Play another Version of the "Place Value" Game

As in previous "Place Value" games, this is a game for two players. Each player should make a layout as shown below.

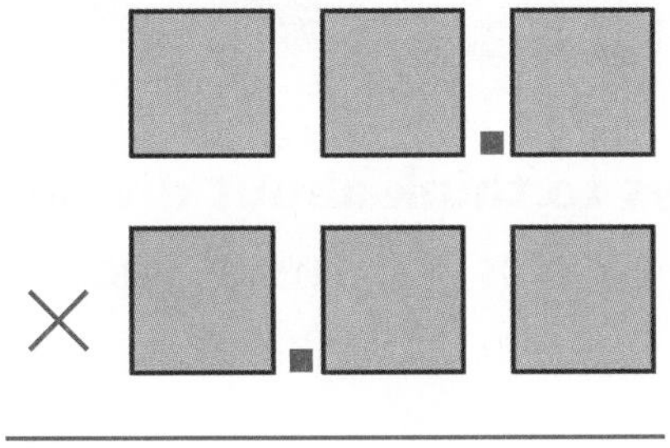

The players take turns rolling a number cube and recording the result of the roll in one of their squares. When each player has filled all of his or her squares, the players compute their products. The player with the greatest product earns a point. Play nine rounds to determine the winner. Then, answer the following questions.

1 What are the greatest and least possible products you could have while playing the game?

2 If you roll 1, 6, 5, 6, 2, and 3, what do you think is the greatest possible product? Compare your answers with those of your classmates.

Why Does It Work?

32.4		324		32.4	←		1 decimal place
× 1.23	⇒	× 123	⇒	× 1.23	←	+	2 decimal places
?????		39852		39.852	←		3 decimal places

When you compute the product of 32.4 and 1.23, you first ignore the decimal points and multiply the whole numbers 324 and 123.

When you ignore the decimal in 32.4, 324 is 10 times greater than 32.4. When you ignore the decimal in 1.23, 123 is 100 times greater than 1.23. So, the product of the whole numbers is 10 times 100, or 1,000, times the product of 32.4 and 1.23. That means that the decimal point is three places too far to the right. You must move the decimal point three places to the left to get the correct answer.

How can you create the greatest product?

hot words product, factor

Homework page 262

8 Patterns and Predictions

DIVIDING DECIMALS

There are many ways to think about dividing decimals. All of these methods involve things you already know how to do.

How do you determine which division problems have the same quotient?

Write Equivalent Division Problems

You can use a calculator to find patterns when dividing decimals.

1 Find the quotient for each set of problems.

a. 24 ÷ 12	**b.** 875 ÷ 125	**c.** 720 ÷ 144
2.4 ÷ 1.2	87.5 ÷ 12.5	72 ÷ 14.4
0.24 ÷ 0.12	8.75 ÷ 1.25	7.2 ÷ 1.44

2 What do you notice about the quotients for each set of division problems? How are the problems in each set similar? How they different?

3 Use a calculator to find each quotient. Then study the results.

a. 495 ÷ 33	**b.** 4,950 ÷ 330	**c.** 49,500 ÷ 3,300
d. 49.5 ÷ 3.3	**e.** 4.95 ÷ 0.33	**f.** 0.495 ÷ 0.033

4 Which of the following quotients do you think will be equal to 408 ÷ 2.4? Explain your reasoning. Then, use a calculator to check your predictions.

a. 4,080 ÷ 24	**b.** 40,800 ÷ 240	**c.** 408 ÷ 24
d. 40.8 ÷ 0.24	**e.** 40.8 ÷ 2.4	**f.** 4.08 ÷ 0.024

5 Rewrite each division problem as an equivalent problem involving whole numbers. Then, find the quotient.

a. $1.2\overline{)36}$ **b.** $1.25\overline{)22.5}$ **c.** $1.45\overline{)261}$

Divide Decimals

What happens if the quotient is a decimal?

If you compute $12 \div 5$, you would get $\frac{12}{5}$ or $2\frac{2}{5}$. Notice that $2\frac{2}{5}$ is the same as $2\frac{4}{10}$ or 2.4. You can get this same result as follows.

$$\begin{array}{r} 2.4 \\ 5\overline{)12.0} \\ \underline{10} \\ 2.0 \\ \underline{2.0} \\ 0 \end{array}$$

Notice that 2.0 is equivalent to 20 tenths. Since $20 \div 5 = 4$, you know that 20 tenths divided by 5 is 4 tenths.

1 Work with a classmate to help each other understand that $19 \div 4 = 4.75$. Two ways of displaying the calculation are shown below.

$$\begin{array}{r} 4.75 \\ 4\overline{)19.00} \\ \underline{16} \\ 3.0 \\ \underline{2.8} \\ 0.20 \\ \underline{0.20} \\ 0.00 \end{array} \qquad \begin{array}{r} 4.75 \\ 4\overline{)19.00} \\ \underline{16} \\ 30 \\ \underline{28} \\ 20 \\ \underline{20} \\ 0 \end{array}$$

2 Copy and complete the division problem below. Then, check your answer by multiplying and by using a calculator.

$$\begin{array}{r} 32.1 \\ 8\overline{)257.000} \\ \underline{24} \\ 17 \\ \underline{16} \\ 10 \\ \underline{8} \\ 2 \end{array}$$

3 Find each quotient. Then, use a calculator to check your answer.

a. $326.7 \div 13.5$ **b.** $323.15 \div 12.5$ **c.** $428.571 \div 23.4$

hot **words** | quotient

Homework

page 263

Write about Dividing Decimals

Write an explanation of how to divide decimals.

9 It Keeps Going and Going

WRITING REPEATING AND TERMINATING DECIMALS

You have already learned to use division to change fractions to decimals. In previous lessons, the decimals could be represented using tenths, hundredths, or thousandths. In this lesson, you will discover decimals that never end.

Write Repeating Decimals

How far can decimals go?

When you use a calculator to find $4 \div 9$, you may get 0.4444444 or 0.44444444 or 0.444444444444. The answer will depend on the number of digits your calculator shows. What is the *real* answer?

1. Find the first five decimal places in the decimal representation of $\frac{4}{9}$ by dividing 4.00000 by 9. Show your work.
2. When Vonda did step **1**, she said, "The 4s will go on forever!" Is she correct? Examine your division and explain.
3. Another repeating decimal is the decimal representation of $\frac{5}{27}$. What part is repeating? How do you know it repeats forever?
4. Every fraction has a decimal representation that either repeats or terminates. Find a decimal representation for each fraction by dividing. Tell if the decimal is *repeating* or *terminating.*

   a. $\frac{5}{16}$ **b.** $\frac{4}{27}$ **c.** $\frac{1}{6}$ **d.** $\frac{4}{13}$ **e.** $\frac{13}{25}$

Repeating Decimals

The decimal equivalent for $\frac{2}{9}$ is 0.2222. . . .This is called a **repeating decimal** because the 2 repeats forever. For 0.3121212 . . . , the 12 repeats. For 0.602360236023. . . , the 6023 repeats. Sometimes a repeating decimal is written with a bar over the part that repeats.

$$0.\overline{2} \qquad 0.3\overline{12} \qquad 0.\overline{6023}$$

A decimal that has a finite number of digits is called a **terminating decimal.**

Use Decimals

What is Celsius?

The United States uses the Fahrenheit scale to measure temperature. Most other parts of the world use the Celsius scale. To change degrees Celsius to degrees Fahrenheit, you can use the following two steps.

- Multiply the Celsius temperature by 1.8.
- Add 32 to the product.

1 Copy the table below. Complete the table by using the two steps.

°C	0	5	7	9	12	15	20	30	40
°F	32								

2 The two steps give an exact conversion. To *approximate* the Fahrenheit temperature given a Celsius temperature, you can double the Celsius temperature and then add 30.

Use this method to approximate each Fahrenheit temperature in the table. Compare the results with the actual temperatures. Which of these temperatures is within 2 degrees of the actual temperature?

3 Work with one or more of your classmates to devise a way to change a Fahrenheit temperature to a Celsius temperature. Use the table in step **1** to check your method. Then, use your method to change each Fahrenheit temperature to Celsius.

a. 50°F **b.** 95°F **c.** 64.4°F

Write about Temperature Conversion

Teresa and Donte's math assignment was to keep track of the daily high temperatures for three days and then find the average high temperature in Celsius. Teresa and Donte only have a thermometer that measures degrees Fahrenheit. They recorded the following temperatures.

Friday: 68°F Saturday: 56°F Sunday: 74°F

Teresa said that they need to average the three temperatures first and then convert to Celsius. Donte said that they could convert all three temperatures to Celsius and then find the average.

- Will both methods work?
- Will one be more accurate than the other?
- Find the average temperature using both methods. Explain which way you think is more efficient and why.

hot words
repeating decimal
terminating decimal
average

HW Homework
page 264

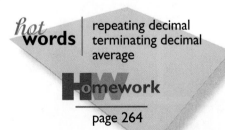

PHASE **THREE**

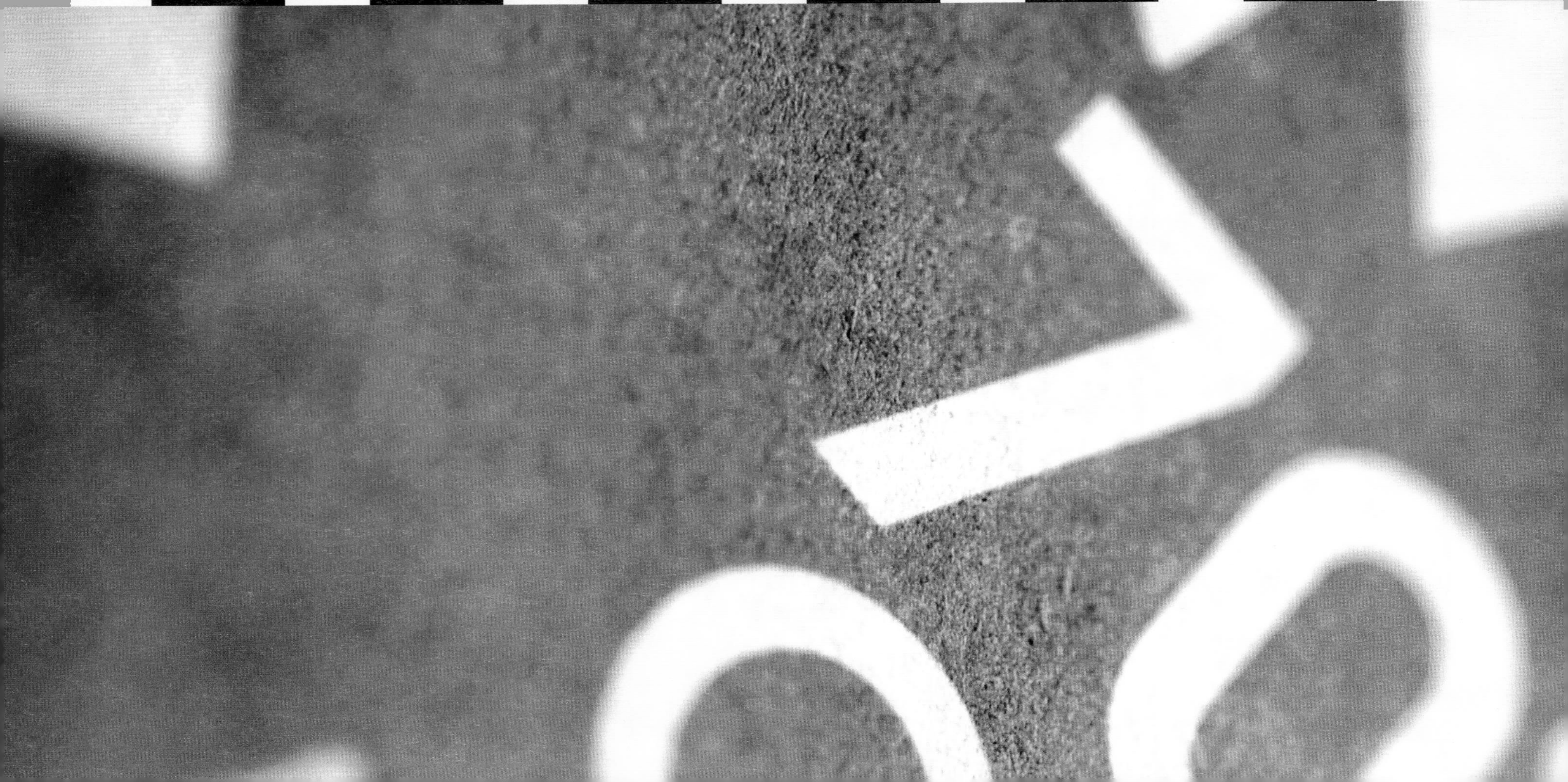

Percents are used everyday. In this phase, you will use mental math and estimation to calculate percents of a number. You will explore the relationship between fractions, decimals, and percents and study how percents are used and misused in the world around us.

Percents

WHAT'S THE MATH?

Investigations in this section focus on:

COMPUTATION

- Understanding the relationships among fractions, decimals, and percents
- Using mental math to get exact answers with percents
- Calculating percent of a number

ESTIMATION

- Using estimation to check whether results are reasonable

MathScape Online
mathscape1.com/self_check_quiz

10 Moving to Percents

WRITING EQUIVALENT FRACTIONS, DECIMALS, AND PERCENTS

When you express a number using hundredths, you can also use percents. The word *percent* means hundredths. The symbol % is used to show the number is a percent.

How can you determine a percent of a whole?

Interpret Percents

Twenty-three hundredths of a whole is twenty-three percent of a whole.

$$\frac{23}{100} = 23\%$$

For each drawing, determine what percent of the whole is shaded.

1

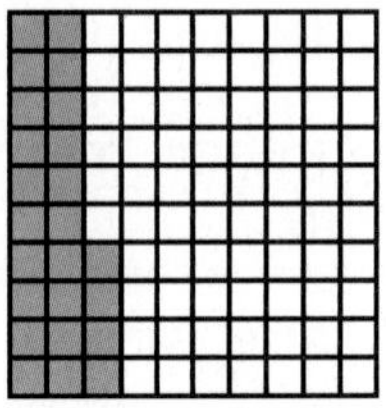

2

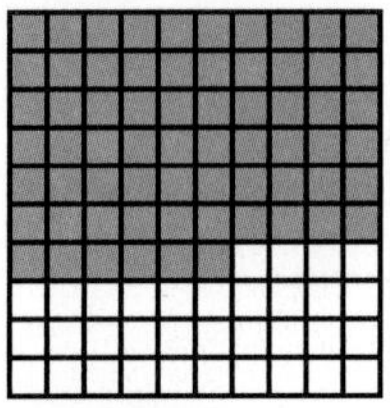

3

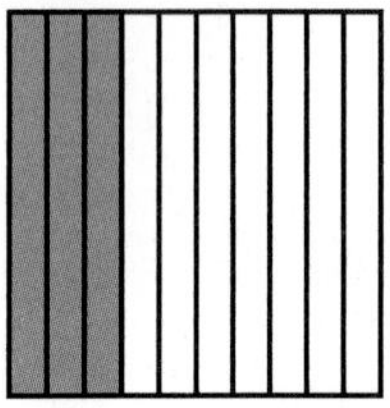

4

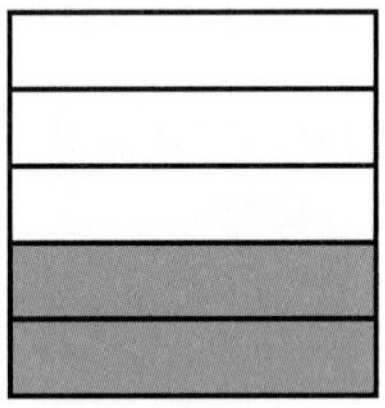

5

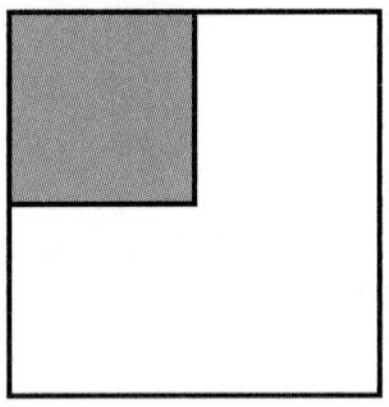

6

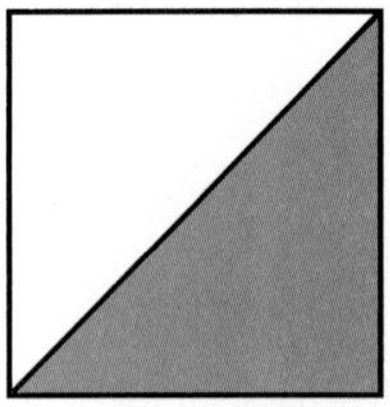

7

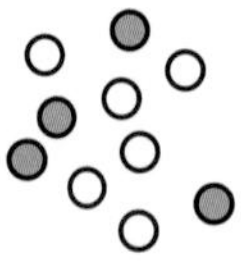

8

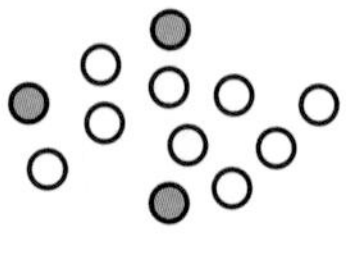

9

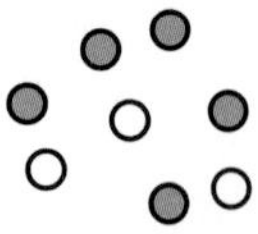

10 Today, the United States has 50 states. Only 13 of the states took part in signing the Declaration of Independence in 1776. What percent of today's states signed the Declaration of Independence?

Write Different Forms for Parts of a Whole

How are fractions, decimals, and percents related?

Here are some examples showing how you can write a percent as a fraction or decimal, or write a fraction or decimal as a percent.

41% can be written as $\frac{41}{100}$ and as 0.41.

0.09 can be written as $\frac{9}{100}$ and as 9%.

$\frac{4}{5}$ can be written as $\frac{8}{10}$, as 0.8, and as 80%.

1 Each of the following numbers can be written as a fraction, as a decimal, and as a percent. Write each number in two other ways.

a. $\frac{3}{10}$ **b.** 80% **c.** 0.75

d. 55% **e.** 0.72 **f.** $\frac{7}{25}$

2 Stephanie collects decks of playing cards. She has 200 decks, and 35 decks are from foreign countries.

a. Use a grid of 100 squares to represent her collection. How many decks does each square represent?

b. Shade the appropriate number of squares to represent the 35 decks from foreign countries.

c. Write a fraction, a decimal, and a percent that represent the part of her collection that are from foreign countries.

3 Write each number in two other ways.

a. 0.755 **b.** 5.9% **c.** $\frac{3}{8}$

d. $\frac{5}{16}$ **e.** 0.003 **f.** 17.4%

Summarize the Conversions

Write instructions someone else can follow to perform each conversion.

- Change a percent to an equivalent decimal.
- Change a percent to an equivalent fraction.
- Change a fraction to an equivalent percent.
- Change a decimal to an equivalent percent.

hot words | percent, ratio

Homework page 265

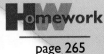

11 Working with Common Percents

REPRESENTING COMMON PERCENTS

Amaze your friends! Impress your parents! With knowledge of just a few percents, you can determine reasonable answers.

How do common fractions and decimals relate to percents?

Model Important Percents

You often see fractions such as $\frac{1}{2}$, $\frac{1}{3}$, $\frac{1}{4}$, and $\frac{3}{4}$. You probably have a good sense of what these numbers mean.

1 Consider the fractions $\frac{1}{2}, \frac{1}{3}, \frac{2}{3}, \frac{1}{4}, \frac{3}{4}, \frac{1}{5}, \frac{2}{5}, \frac{3}{5}, \frac{4}{5}$.

a. Represent each fraction on a 10×10 grid or on a number line. Label each grid or point with the fraction it represents.

b. Label each grid or point with the equivalent decimal and equivalent percent.

c. Represent 0 and 1 with grids or on the number line.

2 For each percent, make a new grid or add a new point to the number line. Label each new grid or point with the percent, fraction, and decimal it represents.

a. 10% b. 30% c. 70% d. 90%

3 Which fraction from step **1** is closest to each percent?

a. 15% b. 85% c. 39% d. 73%

Estimating with Percents

79% is between $\frac{3}{4}$ or 75% and $\frac{4}{5}$ or 80%. It is closer to $\frac{4}{5}$ than $\frac{3}{4}$.

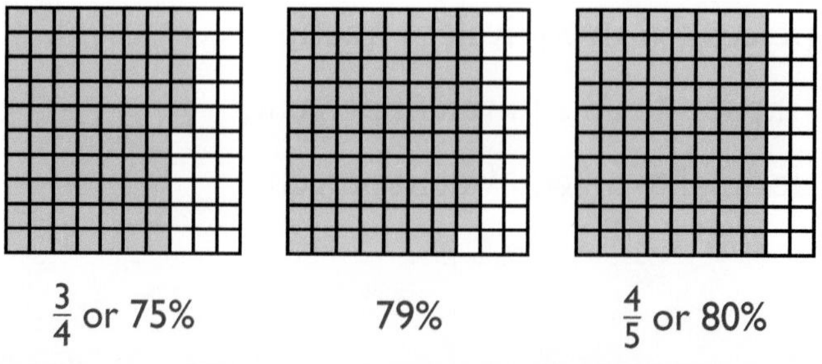

$\frac{3}{4}$ or 75% 79% $\frac{4}{5}$ or 80%

Find Important Percents

Three important percent values are 50%, 10%, and 1%. Fortunately, these are percents that you can calculate quickly.

1 Use the numbers in List A that you compiled in class.

a. Choose a number. On your own, find 50%, 10%, and 1% of the number. Use any strategy that makes sense to you. Record each answer and explain how you found it.

b. On your own, find 50%, 10%, and 1% of another number. See if you can use strategies that are different than those used to find percents of the first number. Look for patterns in your answers.

c. Share your answers and strategies with a partner. Discuss any patterns you see.

d. With your partner, find 50%, 10%, and 1% of each number not already used by you or your partner. Remember to record your answers and strategies.

2 With your partner, choose at least three numbers from List B. Find 50%, 10%, and 1% of each number.

How can you use mental math to find 50%, 10%, and 1% of a number?

Write about Percents

Write a short paragraph explaining how you can quickly find 50%, 10%, and 1% of any number.

hot words | percents

Homework
page 266

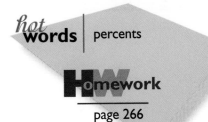

12 Percent Power

CALCULATING PERCENTS

How can you find the percent of a number? In Lesson 11, you found 50%, 10%, and 1% of a number. In this lesson, you will find any percent of a number.

How can you estimate and find the actual percent of a number?

Find a Percent of a Number

In this investigation, you will use fractions to estimate a percent of a number. Suppose you want to estimate 58% of 40.

58% is a little less than 60% or $\frac{3}{5}$.

$\frac{3}{5}$ of 40 is 24.

So, 58% of 40 is a little less than 24.

1 For each problem, find a fraction that you think is close to the percent. Use the fraction to estimate the quantity. Record your answers in the "Estimate" column of the handout Finding Any Percent.

a. 63% of 70 **b.** 34% of 93

c. 12% of 530 **d.** 77% of 1,084

2 Complete the first four rows of the handout. Check the exact value by comparing it to the estimate you found in step **1**.

3 Choose six numbers and six percents from the following list. Write the numbers in the proper columns of the handout. Use at least one three-digit number and at least one four-digit number. Then, complete the table.

Number				Percent Wanted			
25	40	36	90	15%	18%	90%	3%
390	784	135	124	38%	23%	44%	83%
9,300	4,220	3,052	4,040	9%	51%	73%	62%

Use One Hundred to Find Percents

How can you calculate percents using sets of 100?

Liz and her twin brother Nate like trading cards. Some are sports cards (baseball, football, and auto racing), some are comic book cards, and some are game cards. Together, they have 1,000 cards. Liz has separated the cards into 10 sets of 100, so that each set has the same number of each type of card.

As you answer each question, think about how you determined your answer.

1 Each set has 10 auto racing cards. What percent of a set is auto racing cards? What percent of the 1,000 cards is auto racing cards?

2 For each set, 23% are comic book cards. How many comic book cards are in each set? How many comic book cards do Liz and Nate have in all?

3 Of the 1,000 cards, 340 are baseball cards. What percent of the 1,000 cards is baseball cards? What percent of each set is baseball cards?

4 There are 8 football cards in each set of 100, so 8% of each set are football cards. Nate combined five sets into a group of 500, so he could have half of all the cards.

- **a.** How can you find the number of football cards in each set of 500 cards? in all 1,000 cards?
- **b.** How can you find the percent of football cards in each set of 500 cards? in all 1,000 cards?

5 Liz still has sets of 100. In each set, there are 25 game cards.

- **a.** How many sets would she need to combine to have exactly 100 game cards?
- **b.** Of the 25 game cards in each set, 12 cards are one type of game. How many would there be in the combined 100 game cards? What percent of the 100 game cards represents this game?

Write about Using Proportions

When two sets have the same percent or fraction of one item, then the sets are *proportional.* For example, the fraction of baseball cards in each small set is the same as the fraction of baseball cards in the big set. Write a paragraph explaining how to use the idea of proportions to find the percent of a number.

hot words | proportion

Homework

page 267

13 Less Common Percents

UNDERSTANDING PERCENTS LESS THAN 1 AND GREATER THAN 100

Sometimes percents can represent a very small part of a whole or a part greater than a whole. In this lesson, you will study these types of percents.

What does a percent less than one mean?

Understand Percents Less than One

Some percents are less than 1%. To understand the meaning of these types of percents, answer the following questions.

1 Carlos won a gift certificate for $200 at an electronics store. He used his gift certificate to purchase the six items listed below. What percent of the gift certificate was used for each item?

a. a game system: $100 **b.** a game cartridge: $30

c. another game cartridge: $20 **d.** a DVD movie: $24

e. a movie soundtrack: $12 **f.** a rock CD: $13

2 Add the prices in step **1**.

a. How much of the gift certificate remained unspent? What percent of the $200 was left unspent?

b. If all $200 had been spent, what percent of the gift certificate was spent?

3 Write an equivalent decimal and an equivalent fraction for each percent.

a. 0.2% **b.** 0.7% **c.** $0.\overline{3}\%$ **d.** $\frac{1}{10}\%$ **e.** $\frac{2}{5}\%$

4 Which of the following numbers represents a percent less than one? Try to answer without converting the fraction or decimal to a percent. Be prepared to discuss your answers.

a. 0.1 **b.** 0.004 **c.** 0.013 **d.** 0.0082

e. $\frac{1}{200}$ **f.** $\frac{3}{200}$ **g.** $\frac{1}{34}$ **h.** $\frac{10}{1,000}$

Understand Percents Greater than One Hundred

Nichelle is organizing her family's collection of photographs. She has several photo albums that are the same size.

1 Nichelle filled $\frac{3}{4}$ of one photo album with the first box of photographs. The following model represents $\frac{3}{4}$ of the album.

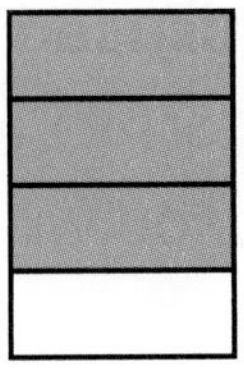

What percent of the album was used?

2 The next box had the same number of photographs in it. The space required for both boxes is twice as much as the first box.

a. Draw a model similar to the one in step **1** to represent how much of the photo albums are filled.

b. What fraction of one album does your drawing represent?

c. What percent of one album does your drawing represent? How did you determine your answer?

3 Each album held 60 photographs.

a. Find the number of photographs in the first box.

b. Find the number of photographs in the second box.

c. When Nichelle was finished, she had filled three albums and $\frac{1}{5}$ of another album. What percent of one album did she fill? How many photographs does Nichelle have?

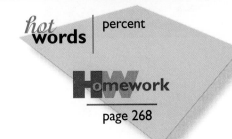

What does a percent greater than 100 mean?

Make Models

Make a model to represent $\frac{1}{2}$%. Then, make a model to represent 250%. Explain why each model represents the appropriate percent.

hot words | percent

Homework
page 268

14 Give It to Me Straight

UNDERSTANDING USES AND MISUSES OF PERCENTS

Sometimes percents are misused in everyday life. Misleading percents may seem correct at first. However, if you use what you know about percents, you will quickly discover which statements are wrong.

Which statements use percents correctly?

Examine Uses of Percents

Study the statements involving percents. For each statement, decide if the percent is used correctly.

- If a statement has been used correctly, describe what the statement means. In some cases, you may want to make up some numbers to give an example of what it means.
- If a statement has not been used correctly, explain why it is incorrect.

1. Jewel said, "I have finished about 50% of my homework."
2. Gracie's Department Store is offering 15% off the regular price on all men's shoes.
3. Of the people who are older than 60 years old and who are living alone, 34% are women and only 15% are men.
4. Central High School enrollment is 105% of last year's enrollment.
5. Ice cream gets its texture by adding air. A few ice cream makers double the volume by having 100% air.
6. One brand of snacks has 13% of the $900 million annual sales for similar snacks.
7. Teenagers represent 130% of the attendance at an amusement park.
8. The rainfall in a particular area is 120% less than it was 25 years ago.

Navigate the Maze

Can you find the correct path through the maze?

For this maze, you want to create a path of marked squares that lead from the lower left corner to the upper right corner using the following rules.

- The lower left square is marked first.
- You can mark an unmarked square to the left or right or above or below the last square marked.
- You can only mark the square whose value is closest to the value in the last square marked.

In the example below, the last square marked was 45%. There are three possibilities for the next square, $\frac{4}{10}$, $\frac{4}{5}$, or $\frac{1}{4}$. Since $\frac{4}{10}$ is closest to 45%, that square must be marked next.

17%	$\frac{4}{5}$	$0.\overline{3}$
$\frac{4}{10}$	~~45%~~	$\frac{1}{4}$
~~0.2~~	~~0.3~~	$\frac{3}{4}$

Try the maze on the handout.

Interpret the Data

Trent and Marika surveyed 150 students in their school. The results are shown below.

Percent of Students Who Own Various Types of Pets

Dogs: $33.\overline{3}$% Cats: 30% Birds: 6% Other: 2% None: 44%

Total Number of Pets Owned by the Students

Dogs: 86 Cats: 95 Birds: 14 Other: 5

1 How many students own each type of pet? How many students own no pets?

2 What percent of all the pets are dogs? cats? birds? other pets?

3 Marika decided they had made a mistake. When she added the number of students from part **1**, she said they had too many students. Why do you think she said that? Is she correct?

hot words | percent

Homework page 269

PHASE FOUR

Negative numbers are used when the temperature is really cold, when someone is losing a game, and when something is missing. All of the numbers you have already used have negative counterparts. If you can handle the whole numbers 0, 1, 2, 3, 4, . . . , you can learn the strategies to handle negative numbers -1, -2, -3, -4, . . . and all integers.

The Integers

WHAT'S THE MATH?

Investigations in this section focus on:

NUMBER and COMPUTATION

- Using negative integers to indicate values
- Placing negative integers on the number line
- Adding and subtracting integers
- Writing equivalent addition and subtraction problems

ALGEBRA FUNCTIONS

- Recognizing and writing equivalent expressions

MathScape Online
mathscape1.com/self_check_quiz

15 The Other End of the Number Line

UNDERSTANDING NEGATIVE NUMBERS

What numbers are less than zero? When the meteorologist says the temperature is 5° below zero, he or she is using a number less than zero.

How can you order numbers less than zero?

Play "How Cold Is It?"

"How Cold Is It?" is a game for two people.

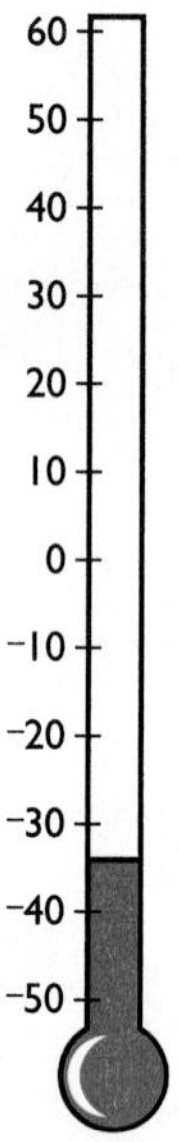

1 Use the following rules to play "How Cold Is It?"

- One person thinks of a cold temperature between 0°F and −50°F and writes it down.
- The other person tries to guess the temperature. After each guess, the first person will say if the guess is warmer or colder than his or her temperature.
- When a person guesses the temperature, record the number of guesses needed. Then, the players switch rolls.
- After four rounds, the player who guessed a temperature using the fewest guesses is the winner.

2 Use the record low temperatures below to determine which city had the coldest record low temperature. Determine which city had the warmest record low temperature. List the cities in order from coldest to warmest record low temperature.

Record Low Temperatures

Nome, Alaska	−54°F	Missoula, Montana	−33°F
Phoenix, Arizona	17°F	El Paso, Texas	−8°F
Apalachicola, Florida	9°F	Burlington, Vermont	−30°F
St. Cloud, Minnesota	−43°F	Elkins, West Virginia	−24°F

Play "Who's Got My Number?"

Play "Who's Got My Number?" with the whole class or with a small group.

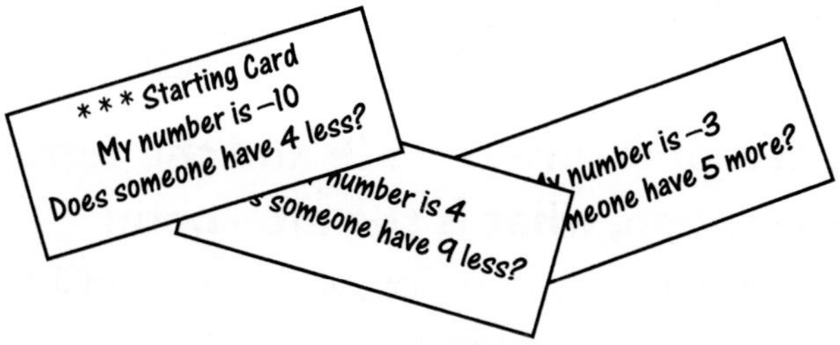

Who's Got My Number?

To play the game, your teacher will give your group a set of cards and a number line recording sheet.

- Shuffle and deal out all of the cards.
- The person with the starred starting card begins by reading the card.
- Locate the number on the number line. Players should use the number line and the clue to determine the next number.
- The person who has that number on one of his or her cards reads it and play continues.

How can you show numbers less than zero on a number line?

Make Your Own Set of Clues

Write a set of your own clues for the game you just played.

- Include clues so that the numbers go as low as −15 and as high as 10.
- Write at least six clues in your set.
- Place a star on the starting card.
- Draw a number line and show the moves to make sure your clues are correct.

hot words | integer

Homework page 270

16 Moving on the Number Line

USING NUMBER LINES TO ADD INTEGERS

If the morning temperature is −5°F and the temperature rises 20°F by afternoon, what is the afternoon temperature? To answer this question, you must add integers. In this lesson, you will use number lines to add integers.

Add Integers on a Number Line

How can you use number lines to add integers?

Adding a positive number corresponds to moving to the right.

$1 + 4 = 5$

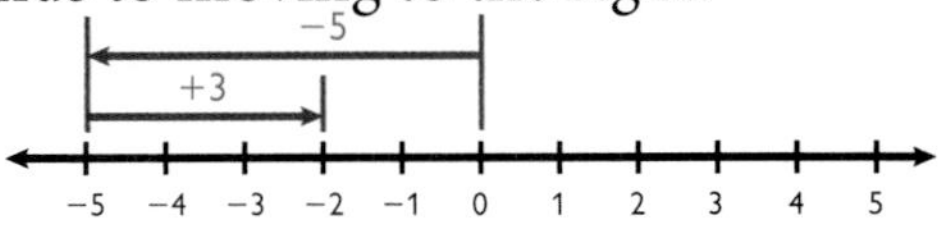

$-5 + 3 = -2$

Adding a negative number corresponds to moving to the left.

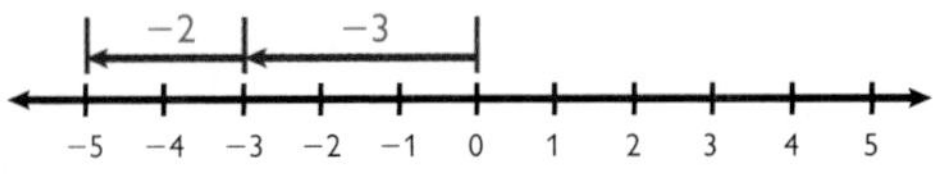

$-3 + (-2) = -5$

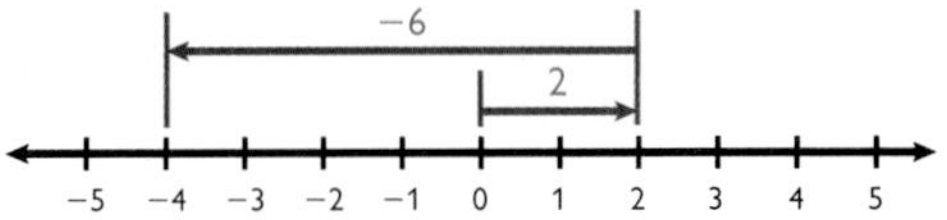

$2 + (-6) = -4$

Play "Number Line Showdown" with a partner.

Number Line Showdown

- Remove the face cards from a regular deck of cards. Shuffle and deal out the cards facedown in front of the players. The black cards will represent positive numbers, and the red cards will represent negative numbers.
- Player 1 turns over the top card from his or her pile. He or she draws an arrow on a number line to represent moving from 0 to the number.
- Player 2 turns over his or her top card. He or she draws an arrow on the number line to show moving from Player 1's number to complete the addition.
- If the sum is positive, Player 1 gets the cards played. If the sum is negative, Player 2 gets the cards played. If the sum is zero, no one gets the cards.
- The players take turns being Player 1 and Player 2.
- The player with the most cards after all the cards have been played wins.

Play Variations of "Number Line Showdown"

Now, you will play a more challenging variation of "Number Line Showdown". Decide which variation of the game you and your partner would like to play.

How can you add more than two integers?

Variations of Number Line Showdown

Variation 1

Player 1 turns over his or her top two cards, uses these numbers to draw an addition problem on the number line, and finds the sum. Then, Player 2 turns over two cards and does the same. The player with the greater sum gets the cards played.

Variation 2

Player 1 turns over his or her top four cards. He or she shows the addition of the four numbers on the number line. Then, Player 2 turns over four cards and does the same. The player with the greater sum gets the cards played.

Write about Addition on a Number Line

You are going to begin to create a handbook that explains how to add and subtract positive and negative numbers. Your handbook should be written for someone who does not know anything about adding and subtracting these numbers.

Today, you will write Part I. This should include a short description of adding positive and negative numbers using a number line.

- Make up three examples of addition problems. Include a positive number plus a negative number, a negative number plus a positive number, and a negative number plus a negative number.
- For each problem, show how to use the number line to solve the problem. Write an addition equation to go with each drawing.
- Include any hints or advice of your own.

hot **words** | signed numbers, number sentence

Homework page 271

17 Taking the Challenge

USING CUBES TO ADD INTEGERS

Cubes can also be used to model addition of integers. In this lesson, you will use several cubes of one color to represent positive integers and several cubes of another color to represent negative integers.

Develop another Model for Addition

What other model can be used to represent addition of integers?

To play "Color Challenge", one partner will need 10 pink cubes to represent positive numbers, and the other partner will need 10 green cubes to represent negative numbers. You will also need a spinner to share. If you and your partner are using cubes of other colors, change the names of the colors on the spinner.

Play several rounds of "Color Challenge".

Color Challenge

- Take turns spinning the spinner. After each spin, place the appropriate cubes as stated on the spinner in the center of the desk or table.
- As you play, work with your partner to develop a method for keeping track of who has more cubes in the center. How many more cubes does the player have?
- The first player to have at least 4 more cubes in the center than the other player wins.

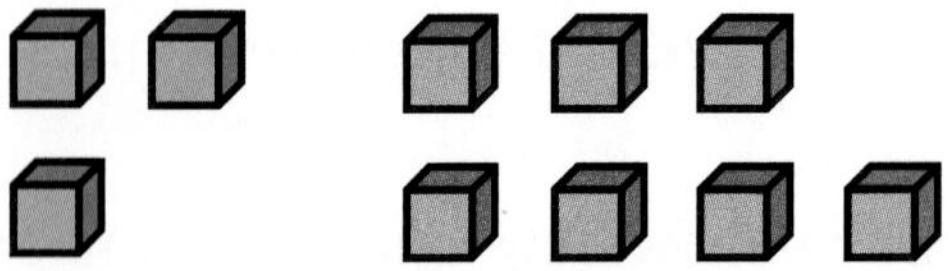

There are 4 more pink cubes than green cubes, so pink has won.

Write Addition Equations

Play "Addition Bingo". As in "Color Challenge", one player will need 10 cubes of one color, and the other player will need 10 cubes of another color. You will also need a new spinner and a game board.

Addition Bingo

- The first player spins the spinner twice, places the appropriate cubes on the desk, and writes the addition problem represented by the cubes.
- The first player finds the sum of the cubes by removing the same number of each color so that only one color remains. The remaining cube(s) represent the sum.
- After finding the sum, the first player crosses off a matching addition problem on his or her game board. If there is no matching addition problem or if it has already been crossed off, the player loses his or her turn.
- The players continue to take turns spinning, writing addition problems, and crossing off squares.
- The first player to have crossed off four squares in a row vertically, horizontally, or diagonally wins.

How can you use cubes to represent addition of numbers?

Write about Addition Using Cubes

Write Part II of your handbook about addition and subtraction of positive and negative numbers. Describe adding positive and negative numbers using cubes.

- Make up three examples of addition problems. Include a positive number plus a negative number with a positive sum, a positive number plus a negative number with a negative sum, and a negative number plus a negative number.
- For each problem, show how to use cubes to solve the problem. Write an addition equation to go with each drawing.
- Include any hints or advice of your own.

hot words: sum, zero pair

Homework
page 272

18 The Meaning of the Sign

USING NUMBER LINES TO SUBTRACT INTEGERS

Mathematical symbols, like words, can sometimes mean more than one thing. One of those symbols is the – sign. Sometimes it means to subtract and sometimes it shows that the number is negative.

How can you use number lines to subtract integers?

Subtract Integers on a Number Line

Subtracting a positive number corresponds to moving to the left on the number line.

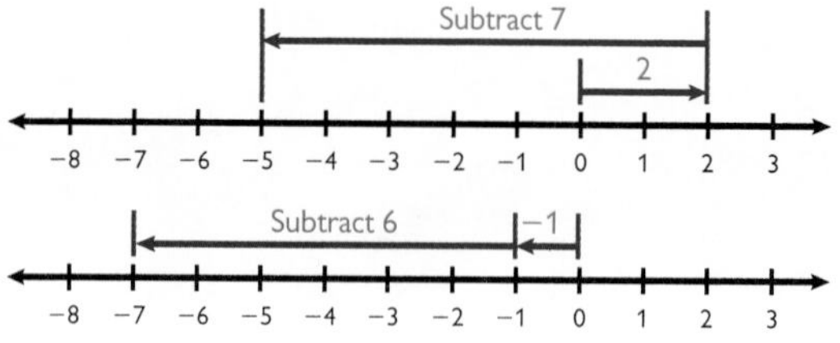

$2 - 7 = -5$

$-1 - 6 = -7$

Your teacher will give you a set of four number cubes and a recording sheet.

1 With a partner, roll the number cubes. Write the numbers in the squares on the number sheet.

2 Each student should use the four numbers and addition and subtraction signs to write an expression. Try to arrange the numbers and signs so that the result is the least possible number. Use at least one addition sign and at least one subtraction sign.

3 On a number line, use arrows to represent the entire expression in step **2**. Write the equation.

4 Write a second expression so that the result is the greatest possible number.

5 On a number line, use arrows to represent the entire expression in step **4**. Write the equation.

6 Compare your answers with your partner's answers. Who got the least answer? Who got the greatest answer?

Solve Subtraction Puzzles

How can you use what you know about subtraction to solve puzzles?

To solve the puzzles on this page, start with the number in the left column and subtract the number in the top row.

1 The example in the puzzle below shows $0 - 4 = -4$. Copy and complete the puzzle.

Second Number (top row); **First Number** (left column)

−	3	6	4	8
2			↓	
−1			↓	
0	→	→	−4	
4				

2 Copy and complete the following two puzzles.

−	2	4		
5				
−2			−3	
		−1		−5
			−7	

−	2	9		
4			−2	
−9				
			−1	
		−3		−5

Write about Subtraction on a Number Line

Write Part III of your handbook about addition and subtraction of positive and negative numbers. Describe subtraction using number lines.

- Make up three examples of subtraction problems. Include a positive number minus a positive number with a positive answer, a positive number minus a positive number with a negative answer, and a negative number minus a positive number.
- For each problem, show how to use the number line to solve the problem. Write a subtraction equation to go with each drawing.
- Include any hints or advice of your own.

hot words | inverse operations

Homework
page 273

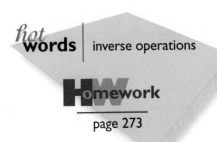

19 The Cube Model

USING CUBES TO SUBTRACT INTEGERS

When you use cubes to model subtraction, sometimes you will not have enough cubes to take away. To solve these problems, you can add zero pairs. A zero pair is one positive cube and one negative cube.

How can you use cubes to subtract integers?

Subtract Integers Using Cubes

When using cubes to model integers, remember that pink cubes represent positive numbers and green cubes represent negative cubes. You may use other colors, but you must first decide which color represents positive numbers and which represents negative numbers.

1 Use cubes to find each difference.

a. $2 - 7$ **b.** $-1 - 5$ **c.** $-2 - 2$

d. $-6 - 3$ **e.** $-3 - 7$ **f.** $4 - 9$

g. $-5 - 1$ **h.** $-4 - 4$ **i.** $-3 - 5$

2 Study the subtraction problems in step **1**. Use what you have learned about subtraction to find each difference.

a. $-\frac{2}{3} - \frac{1}{3}$ **b.** $\frac{3}{4} - \frac{6}{4}$ **c.** $-4.5 - 1.4$

d. $5.2 - 10.1$ **e.** $\frac{1}{2} - \frac{7}{8}$ **f.** $-\frac{4}{5} - \frac{19}{20}$

Subtraction with Cubes

To find $-2 - 1$, use the following steps.

- Use two green cubes to represent -2.
- You need to take away one pink cube, but you do not have any pink cubes.
- Add one zero pair, so that you will have one pink cube to take away.
- Take away one pink cube.
- There are three green cubes left. The answer is -3.

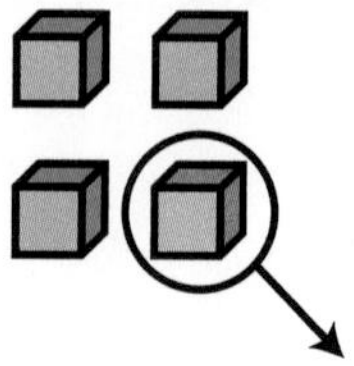

Find the Mistakes

Can you find mistakes in subtraction?

Ms. Parabola collected her students' work and found a number of errors. Each of the following solutions has something wrong with it. Determine why it is incorrect. Write an explanation of what the student did wrong. Then, solve it correctly.

1 I solved the problem $-3 - 3$ by setting out 3 negative cubes. Then, I took away 3 negative cubes. I got 0.

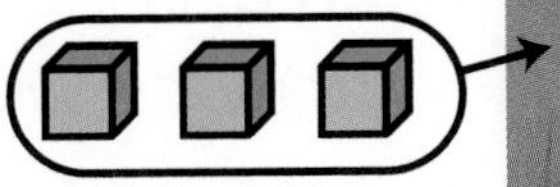

2 I solved the problem $-2 - 4$ by adding two zero pairs to 2 negative cubes. I took away 4 negative cubes. I got $+2$.

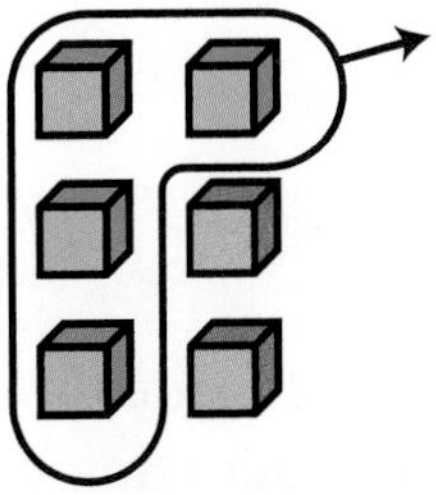

3 I solved the problem $3 - 5$ by adding 5 positive cubes to the 3 positive cubes. Then, I took away 5 positive cubes. I got the answer $+3$, but that cannot be right!

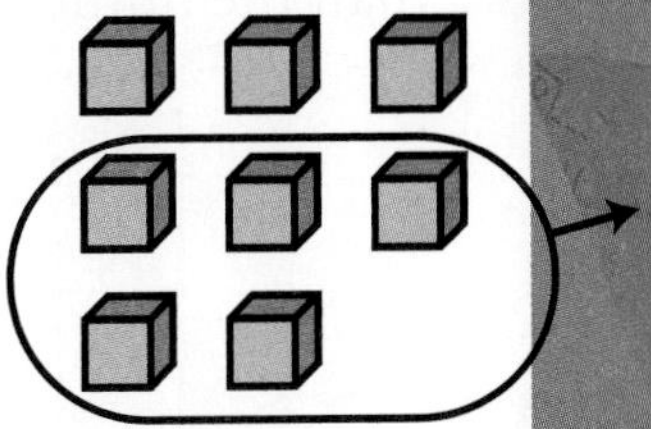

4 I solved the problem $3 - 2$ by adding 2 zero pairs to the 3 positive cubes. Then, I took away 2 negative cubes. I got $+5$.

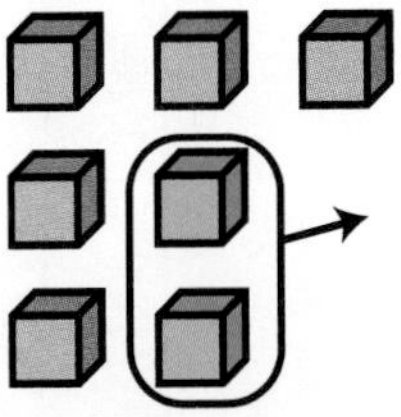

Write about Subtraction Using Cubes

Write Part IV of your handbook about addition and subtraction of positive and negative numbers. Describe subtraction using cubes.

- For each of the following problems, show how to use cubes to solve the problem. Write a subtraction equation to go with each drawing.

  $5 - 8$ $\qquad$ $-4 - 2$ $\qquad$ $-1 - 3$

- Include any hints or advice of your own.

hot words | zero pair

Homework

page 274

20 Write It another Way

EXPLORING EQUIVALENT STATEMENTS

You already know that addition and subtraction are related. In this lesson, you will learn how to write an addition problem as a subtraction problem and a subtraction problem as an addition problem.

How are addition and subtraction related?

Look for Patterns in Addition and Subtraction

Chris made the following list of addition equations. The first number remains the same. The second number is always one less than the number before it.

$4 + 3 = 7$

$4 + 2 = 6$

$4 + 1 = 5$

$4 + 0 = 4$

$4 + (-1) = 3$

$4 + (-2) = 2$

$4 + (-3) = 1$

$4 + (-4) = 0$

$4 + (-5) = -1$

$4 + (-6) = -2$

1. Make a similar list of subtraction equations. Start with $4 - 10$. Make the second number one less than the number before it. Stop when the second number is zero.
2. Compare the two lists. Which problems have the same answers? Describe any patterns you notice.
3. Use what you notice about the patterns to write $3 + (-7)$ as a subtraction problem. Use number cubes to show how you would find the answer for each problem.
4. Write $5 - 8$ as an addition problem. Use a number line to show how you would find the answer for each problem.

Write a Test

Here is your chance! You are going to write a test. Your test should not be so easy that everyone will get all the answers right. It should not be so hard that no one can answer the questions. Be sure to include at least one of each of the following types of problems.

- an integer addition problem with a negative answer
- an integer addition problem with a positive answer
- an integer subtraction problem with a negative answer
- an integer subtraction problem with a positive answer
- a true/false problem about estimating an answer to either an addition or subtraction problem (You may use whole numbers, fractions, or decimals. Make sure the problem shows you whether the test taker understands the order of positive and negative numbers on the number line.)
- a multiple-choice problem that uses the order of operation rules (You must include positive and negative numbers and four answer choices.)
- a writing problem that asks the test taker to tell about a new thing he or she learned about adding and subtracting positive and negative numbers (You might ask the test taker to answer a question, to write an explanation for someone, or to respond to a Dr. Math letter that you create.)

Prepare an answer key for your test. Preparing the answer key will help you to ensure that the test is not too easy or too difficult.

Can you write various types of problems using positive and negative numbers?

hot words | equivalent

Homework
page 275

Homework 1

The Fraction-Decimal Connection

Applying Skills

Write each fraction as a decimal. Then write it as an amount of money using a dollar sign.

1. $\frac{2}{100}$

2. $\frac{48}{100}$

3. $\frac{99}{100}$

4. $\frac{50}{100}$

5. $\frac{2}{10}$

6. $\frac{8}{10}$

Write the value of each group of coins in cents. Write this value as a fraction over 100 pennies. Simplify the fraction. Then, write the value as an amount of money using a dollar sign.

7. 10 pennies

8. 5 nickels

9. 6 dimes

10. 1 quarter, 5 dimes, and 12 pennies

11. 10 dimes, 1 quarter, and 2 nickels

Write the fraction represented by the shaded area of each square or squares. Then, write the corresponding decimal.

12.

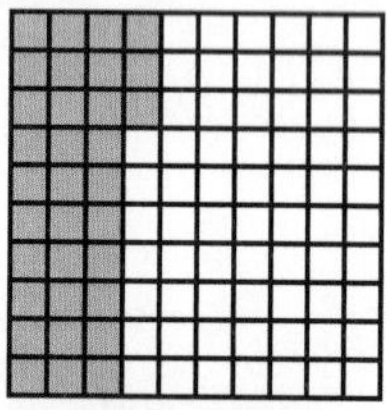

13.

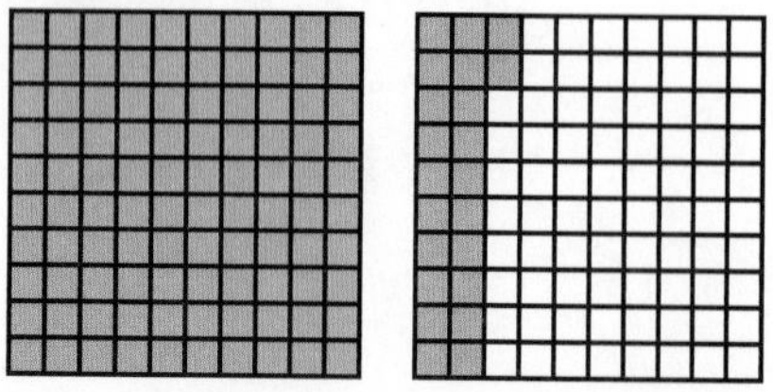

Extending Concepts

Write the fraction represented by the shaded area of each figure. Then, write the corresponding decimal.

14.

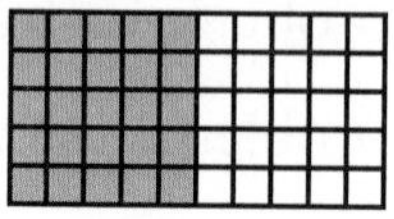

15.

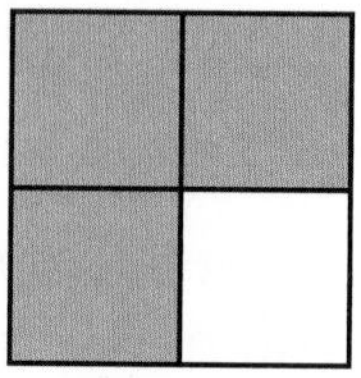

16.

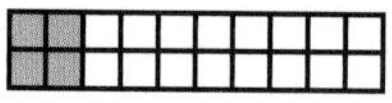

17.

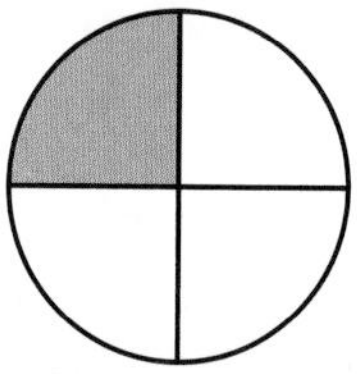

18. This cash register receipt was torn. How much change did the customer get back from the dollar she paid?

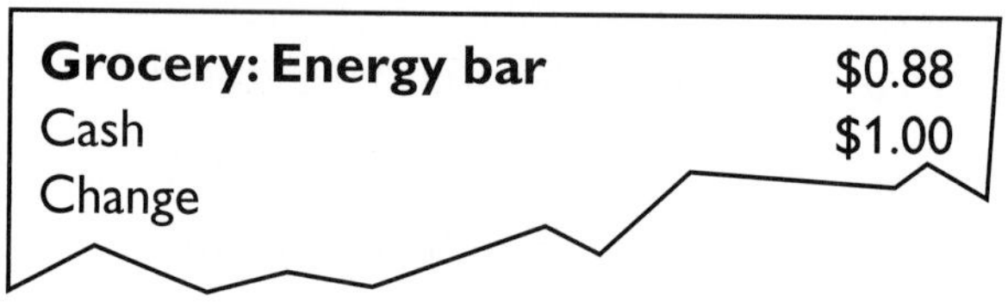

Making Connections

A *millisecond* is a thousandth of a second, and a *millicurie* is a thousandth of a curie (a unit of radioactivity). A *mill* is a monetary unit equal to $\frac{1}{10}$ of a cent.

19. What fraction of a dollar is a mill? Give your answer as a fraction and then as a decimal.

20. Explain why you think this unit is called a mill.

HW Homework 2

What's the Point?

Applying Skills

Write each number.

1. a number with a 3 in the tenths place, a 1 in the hundreds place, a 2 in the tens place, and a 9 in the ones place
2. a number with a 4 in the hundreds place, a 5 in the ones place, a 6 in the tenths place, an 8 in the tens place, and a 3 in the hundredths place
3. a number with a 7 in the thousandths place, a 4 in the ones place, a 1 in the tens place, and zeroes in the tenths and hundredths places
4. a number that is equal to 29.05, but has a digit in its thousandths place

Write each number as a decimal.

5. $5\frac{3}{10}$
6. $137\frac{56}{100}$
7. $10\frac{9}{10}$
8. $6\frac{7}{100}$
9. $12\frac{13}{50}$
10. $4\frac{4}{5}$
11. $5\frac{1}{2}$
12. $12\frac{1}{4}$
13. Complete the pattern by filling in the missing numbers.

$\underline{\ ?\ } \div 4 = 500$ $\qquad \frac{2{,}000}{4} = 500$

$200 \div 4 = 50$ $\qquad \underline{\ ?\ } = 50$

$20 \div 4 = 5$ $\qquad \frac{20}{4} = 5$

$2 \div \underline{\ ?\ } = \underline{\ ?\ }$ $\qquad \frac{2}{4} = \underline{\ ?\ }$

Write each number as a decimal. Use a calculator.

14. $\frac{7}{8}$
15. $\frac{3}{16}$
16. $\frac{25}{200}$
17. $2\frac{3}{4}$
18. $7\frac{4}{5}$
19. $30\frac{23}{40}$
20. $\frac{11}{32}$
21. $\frac{19}{64}$
22. $3\frac{55}{200}$
23. $13\frac{35}{80}$

Extending Concepts

24. Write twelve thousandths as a decimal and as a fraction in simplest form.
25. The egg of the Vervain hummingbird weighs about $\frac{128}{10{,}000}$ ounce. Write this as a decimal. Then, write the decimal in words.

Writing

26. The number 0.52 is read "fifty-two hundredths." The 5 is in the tenths place, and the 2 is in the hundredths place. Explain why these place value positions are named as they are.

Homework 3

Put Them in Order

Applying Skills

Six numbers are graphed on the number line below. Write each number as a fraction, mixed number, or whole number. Then, write each number in decimal notation.

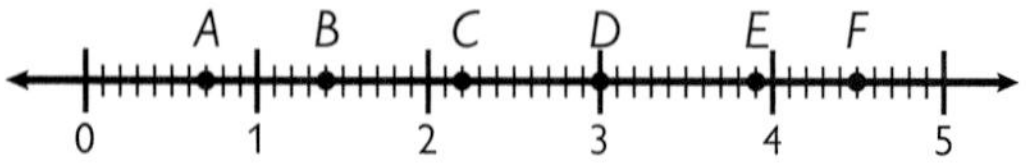

1. point A **2.** point B

3. point C **4.** point D

5. point E **6.** point F

Compare each pair of decimals. Write an expression using <, >, or = to show the comparison.

7. 6.4 and 6.7

8. 5.8 and 12.2

9. 7.02 and 7.20

10. 13.9 and 13.84

11. 16.099 and 160.98

12. 0.331 and 0.303

13. 47.553 and 47.5

Place each set of numbers in order from least to greatest. If two numbers are equal, place an equals sign between them.

14. 14.8, 14.09, 4.99, 14.98, 14.979, 14.099

15. 43, 42.998, 43.16, 42.022, 43.1600, 43.6789

16. 12.3, 12.008, 1.273, 12.54, 120, 12.45

17. 1.2, 4.4. 1.1. 0.9, 17.7, 1.3, 0.95

Extending Concepts

Copy the number line below. Then graph each number.

18. 5.07 **19.** 5.24

20. 5.36 **21.** 5.1

22. 5.17 **23.** 5.51

24. Copy the layout below. Place the digits 0, 2, 3, 4, 5, 6, 7, and 8 in the blanks so that the first number is the greatest possible number and the second number is the least possible number. Each digit can be used only once.

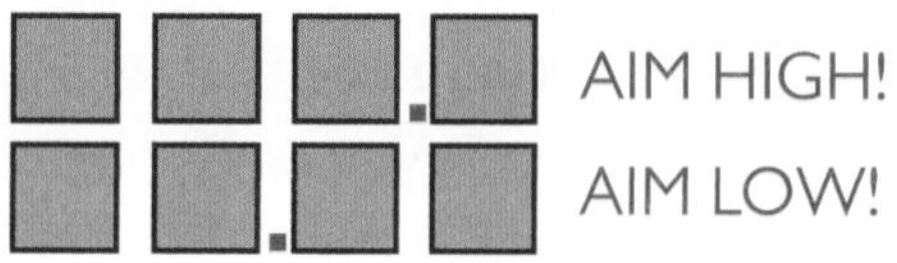

Writing

25. Suppose you are helping a student from another class compare decimals. Write a few sentences explaining why 501.1 is greater than 501.01.

Homework 4

Get It Close Enough

Applying Skills

Copy the number line below. Then graph each number. You may need to estimate some points.

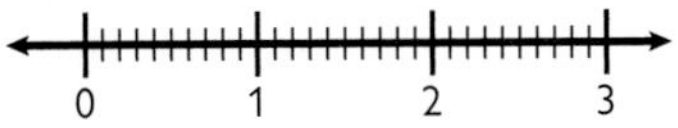

1. 2.5

2. 0.75

3. 2.80

4. 0.07

5. 0.25

6. 1.005

Determine whether each number is between 4.2 and 4.22.

7. 4.06

8. 4.217

9. 4.27

10. 4.022

11. 4.2016

12. 4.2301

13. 4.2099

14. 4.199

15. Round 3.2808 to the nearest tenth.

16. Round 3.2808 to the nearest thousandth.

17. Round 33.81497 to the nearest thousandth.

18. Round 33.81497 to the nearest hundredth.

19. Round 4.6745 to the nearest thousandth.

20. Round 4.6745 to the nearest hundredth.

21. Round 4.6745 to the nearest tenth.

22. Round 0.219 to the nearest hundredth.

23. Round 6.97 to the nearest tenth.

24. Round 19.98 to the nearest tenth.

25. Write each of the following numbers on small pieces of paper so you can move them around. Arrange the numbers from least to greatest. If two numbers are equal, place an equals sign between them. Then, place < between the other numbers.

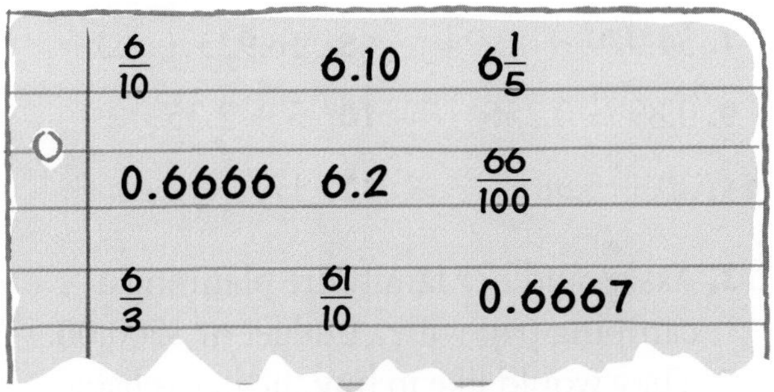

Extending Concepts

26. Write a number greater than 3.8 that rounds to 3.8. Then, write a number less than 3.8 that rounds to 3.8.

27. Write three numbers that each round to 14.37.

Making Connections

In most countries, people do not measure in inches, feet, and yards. They measure in meters. One meter is a little longer than one yard. $\frac{1}{100}$ of a meter is a centimeter (cm). $\frac{1}{1,000}$ of a meter is a millimeter (mm).

Use a metric ruler to draw a line representing each length.

28. 3.2 cm (32 mm)

29. 8.1 cm (81 mm)

30. 12.6 cm (126 mm)

Homework 5

Place the Point

Applying Skills

Find each sum or difference.

1. $1.25 + 0.68$

2. $13.82 - 5.52$

3. $15.3 - 0.92$

4. $16.89 - 2.35$

5. $2.0034 + 25.4$

6. $0.007 + 23.6$

7. $34.079 - 13.24$

8. $16.923 + 2.3$

9. $0.89 - 0.256$

10. $5 + 2.35$

11. $20 - 5.98$

12. $17.9 + 7.41$

13. Asako and her family are planning a camping trip with a budget of \$500.00. They would like to buy the items listed below.

sleeping bag	\$108.36
gas stove	\$31.78
tent	\$359.20
cook set	\$39.42
first-aid kit	\$21.89
Global Positioning System	\$199.99
compass	\$26.14
binoculars	\$109.76
knife	\$19.56
lantern	\$20.88

What different combinations of camping equipment could Asako's family afford with the money they have? Name as many combinations as you can. Show your work.

Extending Concepts

14. When Miguel adds numbers in his head, he likes to use expanded notation. For example, $1.32 + 0.276$ can be interpreted as $1 + 0 = 1$, $0.3 + 0.2 = 0.5$, $0.02 + 0.07 = 0.09$, and $0.000 + 0.006 = 0.006$. The sums are added to give the total of 1.596. Explain what Miguel has done and why it works.

Give the next two numbers for each sequence. Explain how you found the numbers.

15. 2.5, 3.25, 4, 4.75, 5.5, 6.25, . . .

16. 24.8, 23.7, 22.6, 21.5, 20.4, 19.3, . . .

Making Connections

Use the following information from *The World Almanac* about springboard diving at the Olympic Games.

Year	Name/Country	Points
1972	Vladimir Vasin/USSR	594.09
1976	Phil Boggs/U.S.	619.52
1980	Aleksandr Portnov/USSR	905.02
1984	Greg Louganis/U.S.	754.41
1988	Greg Louganis/U.S.	730.80
1992	Mark Lenzi/U.S.	676.53
1996	Xiong Ni/China	701.46
2000	Xiong Ni/China	708.72

17. How many more points did Greg Louganis score in 1984 than in 1988?

18. How many more points did Xiong Ni score in 2000 than in 1996?

Homework 6

More to the Point

Applying Skills

Find each product.

1. 23.62×100

2. 1.876×10

3. $16.8 \times 1{,}000$

4. 78.2×100

5. 125×0.1

6. 56×0.1

7. $7{,}834 \times 0.01$

8. 8×0.01

9. 159×0.001

10. $1{,}008 \times 0.01$

11. Explain how knowing a way to multiply by 0.1, 0.01, and 0.001 can help you multiply by decimals. Use examples if needed.

12. If $670 \times 91 = 60{,}970$, what is the value of 670×9.1?

13. If $1{,}456 \times 645 = 939{,}120$, what is the value of $1{,}456 \times 6.45$?

14. If $57 \times 31 = 1{,}767$, what is the value of 57×0.031?

Find each product.

15. 550×0.3

16. 71×2.2

17. 45×1.1

18. 231×0.12

19. 235×6.2

20. $1{,}025 \times 0.014$

21. 9×25.8

22. 62×0.243

23. $7{,}000 \times 1.8$

24. 41×1.15

25. Choose one of the items 15–24 and explain how you decided where to place the decimal point.

Extending Concepts

26. For Kate's birthday, she and eight of her friends went ice skating. Her mother, father, and grandmother went with them. They took a break for hot chocolate and coffee. The hot chocolate cost $1.35 per cup, and the coffee cost $1.59 per cup. If each of the 9 children had hot chocolate and each of the 3 adults had coffee, how much was the bill?

Give the next two numbers for each sequence. Explain how you found the numbers. You may want to use a calculator.

27. 15.8, 31.6, 63.2, 126.4, 252.8, 505.6, . . .

28. 6.1, 30.5, 152.5, 762.5, 3,812.5, 19,062.5, . . .

29. Write a multiplication sequence of your own. Start with a decimal. Include at least six numbers and the rule for the sequence.

Writing

30. Answer the letter to Dr. Math.

Dear Dr. Math,
What is all the hoopla about the decimal point? I mean, is it really all that important? If it is, why?
Dec-Inez

Homework 7

Decimal Pinpoint

Applying Skills

1. If 382 × 32 = 12,224, what is the value of 38.2 × 0.032?

2. 62 × 876 = 54,312, what is the value of 6.2 × 0.876?

3. If 478 × 52 = 24,856, what is the value of 4.78 × 5.2?

4. If 14 × 75 = 1,050, what is the value of 0.14 × 7.5?

Find each product.

5. 15.2 × 3.4

6. 6.7 × 0.04

7. 587 × 3.2

8. 4.2 × 0.125

9. 0.35 × 1.4

10. 5.2 × 0.065

11. 3.06 × 4.28

12. 0.9 × 0.15

13. 18.37 × 908.44

14. 0.003 × 0.012

15. Choose one of the items 5–14 and explain how you decided where to place the decimal point.

16. Rebecca and Miles are on the decoration committee for the school dance. They need 12 rolls of streamers and 9 bags of balloons. How much will these items cost if a roll of streamers costs $1.39 and a bag of balloons cost $2.09?

17. Pluto's average speed as it travels around the Sun is 10,604 miles per hour. Earth travels 6.28 times faster than Pluto. What is the average speed of Earth?

18. The giant tortoise can travel at a speed of 0.2 kilometer per hour. At this rate, how far can it travel in 1.5 hours?

Extending Concepts

Stacy and Kathryn decided to modify the rules of the "Place Value" game. They used each of the digits 1 to 6 once.

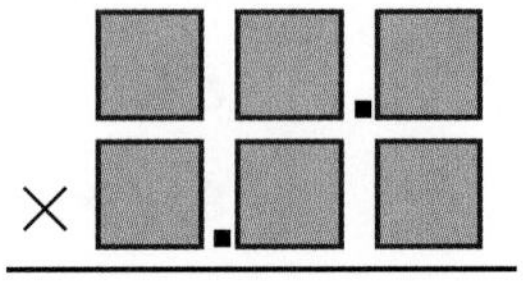

19. How could they place the digits to get the greatest product? What is the greatest product?

20. How could they place the digits to get the least product? What is the least product?

The area of a rectangle equals the length times the width. Find the area of each rectangle.

21.

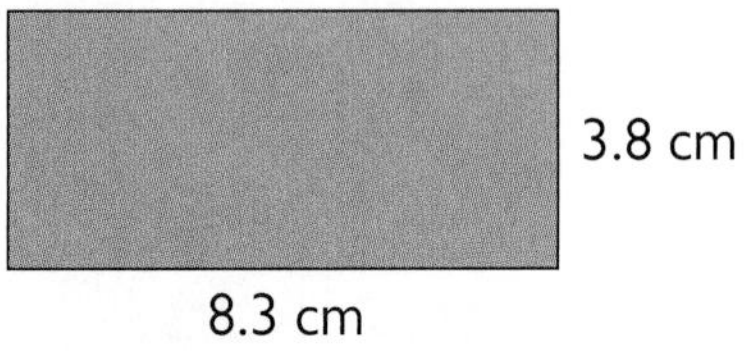

22.

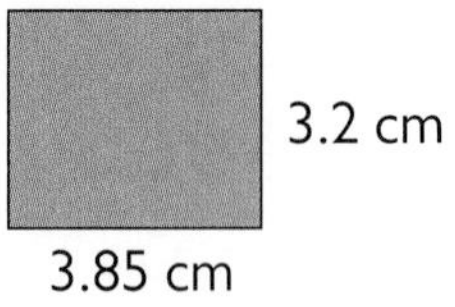

Writing

23. Write some strategies that would help you play the "Place Value" game successfully.

Homework 8

Patterns and Predictions

Applying Skills

1. If $125 \div 5 = 25$, what is the value of $12.5 \div 0.5$?

2. If $288 \div 12 = 24$, what is the value of $2.88 \div 0.12$?

3. If $369 \div 3 = 123$, what is the value of $36{,}900 \div 300$?

4. If $18{,}000 \div 36 = 500$, what is the value of $180 \div 0.36$?

5. Explain how you can change a decimal division problem to make the problem easier to solve.

Find each quotient.

6. $812 \div 0.4$

7. $0.34 \div 0.2$

8. $20.24 \div 2.3$

9. $180 \div 0.36$

10. $23 \div 0.023$

11. $576 \div 3.2$

12. $14.4 \div 0.12$

13. $4.416 \div 19.2$

14. $259.2 \div 6.48$

15. $4.6848 \div 0.366$

16. $97.812 \div 1.1$

17. $38.57 \div 1.9$

18. $199.68 \div 9.6$

19. $131.1 \div 13.8$

20. $5.992 \div 74.9$

21. $39.95 \div 799$

Extending Concepts

22. Vladik's dad filled the gasoline tank in his car. The gasoline cost $1.48 per gallon. The total cost of the gasoline was $22.94. How many gallons of gasoline did Vladik's dad put in the tank?

23. A board is 7.5 feet long. If it is cut into pieces that are each 2.5 feet long, how many pieces will there be?

24. Ann and her mom went grocery shopping. Her mom bought 2 dozen oranges for $2.69 per dozen. Estimate the cost of each orange. Explain your reasoning.

25. Mr. and Mrs. Francisco are buying their first home. During the first year, their total mortgage payments will be $12,159.36. How much will their monthly mortgage payment be? Show your work.

26. Drew wants to invite as many friends as he can to go to the movies with him. He has $20.00.

a. The cost of each ticket is $4.50. Estimate how many friends he can take to the movies.

b. A bag of popcorn costs $1.75. Estimate how many friends he can take to the movies if he is going to buy everyone, including himself, a bag of popcorn.

Writing

27. Find the product of 5.5 and 0.12. Then, write two related division problems. Explain how the division problems support the rules for dividing decimals.

Homework 9

It Keeps Going and Going

Applying Skills

Find a decimal representation for each fraction. Then, tell if the decimal is *repeating* or *terminating.*

1. $\frac{3}{8}$ **2.** $\frac{2}{3}$

3. $\frac{5}{6}$ **4.** $\frac{7}{16}$

5. $\frac{7}{10}$ **6.** $\frac{8}{9}$

7. $\frac{5}{11}$ **8.** $\frac{17}{25}$

9. $\frac{2}{11}$ **10.** $\frac{1}{3}$

11. $\frac{3}{4}$ **12.** $\frac{1}{8}$

13. $\frac{5}{9}$ **14.** $\frac{3}{6}$

15. $\frac{4}{15}$ **16.** $\frac{41}{50}$

17. $\frac{12}{15}$ **18.** $\frac{1}{7}$

19. $\frac{9}{11}$ **20.** $\frac{8}{12}$

Change each Celsius temperature to a Fahrenheit temperature.

21. 25°C **22.** 100°C

23. 55°C **24.** 75°C

Change each Fahrenheit temperature to a Celsius temperature.

25. 95°F **26.** 113°F

27. 131°F **28.** 122°F

Extending Concepts

29. One inch is about 2.54 centimeters. Copy and complete the following table.

Name	Height (inches)	Height (centimeters)
John	53	
Krista		127
Miwa		152.4
Tommy	74	

30. Without doing the division, think of three fractions that would have decimal equivalents that repeat. Write each fraction. Use a calculator to confirm that the decimal equivalents are repeating decimals.

Writing

31. Answer the letter to Dr. Math.

Dear Dr. Math,

My partner and I are putting some numbers in order. My partner says that $3.\overline{3}$ is greater than 3.333333. I say that 3.333333 is greater. In fact, it is obvious. Look how many digits it has! Please, tell us who is right and why.

Sincerely,

Endless Lee Confused

10 Homework

Moving to Percents

Applying Skills

For each drawing, determine what percent of the whole is shaded.

1.

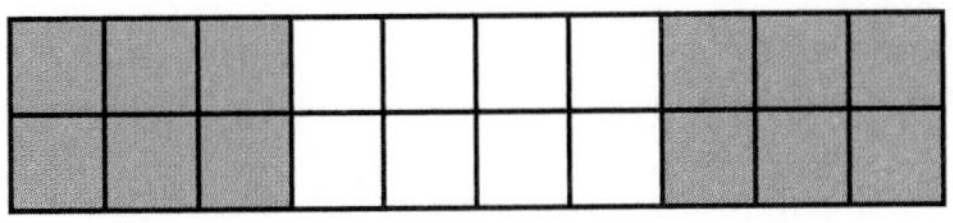

2.

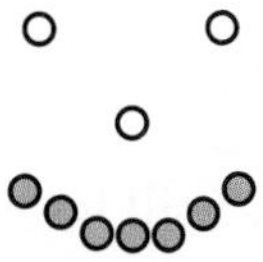

3.

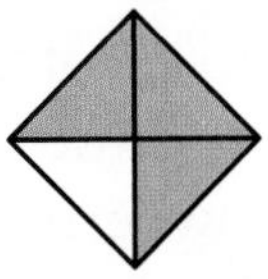

Draw a model showing each percent.

4. 65% **5.** 80% **6.** 30%

7. What percent of the months of the year begin with the letter J?

8. What percent of the months of the year begin with the letter Y?

9. Copy and complete the table.

Fraction	Decimal	Percent
$\frac{4}{10}$		
	0.6	
		90%
		5%
	0.02	
$\frac{1}{8}$		

Extending Concepts

A sixth-grade class is baking cookies for a fund-raiser. They made a graph to show how many of each type were ordered.

10. Write an equivalent fraction for each percent.

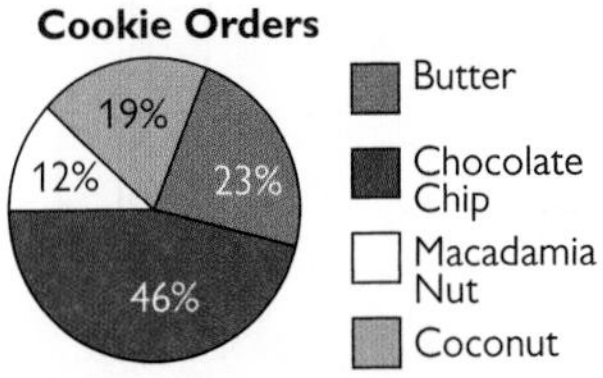

11. If a total of 200 cookies were ordered, how many of each type of cookie will the class need to bake?

12. Because macadamia nuts are expensive, the class would lose money for each macadamia nut cookie they sell. Suppose they substitute coconut cookies for orders of macadamia nut cookies. How many coconut cookies will the class need to bake if a total of 200 cookies were ordered?

Writing

13. Answer the letter to Dr. Math.

Dear Dr. Math,
I understand that I can write a fraction as a decimal and a decimal as a fraction. After all, they are both numbers. But, aren't percents different? They have that funny symbol at the end. How can a fraction, a decimal, and a percent all mean the same thing? Could you help me out?
Sincerely,
P.R. Cent

Homework 11

Working with Common Percents

Applying Skills

Match each percent with the closest fraction.

1. 83%	**A.** $\frac{1}{2}$
2. 35%	**B.** $\frac{1}{3}$
3. 27%	**C.** $\frac{2}{3}$
4. 52%	**D.** $\frac{1}{4}$
5. 65%	**E.** $\frac{3}{4}$
6. 42%	**F.** $\frac{1}{5}$
7. 59%	**G.** $\frac{2}{5}$
8. 73%	**H.** $\frac{3}{5}$
9. 19%	**I.** $\frac{4}{5}$

Find 50%, 10%, and 1% of each number.

10. 100	**11.** 600
12. 60	**13.** 40
14. 150	**15.** 340
16. 18	**17.** 44

Extending Concepts

18. Write the numbers $\frac{672}{900}$, 0.012, $\frac{7}{10}$, 32%, $\frac{1}{10}$, 0.721, and 65% in order from least to greatest.

19. Write the numbers 63%, $\frac{2}{10}$, 0.8, $\frac{8}{29}$, 85%, 0.12, $\frac{55}{90}$, and 89% in order from least to greatest.

20. One day a gardener picked about 20% of the strawberries in his garden. Later that day, his wife picked about 25% of the remaining strawberries. Still later, their son picked about 33% of the strawberries that were left. Even later, their daughter picked about 50% of the remaining strawberries in the garden. Finally, there were only 3 strawberries left. About how many strawberries were originally in the garden? Explain your reasoning.

Writing

21. Explain the connection between fractions, decimals, and percents. Show examples.

Homework 12

Percent Power

Applying Skills

Estimate each value.

1. 19% of 30
2. 27% of 64
3. 48% of 72
4. 73% of 20
5. 67% of 93
6. 25% of 41
7. 65% of 76
8. 81% of 31
9. 34% of 301
10. 41% of 39

Find each value.

11. 25% of 66
12. 13% of 80
13. 52% of 90
14. 16% of 130
15. 37% of 900
16. 32% of 68
17. 66% of 43
18. 7% of 92
19. 42% of 85
20. 78% of 125

Extending Concepts

21. Find the total cost of a $25.50 meal after a tip of 15% has been added.
22. A shirt that sold for $49.99 has been marked down 30%. Find the sale price.
23. Twenty-seven percent of Camille's annual income goes to state and federal taxes. If she earned $35,672, how much did she actually keep?
24. Jared's uncle bought 100 shares of stock for $37.25 per share. In six months, the stock went up 30%. How much were the 100 shares of stock worth in six months?
25. Kyal's family bought a treadmill on sale. It was 25% off the original price of $1,399.95. They paid half of the sale price as a down payment. What was the down payment?
26. Maxine and her family went on a vacation to San Francisco last summer. On the first day, they gave the taxi driver a 15% tip and the hotel bellhop a $2.00 tip for taking the luggage to the room. The taxi fare was $15.75. How much did they spend in all?

Making Connections

For a healthy diet, the total calories a person consumes should be no more than 30% fat. Tell whether each food meets these recommendations. Show your work.

27. Reduced-fat chips
    Calories per serving: 140
    Calories from fat: 70
28. Low-fat snack bar
    Calories per serving: 150
    Calories from fat: 25
29. Light popcorn
    Calories per serving: 30
    Calories from fat: 6
30. Healthy soup
    Calories per serving: 110
    Calories from fat: 25

Homework 13

Less Common Percents

Applying Skills

Write an equivalent decimal and an equivalent fraction for each percent.

1. 0.4%

2. 0.3%

3. 0.25%

4. 0.05%

5. 0.079%

6. 0.008%

7. $\frac{1}{5}\%$

8. $\frac{7}{10}\%$

9. $\frac{1}{4}\%$

10. $\frac{4}{25}\%$

11. $\frac{1}{8}\%$

12. $\frac{1}{25}\%$

State whether each number represents a percent less than one. Write *yes* or *no*.

13. 0.02

14. 0.0019

15. 0.0101

16. 0.7

17. 0.0009

18. 0.0088

19. $\frac{1}{50}$

20. $\frac{1}{2{,}500}$

21. $\frac{7}{300}$

22. $\frac{2}{300}$

23. $\frac{9}{1{,}000}$

24. $\frac{4}{700}$

25. Make a model for 110%.

26. Write an equivalent decimal and an equivalent mixed number for 925%.

27. Order 25%, $\frac{1}{4}\%$, 125%, and 1 from least to greatest.

Extending Concepts

28. Liana makes beaded jewelry. She plans to sell them at the school fair. Liana wants to sell each item for 135% of the cost to make the item. Copy and complete the table below to find the selling price of each item.

Item	Cost to Make	Selling Price
Long Necklace	$22.00	
Short Necklace	$16.00	
Bracelet	$14.50	
Earrings	$9.25	

29. During a recent contest at South Middle School, prizes were awarded to 0.4% of the students. If there are 500 students at the school, how many students won prizes?

30. The diameter of the Sun is 865,500 miles. The diameter of Earth is about 0.9% of the Sun's diameter. Find the diameter of Earth.

Writing

31. A coach wants his players to give 110% of their effort. Explain what is meant by 110%. Is it reasonable for a coach to ask for 110% of their effort? Explain.

Homework 14

Give It to Me Straight

Applying Skills

Examine each statement. Determine whether or not the percent is used correctly. Explain your answer.

1. Unemployment fell 5% in July.

2. Yogi Berra said, "Half the game is 90% mental."

3. Roberto gave away 130% of his stamp collection.

4. 72% of the class prefers snacks that contain chocolate.

5. The team won 70% of the games, lost 25% of the games, and tied 10% of the games.

6. Your height now is 0.1% of your height at one year old.

7. The store's sales for the month of April are 130% of its sales for the month of March.

8. The value of the house is 125% its value 5 years ago.

9. 45% of the class members have a brother, 50% of the class members have a sister, and 20% of the class members have no brother or sister.

10. At the class picnic, the students ate 130% of the cookies prepared by the parents.

Extending Concepts

11. Mark and Fala collected data at lunchtime to determine which types of desserts were purchased most often. The results of the 75 students who went through the lunch line are given below.

Type of Dessert	Percent of Students who Purchased the Dessert
Cookies	24%
Ice Cream	40%
Fruit	4%
Candy	12%
None	20%

How many students bought each type of dessert? How many students bought no dessert?

12. North Middle School has 645 students equally divided into grades six, seven, and eight. During a recent school survey, Dan noticed that only eighth graders played lacrosse. After analyzing the data, he said, "86 students play lacrosse." Paula said, "No way! 40% of the eighth graders play lacrosse." Explain how both students can be correct.

Writing

13. Throughout this unit, you have used fractions, decimals, and percents. Explain how the three are connected. Describe the advantage of each representation. Use examples to help to clarify your explanation.

15 Homework: The Other End of the Number Line

Applying Skills

Determine whether each sentence is *true* or *false.*

1. $4 > -2$
2. $3\frac{1}{2} < 3\frac{3}{4}$
3. $0.3 > 0.25$
4. $0 > -1$
5. $-\frac{1}{4} < -\frac{1}{2}$
6. $-5 > -4$
7. $-6.5 > -6$
8. $-107 < -106$

Use $<$ or $>$ to compare each pair of numbers.

9. 37 and 42
10. -25 and -37
11. -12 and 12
12. -144 and -225
13. -512 and -550
14. -300 and -305
15. -960 and -890
16. 385 and 421

Evan and Alison are playing "Who's Got My Number?" They are reading cards to each other. Determine the next number for each card.

17. My number is -13. Does anyone have 4 more?
18. My number is -21. Does anyone have 25 more?
19. My number is -9. Does anyone have 9 more?
20. My number is -14. Does anyone have 7 less?
21. My number is 4. Does anyone have 17 less?
22. Arrange the answers from items 17–21 in order from least to greatest.

Extending Concepts

Order each set of numbers from least to greatest.

23. $2, -2.5, -6.8, 6.7, 4, 0.731, -3$
24. $-\frac{1}{3}, 4, -\frac{3}{4}, \frac{7}{8}, 1, -\frac{7}{8}, -2, 5$
25. $-1\frac{2}{3}, -\frac{6}{4}, 3, 0, 4\frac{1}{5}, \frac{5}{3}, 4, -2, -1\frac{4}{5}$

Making Connections

26. Consider the elevations listed below. List the elevations from greatest to least.

Location	Elevation (meters)
Dead Sea, Israel	−408
Chimborazo, Ecuador	6,267
Death Valley, California	−86
Challenger Deep, Pacific Ocean	−10,924
Zuidplaspoldor, Netherlands	−7
Fujiyama, Japan	3,776

27. Find the elevations of three other interesting places around the world. Add these elevations to the list in item **26**.

Homework 16
Moving on the Number Line

Applying Skills

Write an addition equation for each number line.

1.

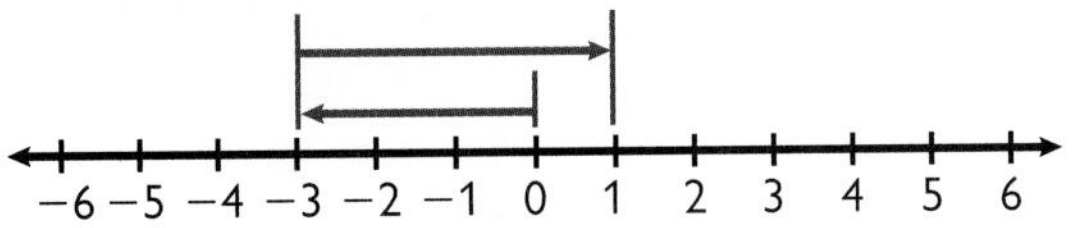

2.

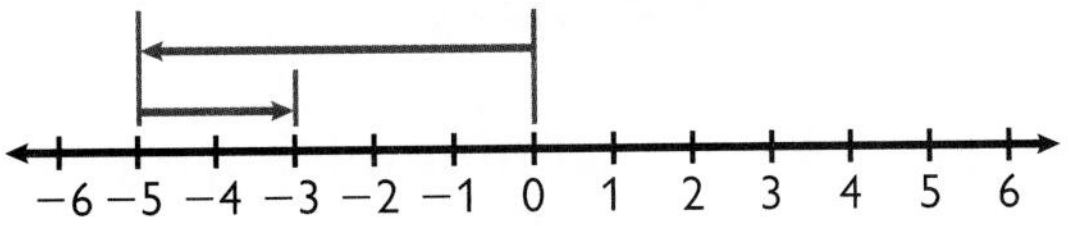

3.

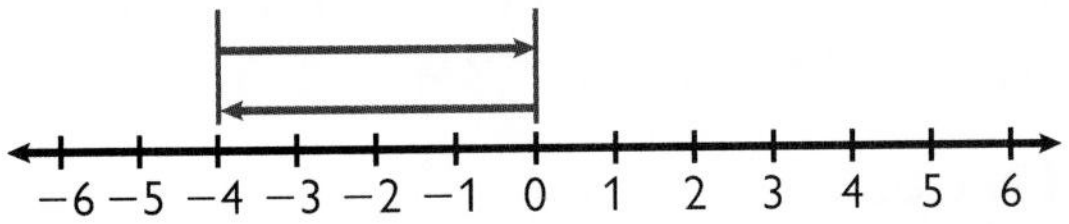

4.

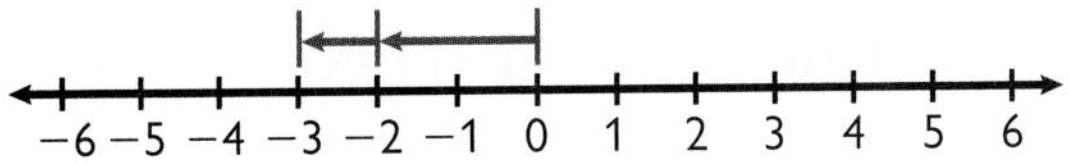

5.

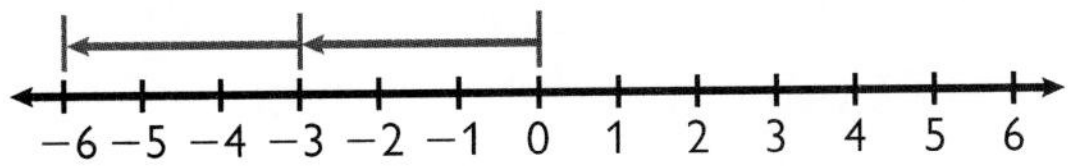

Use a number line to find each sum.

6. $-4 + 7$

7. $6 + (-3)$

8. $-3 + (-2)$

9. $-2 + (-2)$

10. $1 + (-5)$

11. $-8 + 5$

12. $-5 + 5$

13. $-5 + 7$

14. $6 + (-9)$

15. $-2 + (-5)$

Extending Concepts

Fractions and decimals also have negatives. Use what you have learned to find each sum.

16. $1\frac{3}{5} + (-5)$

17. $-\frac{3}{7} + (-\frac{6}{7})$

18. $5.2 + (-7.3)$

19. $-12.352 + 4.327$

Writing

20. Answer the letter to Dr. Math.

Dear Dr. Math,

The other day in math class, I was finding -3 + 7 + (-5) + 1. I started drawing arrows on the number line. Then, my friend Maya walked by and said the answer was zero. I asked how she got the answer so fast. She told me that she added the two positive numbers (7 + 1) and got 8. Then, she added the two negative numbers (-3 + (-5)) and got -8. She said 8 + (-8) is zero. I thought she was wrong, but when I checked it on the number line, the answer was zero. Why is it OK to add the numbers out of order?

Signed,
Addled About Addition

Homework 17

Taking the Challenge

Applying Skills

Margo (pink) and Lisa (green) are playing "Color Challenge." For each set of spins, determine who is ahead. Then determine how much the person is ahead.

1. Add 2 green cubes.
Add 1 pink cube.
Add 1 green cube.
Add 2 pink cubes.

2. Add 1 green cube.
Add 1 green cube.
Add 3 pink cubes.
Add 2 green cubes.
Add 1 pink cube.
Add 2 pink cubes.

3. Add 2 pink cubes.
Add 1 pink cube.
Add 1 pink cube.
Add 2 green cubes.
Add 1 green cube.
Add 2 pink cubes.
Add 3 green cubes.
Add 1 green cube.

Write an addition equation for each drawing.

4.

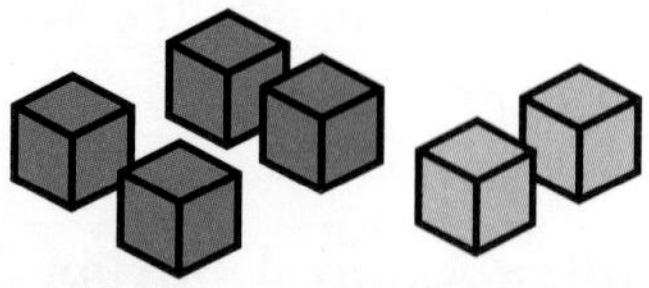

5.

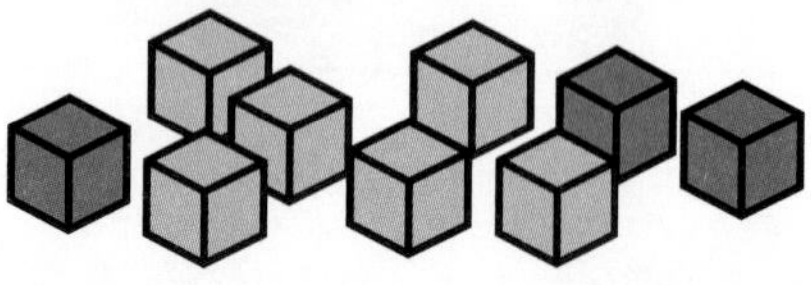

Use cubes to find each sum.

6. $-2 + 6$

7. $-5 + 3$

8. $4 + (-5)$

9. $7 + (-4)$

10. $-5 + (-6)$

11. $-1 + (-8)$

Extending Concepts

You may know that multiplying is just a fast way of adding. For example, you can use 4×3 to find $3 + 3 + 3 + 3$.

12. Consider $-3 + (-3) + (-3) + (-3)$.

a. Draw a collection of cubes that represent the addition problem.

b. Write a corresponding multiplication problem and its product.

13. Consider $-2 + (-2) + (-2) + (-2) + (-2)$.

a. Draw a collection of cubes that represent the addition problem.

b. Write a corresponding multiplication problem and its product.

14. What is $4 \times (-5)$? Try to find the answer without using drawings or cubes.

Making Connections

15. At the end of last month, Keston had $1,200 in his savings account. His next bank statement listed a deposit of $100, a withdrawal of $300, and another deposit of $150. Write an addition problem to represent the situation. How much money does he have in his account now?

18 Homework: The Meaning of the Sign

Applying Skills

Write a subtraction equation for each number line.

1.

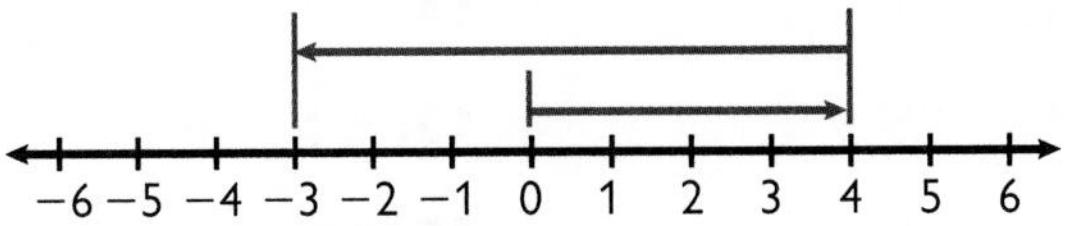

2.

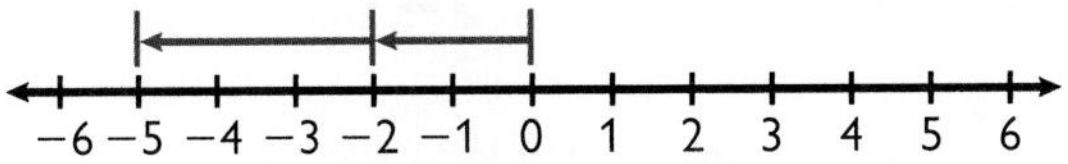

3.

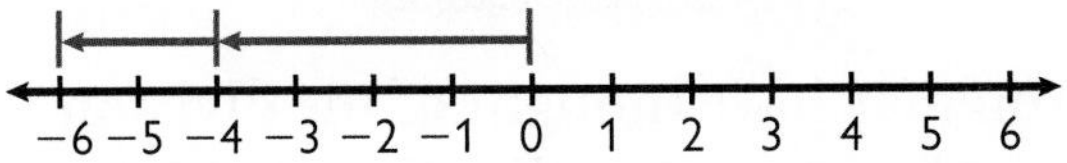

4.

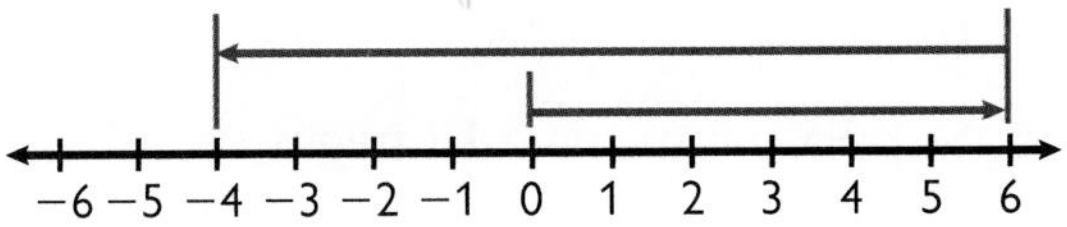

5.

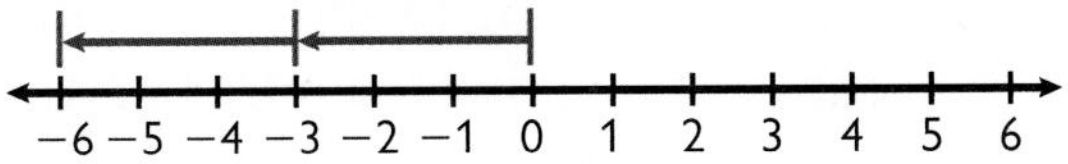

Use a number line to find each difference.

6. $3 - 8$

7. $-1 - 5$

8. $-3 - 2$

9. $-2 - 2$

10. $5 - 9$

11. $-2 - 9$

12. $-5 - 4$

13. $6 - 8$

14. $-7 - 8$

15. $4 - 10$

Extending Concepts

You must add and subtract integers according to the order of operations. For each problem, two students got different answers. Decide which student has the correct answer.

16. Ellen: $-3 - 7 + 1 = -11$
Tia: $-3 - 7 + 1 = -9$

17. Robert: $4 - 3 + 6 = 7$
Josh: $4 - 3 + 6 = -5$

18. Emma: $-2 + (-4) - 7 + (-1) = -12$
Lucas: $-2 + (-4) - 7 + (-1) = -14$

19. HaJeong: $3 - 5 + (-3) - 1 = 0$
Janaé: $3 - 5 + (-3) - 1 = -6$

Making Connections

20. Opal and Dory are piloting a deep sea submersible at the depth of 200 meters below sea level. To retrieve some dropped equipment, they must descend another 2,400 meters. Write a subtraction equation that describes the situation. What is the final depth?

21. The afternoon temperature was 15°F. During the night, the temperature dropped 17°F. Write a subtraction equation that describes the situation. What was the nighttime temperature?

The Cube Model

Homework 19

Applying Skills

Write a subtraction equation for each drawing.

1.

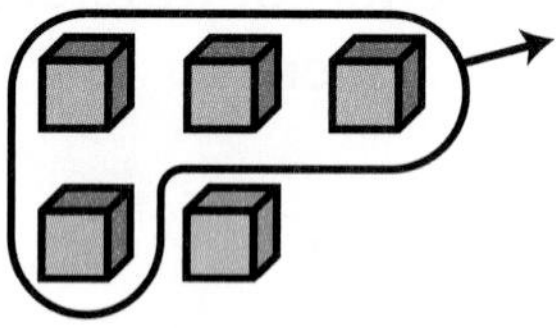

2.

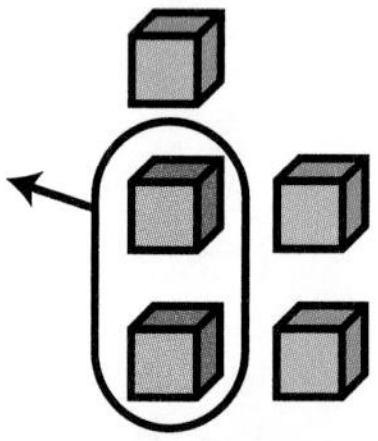

3.

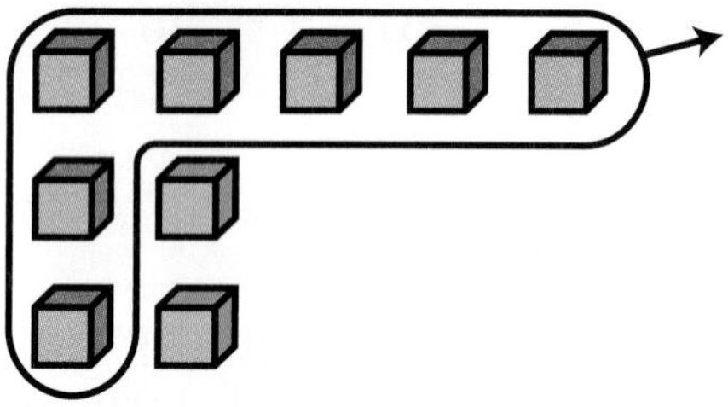

4.

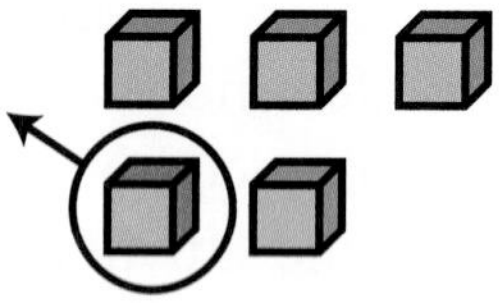

5.

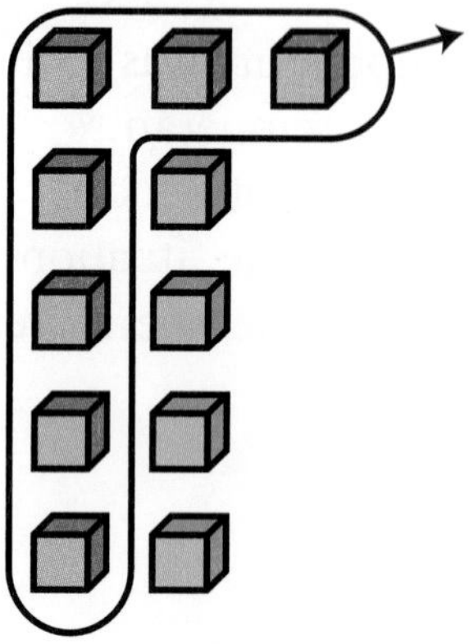

Use cubes to find each difference.

6. $-2 - 6$

7. $5 - 7$

8. $-5 - 2$

9. $4 - 7$

10. $-1 - 1$

11. $-1 - 3$

12. $6 - 9$

13. $-5 - 6$

14. $-8 - 2$

15. $8 - 9$

Extending Concepts

You may have thought of the division problem 15 ÷ 5 as dividing 15 things into 5 groups.

16. Suppose you have 12 negative cubes.

a. Make a drawing to show how the cubes can be divided into 3 groups.

b. Write a division problem for the situation.

c. What is the quotient?

17. Suppose you have 10 negative cubes.

a. Make a drawing to show how the cubes can be divided into 5 groups.

b. Write a division problem for the situation.

c. What is the quotient?

18. What is $-24 \div 6$? Try to find the answer without using drawings or cubes. Explain your answer.

Writing

19. Use words and/or pictures to describe how you would use zero pairs to find $-4 - 6$.

20 Homework

Write It another Way

Applying Skills

Write each subtraction problem as an addition problem. Then, solve the problem.

1. $-5 - 1$

2. $4 - 11$

3. $-2 - 7$

4. $8 - 6$

5. $-1 - 8$

6. $0 - 7$

Write each addition problem as a subtraction problem. Then, solve the problem.

7. $-4 + (-5)$

8. $8 + (-7)$

9. $3 + (-11)$

10. $-7 + (-8)$

11. $-1 + (-5)$

12. $6 + (-6)$

Find each sum or difference.

13. $-3 + 8$

14. $-3 - 8$

15. $-4 + (-6)$

16. $5 - 9$

17. $0 - 2$

18. $-3 + 10$

Extending Concepts

19. Use the subtraction pattern you started at the beginning of the lesson.

a. Extend the list of subtraction problems to include the following:

$4 - (-1)$

$4 - (-2)$

$4 - (-3)$

$4 - (-4)$

$4 - (-5)$

b. Use the pattern to find the solutions to the subtraction problems.

c. Copy and complete the following sentence.

Subtracting a negative number gives the same answer as if you __?__.

Use what you have learned from the pattern to find each difference.

20. $5 - (-3)$

21. $-3 - (-7)$

22. $-8 - (-4)$

23. $1 - (-8)$

24. $0 - (-3)$

25. $-1 - (-6)$

26. $9 - (-5)$

27. $7 - (-7)$

Making Connections

28. Death Valley, California, is the lowest elevation in the United States. It is 280 feet below sea level, or -280 feet in elevation. Mt. McKinley, Alaska, has the highest elevation at about 20,000 feet above sea level. Write a number sentence that shows the difference in elevation between Mt. McKinley and Death Valley. What is the difference?

29. The highest temperature recorded in North America is 134°F in Death Valley, California. The lowest recorded temperature is -87°F in Northice, Greenland. What is the difference in temperatures?

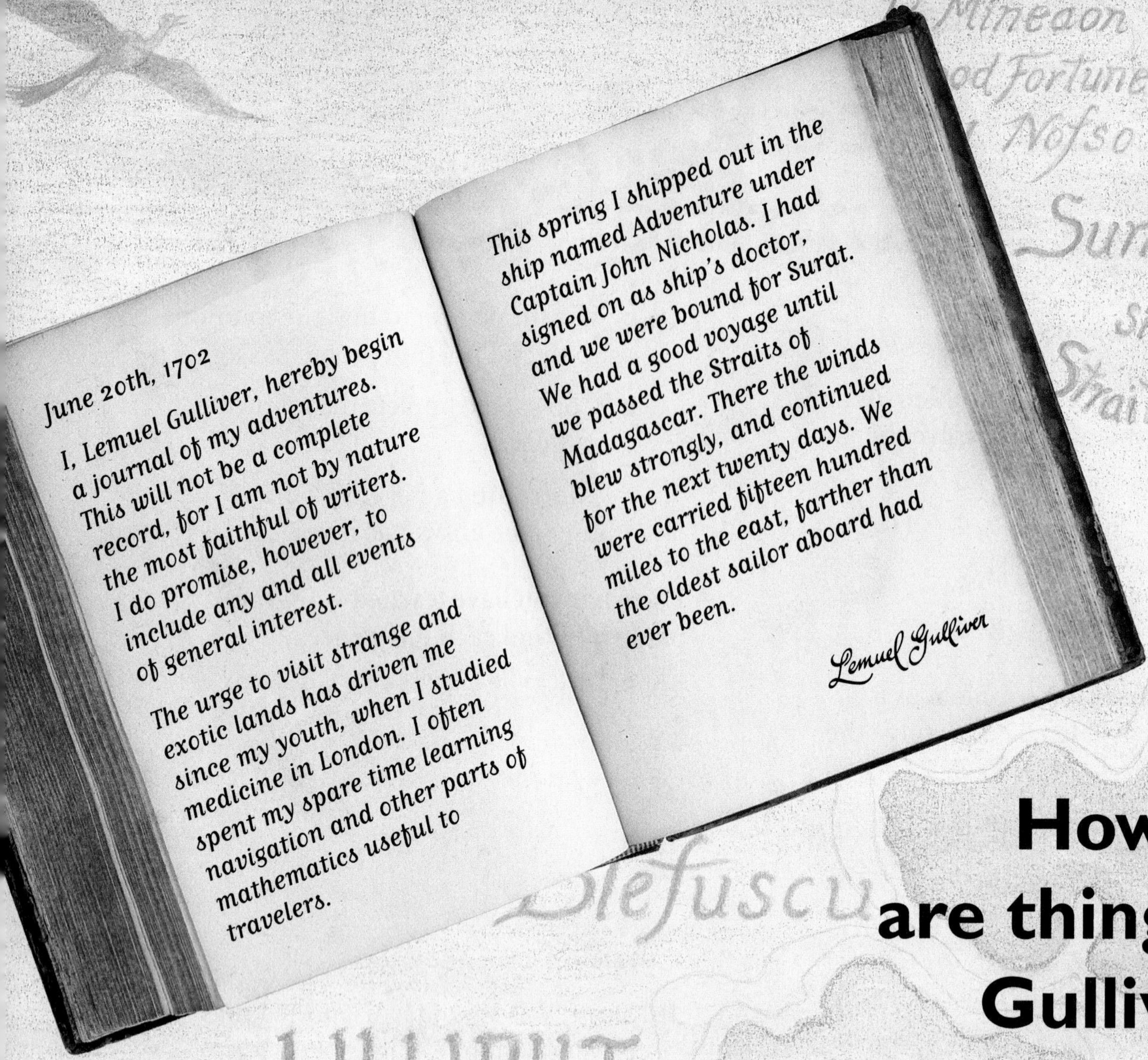
June 20th, 1702

I, Lemuel Gulliver, hereby begin a journal of my adventures. This will not be a complete record, for I am not by nature the most faithful of writers. I do promise, however, to include any and all events of general interest.

The urge to visit strange and exotic lands has driven me since my youth, when I studied medicine in London. I often spent my spare time learning navigation and other parts of mathematics useful to travelers.

This spring I shipped out in the ship named Adventure under Captain John Nicholas. I had signed on as ship's doctor, and we were bound for Surat. We had a good voyage until we passed the Straits of Madagascar. There the winds blew strongly, and continued for the next twenty days. We were carried fifteen hundred miles to the east, farther than the oldest sailor aboard had ever been.

Lemuel Gulliver

How big are things in Gulliver's Worlds?

GULLIVER'S WORLDS

PHASE**ONE**
Brobdingnag

Gulliver's journal holds clues to sizes of things in Brobdingnag, a land of giants. Using these clues, you will find ways to predict the sizes of other things. Then you will use math to create a life-size drawing of a giant object. You will also compare sizes in the two lands. Finally, you will use what you know about scale to write a story set in Brobdingnag.

PHASE**TWO**
Lilliput

Lilliput is a land of tiny people. Gulliver's journal and drawings will help you find out about the sizes of things in Lilliput. You will compare the measurement system in Lilliput to ours. Then you will explore area and volume as you figure out how many Lilliputian objects are needed to feed and house Gulliver. Finally, you will write a story set in Lilliput.

PHASE**THREE**
Lands of the Large and Lands of the Little

Clues from pictures will help you write a scale factor that relates the sizes of things in different lands to the sizes of things in Ourland. You will continue to explore length, area, and volume, and see how these measures change as the scale changes. Finally, you will put together all you have learned to create a museum exhibit about one of these lands.

PHASE ONE

August 29, 1702

We finally sighted land again today. We went ashore near a small creek.

I was gone only a short time. Yet when I headed back toward the landing site, the sailors were already rowing frantically out to sea. I could see a huge creature chasing them through the water. It stopped, though, at a sharp reef, and so the sailors escaped.

This was, I admit, of small comfort to me, because I was now alone. Fearing for my safety, I scampered inland. Beyond a steep hill, I discovered tall stalks, about eighteen feet high. They appeared to be wheat. I reached a stone stairway, but finding each step to rise six feet, I was unable to climb it. The trees along its edge were so tall I could not guess their height.

Lemuel Gulliver

Imagine a world in which everything is so large that you would be as small as a mouse. How can you predict how large things will be in this land?

In this phase you will learn to figure out a scale factor that describes how sizes of things are related. You will use the scale factor to create life-size drawings, solve problems, and write stories.

Brobdingnag

WHAT'S THE MATH?

Investigations in this section focus on:

DATA COLLECTION

- Gathering information from a story
- Organizing data to find patterns

MEASUREMENT and ESTIMATION

- Measuring with inches, feet, and fractions of inches
- Estimating the sizes of large objects

SCALE and PROPORTION

- Finding the scale factor that describes the relationship between sizes
- Applying the scale factor to predict sizes of objects
- Creating scale drawings
- Exploring the effect of rescaling on area and volume

MathScape Online
mathscape1.com/self_check_quiz

1 The Sizes of Things in Brobdingnag

DETERMINING THE SCALE FACTOR

How well can you picture in your mind the events described in Gulliver's journal entry? Here you will gather clues from the journal entry about the sizes of things in Brobdingnag. As you compare sizes of things in Brobdingnag to sizes in Ourland, you will learn about scale.

August 29, 1702
I had not a moment to rest, as another
monster was approaching. I now saw
that in form he resembled a human
being. It was his size—as tall as a ship's
mast—that made him appear to be a
monster. Scared and confused, I backed
away, tripping over an apple core that
lay like a log behind me. As I stood up
again, the giant began cutting wheat
with a great scythe. With every stride he
traveled about ten yards closer to me,
and I was faced with either being
trampled on or cut in two. Therefore,
I gave up my hiding place and shouted
for his attention.

Compare Sizes to Determine a Scale Factor

How are sizes of things in Brobdingnag related to sizes in Ourland?

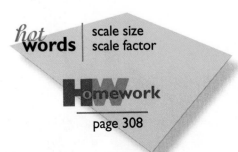

A scale factor is a ratio that tells how the sizes of things are related. For example, some model trains use a 20:1 scale factor. This means that each part on the real train is 20 times as large as the same part on the model train. Follow these steps to find the scale factor that relates sizes in Brobdingnag to sizes in Ourland.

1. Make a chart with three columns. Column 1 is for the name of each object. Column 2 is for the size of the object in Brobdingnag. Column 3 is for the size of the corresponding object in Ourland.

2. Fill out column 1 and column 2 with clues you found in the story about the sizes of objects in Brobdingnag. Measure or estimate how big each of the objects would be in Ourland. Enter that information in column 3.

3. Use the information in your chart to figure out a scale factor that tells how sizes of things in Brobdingnag are related to sizes in Ourland.

Object	Brobdingnag	Ourland
Stalk of wheat	About 18 feet	

How big would an Ourland object be in Brobdingnag?

hot words scale size
scale factor

Homework page 308

2 A Life-Size Object in Brobdingnag

RESCALING THE SIZES OF OBJECTS

The story continues as Gulliver describes more events from his life in Brobdingnag. In the last lesson, you figured out how sizes in Brobdingnag relate to sizes in Ourland. In this lesson you will use what you know to create a life-size drawing of a Brobdingnag object.

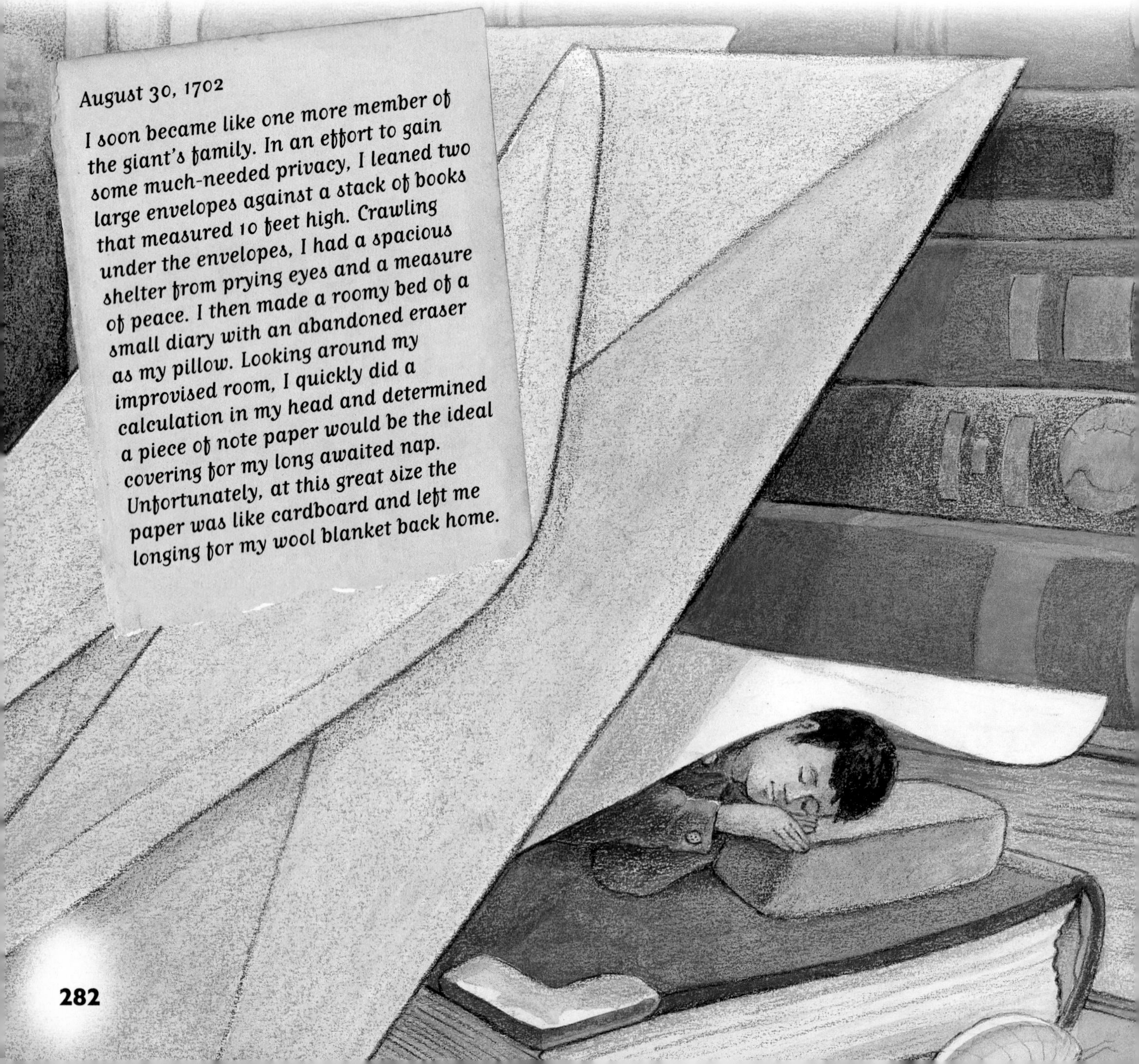

August 30, 1702
I soon became like one more member of the giant's family. In an effort to gain some much-needed privacy, I leaned two large envelopes against a stack of books that measured 10 feet high. Crawling under the envelopes, I had a spacious shelter from prying eyes and a measure of peace. I then made a roomy bed of a small diary with an abandoned eraser as my pillow. Looking around my improvised room, I quickly did a calculation in my head and determined a piece of note paper would be the ideal covering for my long awaited nap. Unfortunately, at this great size the paper was like cardboard and left me longing for my wool blanket back home.

Make a Life-Size Drawing

Choose an object from Ourland that is small enough to fit in your pocket or in your hand. What would be the size of your object in Brobdingnag? Make a life-size drawing of the Brobdingnag object.

1. Make the size of your drawing as accurate as possible. Label the measurements.
2. After you finish your drawing, figure out a way to check that your drawing and measurements are accurate.
3. Write a short description about how you determined the size of your drawing and how you checked that the drawing was accurate.

How can you figure out the size of a Brobdingnag object?

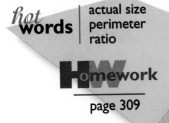

Investigate the Effect of Rescaling on Area

How many of the Ourland objects does it take to cover all of the Brobdingnag object? Use your drawing and the original Ourland object to investigate this question. Write about how you figured out how many Ourland objects it took to cover the Brodingnag object.

hot words
actual size
perimeter
ratio

Homework
page 309

3 How Big Is "Little" Glumdalclitch?

ESTIMATING LENGTHS, AREAS, AND VOLUMES

What if Glumdalclitch visited Ourland? Would she fit in your classroom? Drawing a picture of "little" Glumdalclitch in actual Ourland measurements would take a great deal of paper. To get a sense of the size of a very large object, it is sometimes easier to use estimation.

November 25, 1702
My first impression of the girl in the family proved to be correct. She was very good-natured, kind in spirit, and patient in teaching me her language. She was considered small for her age, being just under forty feet tall. Therefore I called her Glumdalclitch, which I learned means "Little Nurse" in her language. She called me Grildrig, meaning "Little Puppet."

Use Estimation to Solve Problems

How do objects from Ourland and Brobdingnag compare in length, area, and volume?

Estimate the size of each Brobdingnag object. Answer the questions about how the size of the Brobdingnag object compares to size of the same Ourland object and explain how you found each answer.

1. Could a mattress that would fit Glumdalclitch fit in the classroom? How much of the floor would it cover? How many Ourland mattresses would it take to cover the same amount of floor?
2. How big would Glumdalclitch's notebook be? How many sheets of our notebook paper would we need to tape together to make one sheet for her notebook?
3. How big a shoe box do you think Glumdalclitch might have? How many of our shoe boxes would fit inside hers?
4. How many slices of our bread would it take to make one slice of bread big enough for Glumdalclitch to eat?

Why are so many Ourland objects needed to cover or to fill a Brobdingnag object?

hot words | area

Homework page 310

4 Telling Tales in Brobdingnag

USING ACCURATE SIZES IN CREATIVE STORIES

Imagine how it would be for you to visit Brobdingnag. By now you have a good understanding of the scale factor in Brobdingnag. You can use what you know to write your own story. You will see that good mathematical thinking is important in writing a believable story.

June 12, 1704

My size led to some frightening situations. One morning, I was sitting by the window when twenty giant wasps came flying into the room. Some of them carried off the sweet cake I was about to eat for breakfast. Others flew around my head, confusing me with the noise and threatening me with their stings. I killed four of them with my sword and drove the rest off. In other situations, my small size proved very useful. For example, I was once lowered in a bucket down the well to retrieve a ring that the princess had dropped accidentally. She was so happy when they pulled me back up and she saw her prized ring, which I had placed over my head and around my neck for safekeeping.

Write a Story Using Accurate Dimensions

Choose one place in Brobdingnag. Imagine what it would be like to visit that place. Describe in detail the place and at least one adventure that happened to you there.

1. Write a story about Brobdingnag. Make the story believable by using accurate measurements for the objects you describe.
2. Include a size description of at least three objects found in that place.
3. Write a believable title. The title should include at least one size comparison between Brobdingnag and Ourland.
4. Record and check all of your measurements.

How can you use rescaling to write a story about Brobdingnag?

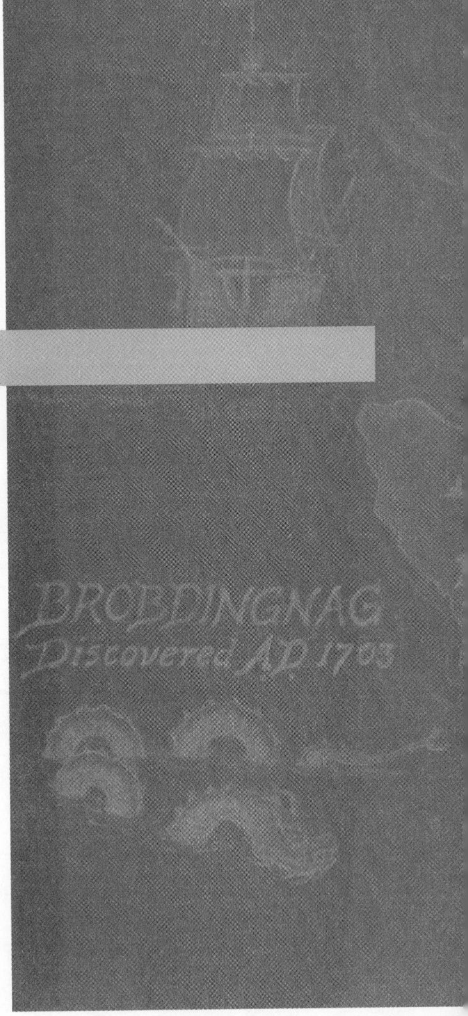

Summarize the Math Used in the Story

After you write your story, summarize how you used math to figure out the sizes of things in the story. Include the following in your summary:

- Make a table, list, or drawing showing the sizes of the three objects in both Ourland and Brobdingnag.
- Explain how you used scale, estimation, and measurement to figure out the sizes of these objects.

hot words | linear measure
scale drawing

Homework
page 311

PHASE TWO

August 5, 1706

Having been condemned by nature and fortune to a restless life, I took a post on the Antelope, a trading ship.

I had been spending much time on deck, until a storm suddenly came upon us.

The storm grew steadily worse. The ship sank on a reef, and I became separated from my crew. Pushed toward shore by the tide, I fell exhausted upon the softest and shortest grass I had ever seen. But before I could examine it further, I fell into a deep sleep.

When I awoke, I found my arms and legs strongly fastened to the ground. Even my hair was tied down. My frustration gave way to surprise as I saw who had done this. Hundreds of them were scattered on the ground around me. They looked like men in every way but one—they were a mere six inches tall.

Lemuel Gulliver

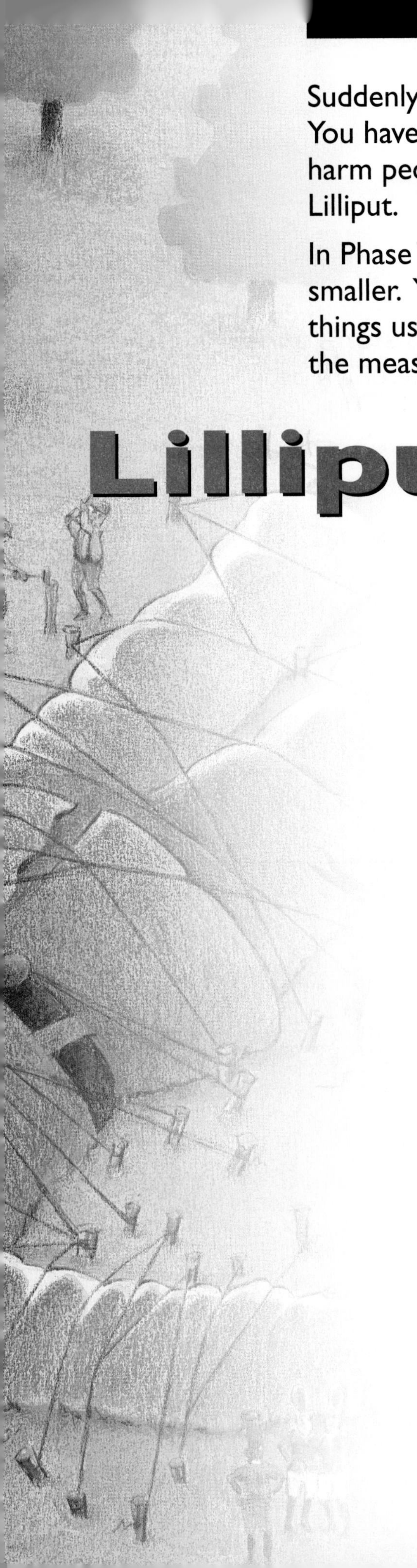

Suddenly you are in a world in which everything is tiny. You have to be careful where you step, so that you don't harm people or destroy houses. The name of this land is Lilliput.

In Phase Two, you will use a scale factor that makes things smaller. You will compare the different ways of measuring things using inches and feet, centimeters and meters and the measurement units used in Lilliput.

Lilliput

WHAT'S THE MATH?

Investigations in this section focus on:

DATA COLLECTION

- Gathering information from a story and pictures
- Organizing data to find patterns

MEASUREMENT and ESTIMATION

- Measuring accurately using fractions
- Comparing the U.S. customary and metric systems of measurement
- Estimating the sizes of objects
- Exploring area and volume measurements

SCALE and PROPORTION

- Working with a scale factor that reduces the sizes of objects
- Applying the scale factor to predict sizes of objects and to create a three-dimensional scale model
- Exploring the effect of rescaling on area and volume

MathScape Online
mathscape1.com/self_check_quiz

5 Sizing Up the Lilliputians

DETERMINING
A SCALE FACTOR
LESS THAN ONE

Gulliver is swept overboard in a storm at sea and wakes up in a new land. He is the captive of tiny people in the land of Lilliput. How are the sizes of things in Lilliput related to sizes in Ourland? Clues in the journal will help you find out how small things are in Lilliput.

September 5, 1706

One afternoon I happened across a country school yard. The children swarmed over me, begging me to play. I let them tumble through my hair and slide down my hand. One brave boy, about twelve years old, let me make a tracing of him under his schoolmaster's supervision. He lay down on my journal while I traced his outline with my quill. He then stamped his hand print and footprint on the paper. The schoolmaster agreed to let me draw his outline as well, along with his spectacles and belt.

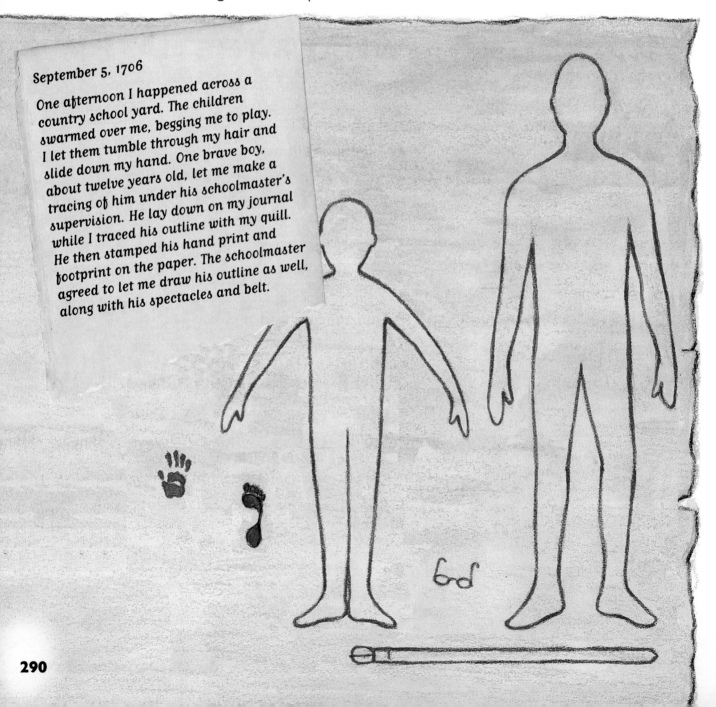

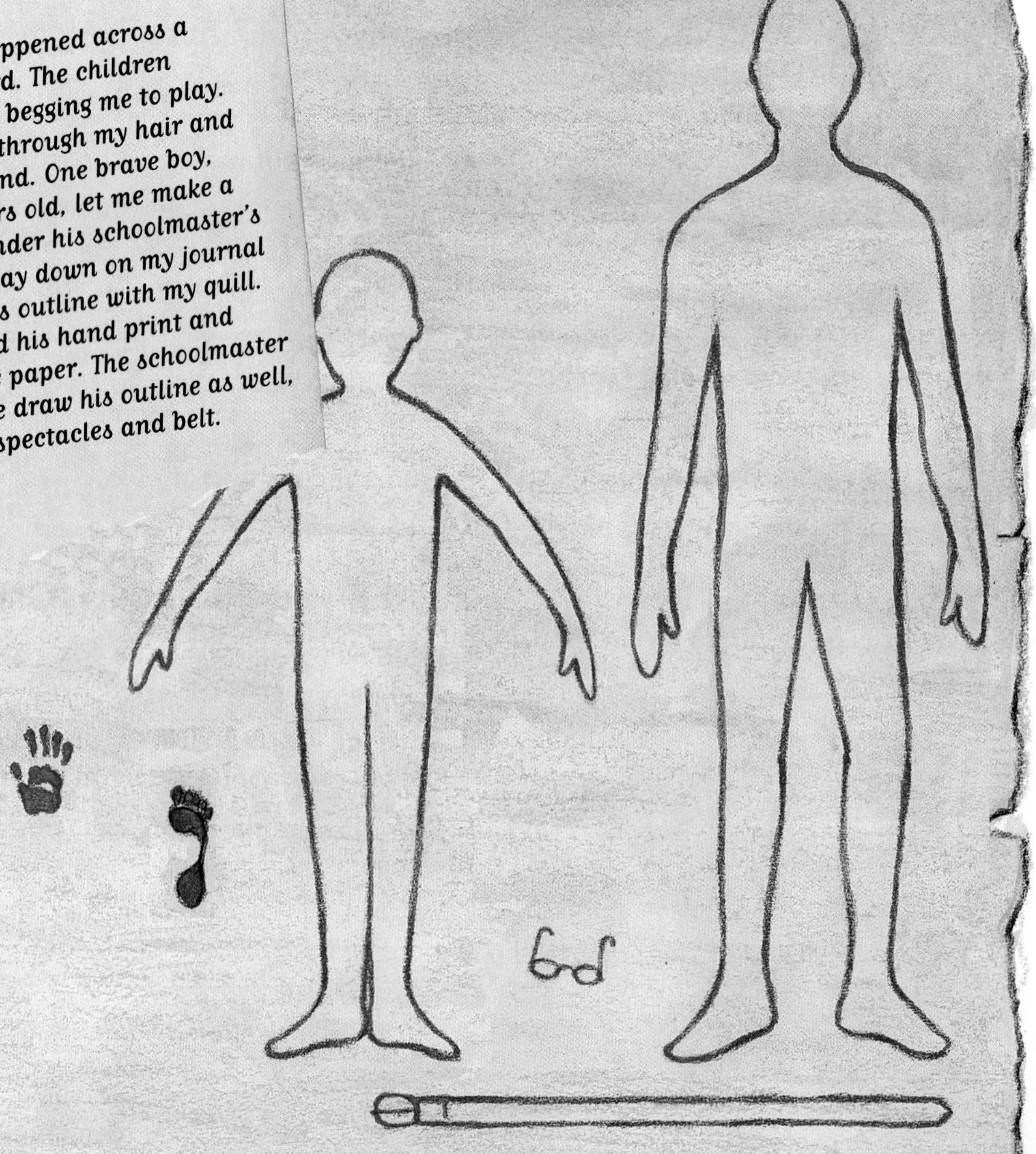

Create a Chart to Compare Sizes

Make a scale chart with three columns. Column 1 is for the name of an object. Column 2 is for the size of the object in Lilliput. Column 3 is for the size in Ourland. Use a ruler to measure the tracings.

1. Use the words and tracings in the story to record the name of the object and its Lilliputian measurements on the chart. Measure or estimate the size of the same object in Ourland.
2. Use the information in your scale chart to find a scale factor that shows how sizes in Lilliput are related to sizes in Ourland.
3. Estimate or measure the sizes of some more objects in Ourland. Add these objects and their Ourland measurements to the chart. Find the size each object would be in Lilliput and add that information to the chart.

Do you think the Lilliputian student in the tracing is tall, short, or average-size in a Lilliputian sixth-grade class?

How are sizes of things in Lilliput related to sizes in Ourland?

Write About Estimation Strategies

Write about what you did and learned as you investigated sizes of things in Lilliput.

- Describe the measurement and estimation strategies you used to find the sizes of things in Ourland and Lilliput. Show how you used the scale factor to complete your chart.
- What did you discover about finding the average size of an object?

hot words mean, median, mode

Homework

page 312

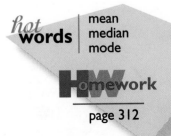

6 Glum-gluffs and Mum-gluffs

MEASURING WITH NONSTANDARD UNITS

The same object can be measured in different units of measurement. Inches and feet are units in the U.S. customary system of measurement. Centimeters and meters are units in the metric system. How do these units compare to the units used in Lillilput?

October 5, 1706

During dinner, the King and Queen told stories about their country and people, and I told stories of mine. The King found it especially hard to believe that he, one of the tallest men in his land, would be no bigger than a child's doll in mine. He informed me that he was $8\frac{1}{2}$ glum-gluffs tall, and that the Queen was 6 glum-gluffs tall. When I inquired what a glum-gluff was, he replied that it was $\frac{1}{20}$ of a mum-gluff. He then kindly agreed to have his steward mark the length of 1 glum-gluff in my journal.

——— 1 glum-gluff

Measure an Object in Different Systems

Choose an object that you added to the Lilliput scale chart you made in Lesson 5.

1. Use the measurements recorded in the chart to make an accurate, life-size drawing of the object in Lilliput.
2. Use a metric ruler to measure the drawing in metric units (centimeters). Write the metric measurements on the drawing.
3. Calculate what the measurements of the drawing would be in the Lilliputian units of glum-gluffs. Write the Lilliputian measurements on the drawing.
4. Compare the object in the drawing to any object in Ourland that would be about the same size. Write the name of the Ourland object on the drawing.

How do the units used in different measurement systems compare?

Compare Measurement Systems

Write a letter to the King and Queen of Lilliput. Compare the measurement systems of Ourland and Lilliput. Make sure your letter answers the following questions:

- When would you prefer to use the U.S. customary system of measurement? When would you prefer to use the metric system?
- Would you ever prefer to use glum-gluffs and mum-gluffs? Why?
- Suppose the people in Lilliput were going to adopt one of our measurement systems. Which one would you recommend to them? Why or why not?

hot words | standard measurement
measurement units

Homework
page 313

7 Housing and Feeding Gulliver

SOLVING PROBLEMS INVOLVING AREA AND VOLUME

Gulliver's needs for food and shelter in Lilliput present some interesting problems. These problems involve area and volume. In the last phase, you solved problems in one dimension. Now you will extend your work with the Lilliputian scale factor to solve problems in two and three dimensions.

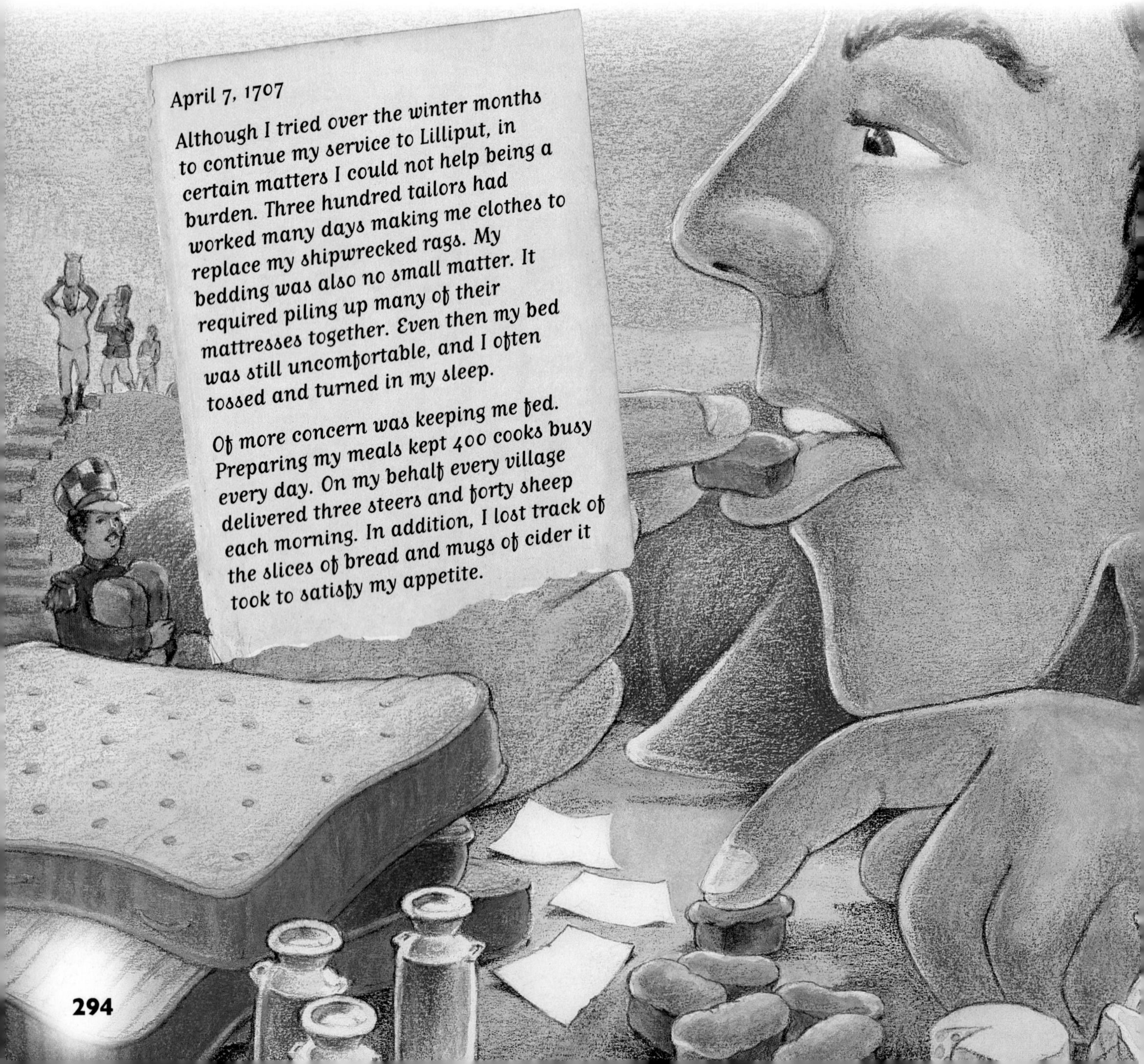

April 7, 1707
Although I tried over the winter months to continue my service to Lilliput, in certain matters I could not help being a burden. Three hundred tailors had worked many days making me clothes to replace my shipwrecked rags. My bedding was also no small matter. It required piling up many of their mattresses together. Even then my bed was still uncomfortable, and I often tossed and turned in my sleep.
Of more concern was keeping me fed. Preparing my meals kept 400 cooks busy every day. On my behalf every village delivered three steers and forty sheep each morning. In addition, I lost track of the slices of bread and mugs of cider it took to satisfy my appetite.

Estimate to Solve Area and Volume Problems

How can you use estimation to solve problems in two and three dimensions?

Estimate how many of the Lilliputian objects Gulliver needs. Then make a Lilliputian-size model of one of the four objects.

1. How many Lilliputian-size mattresses would Gulliver need to make a bed? How should Gulliver arrange those mattresses to make a comfortable-size bed?
2. How many sheets of Lilliputian paper would need to be taped together to made one sheet of writing paper for Gulliver?
3. At home, Gulliver would eat two loaves of bread each week. How many Lilliputian loaves of bread would Gulliver need each week?
4. At home, Gulliver drank 3 cups of milk each day. How many Lilliputian-size quarts of milk would Gulliver need each day?

Use a Model to Check an Estimate

Describe in writing how you can use your Lilliputian-size model to check your estimate. Include a sketch with measurements to show your thinking.

hot words | volume

Homework page 314

8 Seeing Through Lilliputian Eyes

WRITING ABOUT AREA AND VOLUME

Imagine yourself in Lilliput. What objects would you bring with you? How would the Lilliputians describe these objects? You will use what you have learned about scale in one, two, and three dimensions when you write a story describing your own adventures in Lilliput.

Gulliver's Pocket Contents:
1. One great piece of coarse cloth, large enough to be a carpet for your Majesty's chief Room of State
2. A great bundle of white thin substances, folded one over another, about the thickness of three men, tied with a strong cable and marked with black figures, with every letter almost half as large as the palm of our hands
3. A long pole from the back of which extended 20 shorter poles, resembling the palace railings
4. Several round flat pieces of yellow and silver metal, of different bulk, some so large and heavy that my comrade and I could hardly lift them
5. Some wonderful kind of globe-like engine, part silver and part transparent metal, with a loud noise like the sound of a water-mill, attached by a great silver chain

Write a Story Using 3-D Measurements

Write a believable story using three-dimensional measurements. You will need to figure out the correct length, width, and height of the objects you describe.

1. Imagine a place in Lilliput. Describe at least one adventure that could happen to you there.
2. Describe the measurements of at least three objects found in the place. You could include an Ourland object in your story for comparison.
3. Include a conversation with a Lilliputian that compares the sizes of the objects in the story to the same objects in Ourland.
4. Record and check all of your measurements.

How can you describe a three-dimensional Lilliputian world?

Describe Rescaling Strategies

Summarize how you determined the length, width, and height of the three objects described in your story.

- Make a table, list, or drawing showing the length, width, and height of each object in both Ourland and Lilliput.
- Explain the methods you used to estimate or measure each object. Show how you rescaled it using the scale factor.

hot words | scale
metric system

Homework
page 315

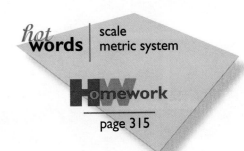

PHASE THREE

September 13, 1708

It has now been some months since my journal was published. News of my travels has prompted a greater response than I expected. I must confess, too, a certain pleasure at the attentions I have received from my publisher. He has pronounced himself well-pleased with the sales figures, and credits both my nature and my writing for this success.

Although I did not anticipate it, I have recently received much correspondence from my readers. Pictures and letters have come from nearby countries—France, Spain and Holland—as well as more distant lands. It is my dream that someday all my treasures and adventures may be shared with the public as part of a glorious exhibit, but, alas, I fear it will not happen in my lifetime.

Your friend,

Lemuel Gulliver

Imagine you are in charge of a special exhibit about *Gulliver's Worlds*. How would you show the sizes of the different lands he visited? In this final phase, you will explore size relationships in different lands, both big and small. You will find ways to show how the sizes compare in length, area, and volume. Finally, your class will create displays of life-size objects from one of *Gulliver's Worlds*.

Lands of the Large and Lands of the Little

WHAT'S THE MATH?

Investigations in this section focus on:

DATA COLLECTION

- Gathering information from pictures
- Creating displays to show size relationships

MEASUREMENT and ESTIMATION

- Measuring accurately using fractions
- Exploring area and volume measurements

SCALE and PROPORTION

- Finding scale factors that describe relationships among sizes
- Enlarging and reducing the sizes of objects according to scale factors
- Creating 2-D scale drawings
- Creating a 3-D scale model
- Exploring the effects of rescaling on area and volume

MathScape Online
mathscape1.com/self_check_quiz

9 Lands of the Large

REPRESENTING SIZE RELATIONSHIPS

Ourland Museum needs a display that compares the sizes of objects from Lands of the Large to Ourland. Can you figure out the scale factor for each of the Lands of the Large? Can you find a way to show the size relationships among the different lands?

How large is a life-size face in each of the Lands of the Large?

Investigate Proportions of Faces

Compare the objects in the photos to find the scale factor. The smaller object is always from Ourland. When you are finished, check to make sure your scale factor is correct before you do the following group investigation.

1. As a group, select one of the Lands of the Large for this investigation. Have each member of your group draw a different feature of a face from your group's land.
2. As a group, arrange the features to form a realistic face. Check to see if your measurements are correct and the features are in proportion. Work together to draw the outline of the face.

How tall would a person in your group's Land of the Large be?

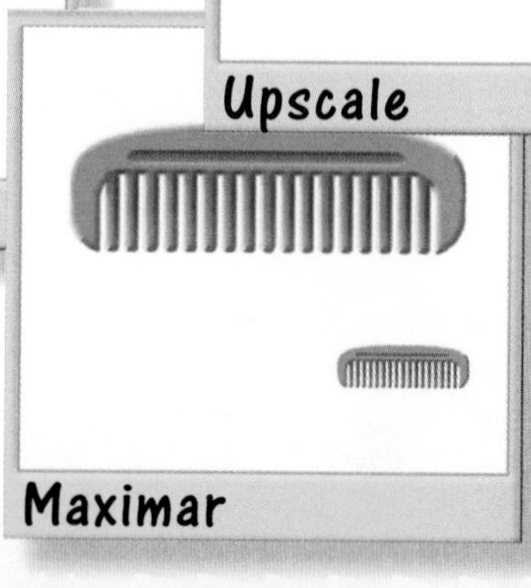

Represent Size Relationships

Use the scale drawing of an Ourland face on this page to make a simple scale drawing of a face from each of the Lands of the Large. Organize your drawings into a visual display to show size relationships.

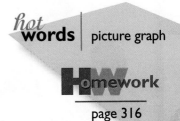

1. Measure the scale drawing of the Ourland face.
2. Use the scale factors from the Lands of the Large to make a scale drawing of a face from each land. You do not need to draw in the features.
3. Write the name of the land and the scale factor compared to Ourland next to each drawing.
4. Organize your drawings into a display of size relationships that compares the sizes of faces from different lands and shows how they are related.

How do things in the Lands of the Large compare in size to things in Ourland?

Describe a Scale Factor for Brobdingnag

Compare the sizes of things in each of the Lands of the Large to the sizes of things in Brobdingnag. Use this to explain how the scale factor describes size relationships.

- Figure out the scale factor for each land compared to Brobdingnag. Write it next to the scale drawing from that land.
- Describe in writing how you figured out the scale factor. Tell why it is different from the Ourland scale factor.

hot words | picture graph

Homework

page 316

10 Lands of the Little

PREDICTING SCALE SIZE

Can you find the mistakes in the Lands of the Little display? Here you will correct the scale drawings and create a chart that you can use to find the size of any object in a Land of the Little.

How do things in the Lands of the Little compare in size to things in Ourland?

Compare Objects in the Lands of the Little

Measure each pair of scale drawings on this page. The larger object is always from Ourland. Does the size relationship for each Land of the Little object match the scale factor below it?

1. As a group, choose an Ourland object from your classroom. Draw your object on a piece of paper.
2. Use each of the four scale factors below to draw a new picture of your object. Now, which scale drawings below do you think are incorrect?

Quarterville 1:4 (.25:1)

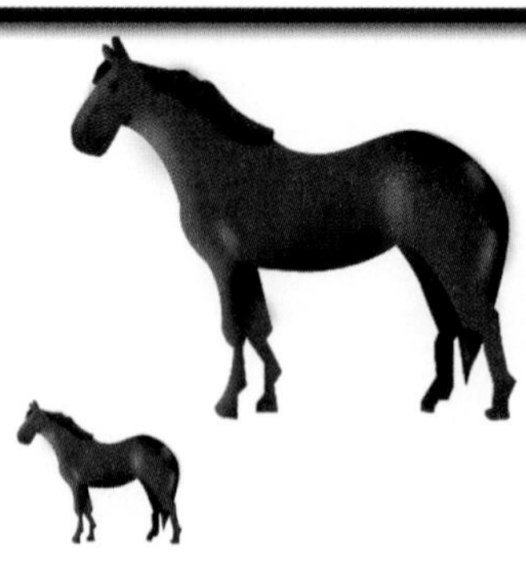

Micropolus 3:8 (.375:1)

Small Town 2:3 (.67:1)

Dimutia 1:6 (.167:1)

Create a Table to Show Size Relationships

Make a table that shows how objects in the Lands of the Little compare in size to objects in Ourland. Record the size an object would be in other lands if you knew its size in Ourland.

1. Fill in the names of the Lands of the Little at the top of each column. Mark the Ourland measurements (100 inches, 75 inches, 50 inches, 25 inches, 10 inches) in the Ourland column.
2. Figure out how big an object would be in the Lands of the Little for each of the Ourland measurements. Mark the Lands of the Little measurements in the appropriate columns.
3. Find a way to use your group's scale drawings to check that your table is correct.

Ourland	Lilliput	Dimutia	Quarterville	Micropolus	Small Town
100 inches					
75 inches					
50 inches					
25 inches					
10 inches					

How could you show the size relationships of things in different lands?

Write a Guide for Using the Table

Explain how you can use your table to answer each question.

- If an object is 80 inches long in Ourland, how long would it be in each of the other lands?
- If an object is 5 inches long in Lilliput, how long would it be in Ourland?
- If an object is 25 inches long in Small Town, how long would it be in the other lands?

hot words | coordinate graph

Homework
page 317

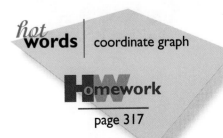

11 Gulliver's Worlds Cubed

RESCALING IN ONE, TWO, AND THREE DIMENSIONS

The *Gulliver's Worlds* group exhibit needs a finishing touch. The exhibit needs to show how rescaling affects area and volume. How do length, area, and volume change when the scale of something changes? Can you create a display that will help visitors understand this?

How does a change in scale affect measurements of length, area, and volume?

Investigate Cube Sizes in Different Lands

Use the information on this page to figure out a way to use Ourland cubes to build a large cube at each of the following scale factors: 2 : 1, 3 : 1, 4 : 1.

1. Record how many Ourland cubes make up each large cube.
2. Estimate how many cubes it would take to make a Brobdingnag cube (scale factor = 12 : 1)

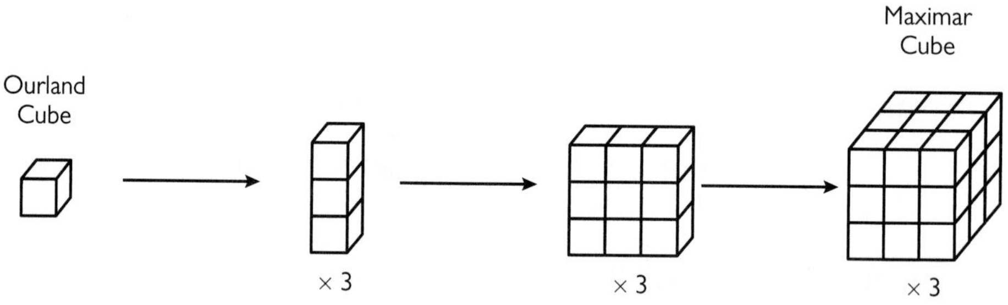

Comment Card

If Maximar is 3 times bigger than Ourland, why does it take more than 3 cubes from Ourland to make a cube in Maximar?

Collect Data for Two and Three Dimensions

Use your cubes to collect size information.

1 Organize the information into a table that answers the following questions:

a. What is the scale factor of the cube?

b. How many Ourland cubes high is one edge of this cube?

c. How many Ourland cubes are needed to cover one face of this cube completely?

d. How many Ourland cubes are needed to fill this cube completely?

Scale Factor	How Many Cubes Long Is an Edge? (length)	How Many Cubes Cover a Face? (area)	How Many Cubes Fill the Cube? (volume)
2:1			
3:1			
4:1			
5:1			
10:1			
25:1			

2 Find a rule that can predict how big a cube would be for each of the following scales. Then add the information to your table:

2.5:1   6:1   20:1   100:1

How can you predict how much each measurement will change when you rescale?

Write About Scale, Area, and Volume

Write down the set of rules you used to complete your table. Make sure that someone else could apply your rules to any scale factor.

1 Explain how your rules work.

2 Make a diagram showing how to use the rules to predict the following:

a. The length of one edge of a cube

b. The area of one face of a cube

c. The volume of a cube

hot words
exponent
cubic centimeter

Homework
page 318

12 Stepping into Gulliver's Worlds

FINAL PROJECT

A life-size display in the correct scale and proportion can make you feel like you have stepped into another world. You will help the Ourland Museum create a life-size display of one of the lands in *Gulliver's Worlds.* The goal is for museum visitors to get involved with your display.

What would it look like if you stepped into one of the lands in *Gulliver's Worlds*?

Create a Display Using Accurate Dimensions

Choose one of the lands in *Gulliver's Worlds.* Create a display and tour that compares the sizes of things in that land to those in Ourland.

1. Create at least three objects in one, two, or three dimensions to use in the display.
2. Write a short tour that describes the measurements of the objects in the display and compares them to sizes in Ourland. Describe and label the areas and volumes of the objects.
3. Find a way for visitors to get involved with the display.
4. Include your writing from previous lessons and charts to help museum visitors understand the scale of the land.

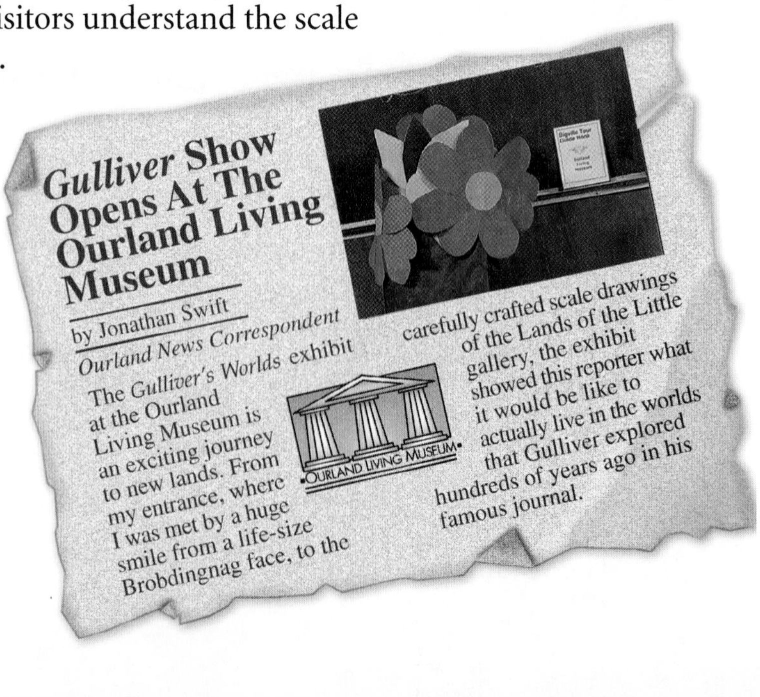

Gulliver Show Opens At The Ourland Living Museum

by Jonathan Swift
Ourland News Correspondent

The *Gulliver's Worlds* exhibit at the Ourland Living Museum is an exciting journey to new lands. From my entrance, where I was met by a huge smile from a life-size Brobdingnag face, to the carefully crafted scale drawings of the Lands of the Little gallery, the exhibit showed this reporter what it would be like to actually live in the worlds that Gulliver explored hundreds of years ago in his famous journal.

Review a Display

You will be reviewing a classmate's display and presentation. As you review the exhibit, write down the scale factor and as many measurements as you can. Use the following questions to help write your review:

1. What parts of the display look life-size?
2. How did you check that the sizes of the objects in the display were correct?
3. How does the presentation describe linear, area, and volume measurements?
4. How does the presentation compare sizes to those in Ourland?
5. Would you add or change anything to make the display more believable?

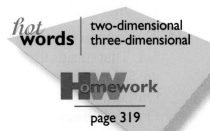

How would you evaluate your own display?

hot **words** | two-dimensional
three-dimensional

Homework
page 319

The Sizes of Things in Brobdingnag

Homework 1

Applying Skills

Fill in the missing height conversions to complete the chart.

	Name	Height (in.)	Height (ft and in.)	Height (ft)
	Marla	49″	4′1″	$4\frac{1}{12}'$
1.	Scott	56″		
2.	Jessica		4′7″	
3.	Shoshana	63″		
4.	Jamal	54″		
5.	Louise		4′11″	
6.	Kelvin	58″		
7.	Keisha		5′2″	
8.	Jeffrey		4′2″	

9. List the names in height order from tallest to shortest.

10. The scale factor of Giantland to Ourland is 11 : 1. That means that objects in Giantland are 11 times the size of the same objects in Ourland. Figure out how large the following Ourland objects would be in Giantland:

a. a tree that is 9 ft tall

b. a man that is 6 ft tall

c. a photo that is 7 in. wide and 5 in. high

11. The scale factor of Big City to Ourland is 5 : 1. That means that objects in Big City are 5 times the size of the same objects in Ourland. How large would each of the Ourland objects from item **10** be in Big City?

Extending Concepts

12. Duane made an amazing run at the football game Friday night.

Examine the diagram below and give the distance of the play in:

a. yards **b.** feet **c.** inches

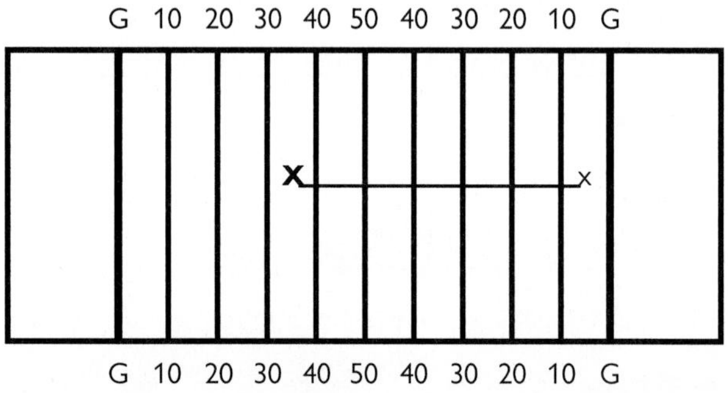

HINT: From goal line to goal line in a football field is 100 yards.

Making Connections

13. Answer this Dr. Math letter:

Dear Dr. Math,
Today in science class we were using microscopes. The lenses were 10×, 50×, and 100×. I think there is a way scale factor applies to what I see and what the actual size is. Is that true? If so, could you please explain?
William Neye

A Life-Size Object in Brobdingnag

Applying Skills

Reduce these fractions to lowest terms.

1. $\frac{21}{49}$ **2.** $\frac{33}{126}$ **3.** $\frac{54}{81}$

4. $\frac{28}{48}$ **5.** $\frac{15}{75}$ **6.** $\frac{10}{18}$

7. $\frac{126}{252}$ **8.** $\frac{8}{24}$ **9.** $\frac{16}{12}$

10. $\frac{64}{6}$

Follow the instructions to describe each relationship in a different way.

11. Write $\frac{10}{1}$ as a ratio.

12. Write 4:1 as a fraction.

13. Write "2 to 1" as a fraction.

14. Write out 6:1 in words.

15. Write $\frac{8}{1}$ as a ratio.

Extending Concepts

16. The height of a blade of grass in a giant-size display is $4\frac{2}{3}$ ft. The blade of grass in your yard is 4 in. high. What is the scale factor?

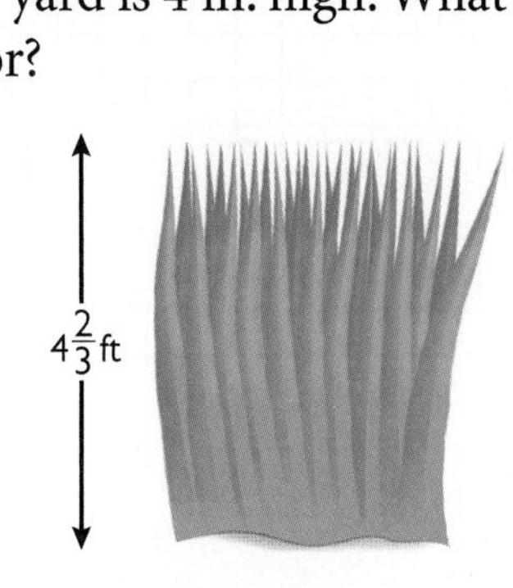

17. The scale factor of Vastland to Ourland is 20:1. That means that objects in Vastland are 20 times the size of the same objects in Ourland. Figure out how large the following Ourland objects would be in Vastland:

a. a car that is $4\frac{1}{2}$ ft high and 8 ft long

b. a building that is 23 yd high and 40 ft long

c. a piece of paper that is $8\frac{1}{2}$ in. wide and 11 in. long

Making Connections

18. The science class is creating insects that are larger than life. First they will study the ant. The queen ant that they have to observe is $\frac{1}{2}$ in. long. The large model they create will be 5 ft long. Mr. Estes wants to have a ladybug model created too. Tasha found a ladybug and measured it at $\frac{1}{8}$ in. long.

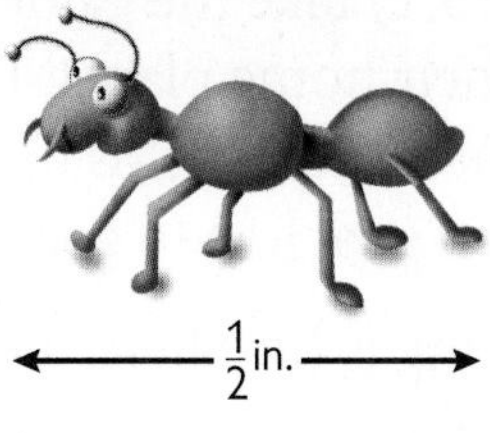

a. What is the scale factor for the ant model?

b. How big will the ladybug model be if the same scale factor is used?

Homework 3

How Big Is "Little" Glumdalclitch?

Applying Skills

Reduce these fractions to the lowest terms.

1. $\frac{25}{75}$ **2.** $\frac{69}{23}$ **3.** $\frac{1,176}{21}$

4. $\frac{16}{4}$ **5.** $\frac{36}{30}$ **6.** $\frac{49}{14}$

7. $\frac{24}{3}$ **8.** $\frac{54}{18}$ **9.** $\frac{8}{12}$

Convert these fractions to like measurement units and then reduce each fraction to show a size relationship. See if you can make each fraction into a scale factor.

10. $\frac{3 \text{ in.}}{4 \text{ ft}}$ **11.** $\frac{8 \text{ in.}}{2 \text{ yd}}$ **12.** $\frac{440 \text{ yd}}{\frac{1}{2} \text{ mi}}$

13. $\frac{18 \text{ in.}}{1 \text{ yd}}$ **14.** $\frac{2 \text{ yd}}{12 \text{ ft}}$

Extending Concepts

15. Ali is writing a script for a new movie in which aliens that are 3 times the size of humans (3:1) take the game of football back to their home planet. HINT: U.S. regulation football fields measure 100 yards from goal line to goal line.

a. How long is the aliens' field in yards?

b. How long is the aliens' field in inches?

c. How long is the aliens' field in feet?

16. The scale factor of Jumbolia to Ourland is 17:1. That means that objects in Jumbolia are 17 times the size of the same objects in Ourland. Figure out how large the following Ourland objects would be in Jumbolia:

a. a radio that is $4\frac{1}{2}$ in. high, 8 in. long, and $3\frac{1}{2}$ in. wide

b. a rug that is $6\frac{1}{2}$ ft wide and $9\frac{3}{4}$ ft long

c. a desk that is $2\frac{1}{3}$ ft high, 3 ft long, and $2\frac{1}{2}$ ft wide

Making Connections

17. Provide the scale factor for the following map by measuring the distance with a ruler. The distance from the library to the school is $2\frac{1}{2}$ miles.

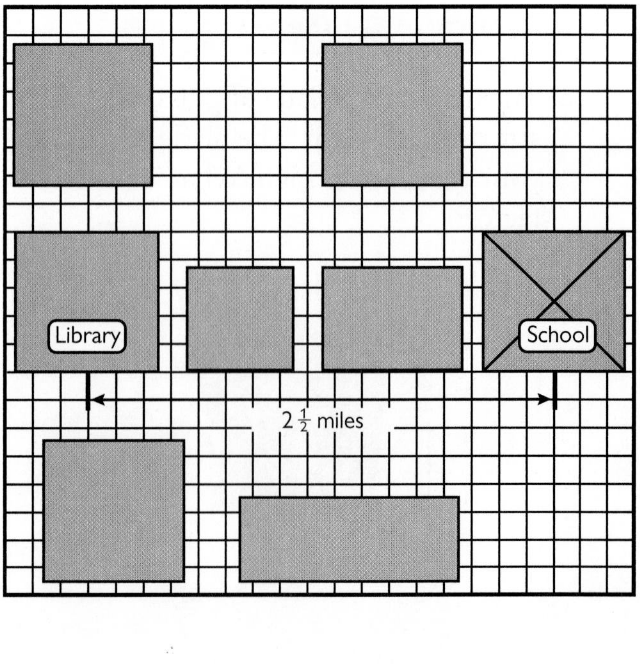

Homework 4

Telling Tales in Brobdingnag

Applying Skills

Convert these ratios to like measurement units and then reduce the fraction to create a scale factor.

Example $\frac{1}{2}$ yd: 6 in. $= \frac{\frac{1}{2}\text{ yd}}{6\text{ in.}} = \frac{18\text{ in.}}{6\text{ in.}} = \frac{3\text{ in.}}{1\text{ in.}}$ = scale factor 3:1

1. $\frac{3}{4}$ ft:3 in.

2. 6 ft:$\frac{1}{3}$ yd

3. $\frac{1}{2}$ mi : 528 ft

4. $2\frac{1}{2}$ yd : $\frac{1}{4}$ ft

5. 14 in : $\frac{7}{12}$ ft

6. $\frac{1}{6}$ yd:2 in.

7. $\frac{1}{15}$ mi:16 ft

8. $4\frac{1}{12}$ ft:7 in.

9. 4 mi:1,760 yd

10. 4,392 in:6 ft

Extending Concepts

11. The scale factor of Mammothville to Ourland is 1 yd:1 in. That means if an object in Mammothville is one yard long, then the same object in Ourland would be only one inch long. Figure out how large the following Ourland objects would be in Mammothville:

a. a soda can that is 5 in. tall and $2\frac{1}{2}$ in. wide

b. a football field that is 100 yd long

c. a table that is $3\frac{1}{2}$ ft high, 4 ft wide, and 2 yd long

12. The scale factor of Colossus to Ourland is $\frac{1}{4}$ yd:3 in. That means if an object in Colossus is $\frac{1}{4}$ of a yard long, then the same object in Ourland would be only 3 inches long. How large would each of the Ourland objects from item **11** be in Colossus?

13. Match the following scale factors to the correct measurement units:

a. 3:1	**i.** 1 ft:1 in.
b. 5,280:1	**ii.** 1 mi:1 yd
c. 12:1	**iii.** 1 yd:1 ft
d. 1,760:1	**iv.** 1 mi:1 ft

Making Connections

14. In Humungoville the scale factor to Ourland is 4:1. Use the postage stamp from Ourland pictured below to draw a postage stamp for Humungoville. Make sure the length and width are at a scale factor of 4:1. You can be creative with the picture inside.

Homework 5

Sizing Up the Lilliputians

Applying Skills

Write each of the following decimals as a fraction.

Example $0.302 = \frac{302}{1,000}$

1. 0.2 **2.** 0.435

3. 0.1056 **4.** 0.78

5. 0.44 **6.** 0.025

7. 0.9 **8.** 0.5002

9. 0.001 **10.** 0.67

Write each decimal in words.

Example 0.5 = five tenths

11. 0.007 **12.** 0.25

13. 0.3892 **14.** 0.6

15. 0.04

16. The scale factor of Pint-Size Place to Ourland is 1:11. That means that objects in Ourland are 11 times the size of the same objects in Pint-Size Place. Figure out about how large the following Ourland objects would be in Pint-Size Place:

a. a house that is 15 ft high, 33 ft wide, and 60 ft long

b. a train that is 363 ft long and 20 ft high

c. a woman who is 5 ft 6 in. tall

Extending Concepts

17. Measure the height of each picture. Compare the sizes of the pictures and determine the scale factor. What is the scale factor when:

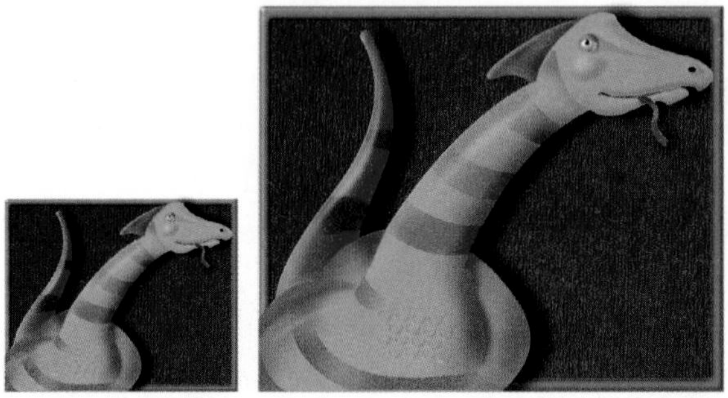

a. the larger picture is 1?

b. the smaller picture is 1?

Making Connections

18. The regulation size of a soccer field varies from the largest size, 119 m × 91 m, to the smallest size allowed, 91 m × 46 m. What is the difference in the perimeters of the two field sizes? How do you think the difference in perimeters affects the game?

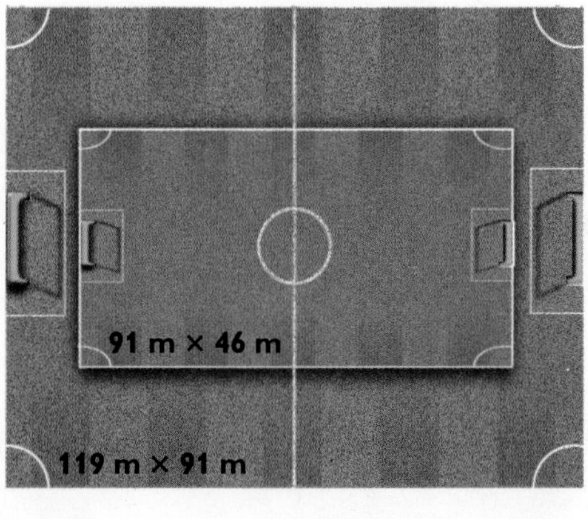

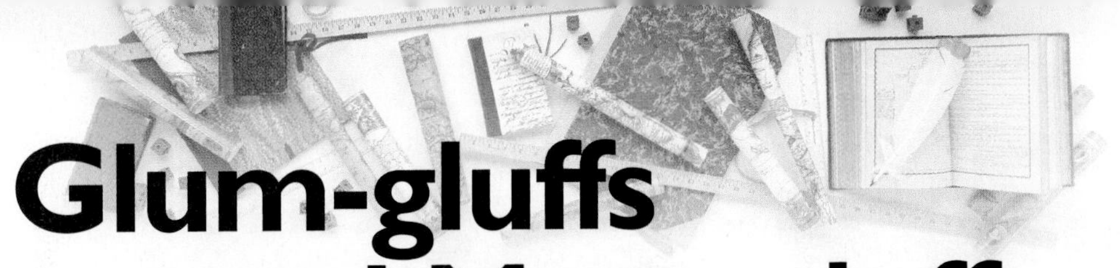

Homework 6

Glum-gluffs and Mum-gluffs

Applying Skills

Complete the following table showing equivalencies in the metric system.

	mm	cm	dm	m	km
1.				1,000	1
2.	1,000	100	10	1	
3.			1		
4.		1			
5.	1				

Supply the missing equivalent.

6. 42 dm = ____ m

7. 5 cm = ____ m

8. 0.5 m = ____ cm

9. 0.25 cm = ____ mm

10. 0.45 km = ____ m

11. 1.27 m = ____ dm

12. 24.5 dm = ____ cm

13. 38.69 cm = ____ m

14. 0.2 mm = ____ cm

15. 369,782 mm = ____ m

16. 0.128 cm = ____ mm

17. 7.3 m = ____ dm

Extending Concepts

18. Place the following measurements in height order from shortest to tallest.

- 1967 mm
- 0.0073 km
- 43.5 cm
- 0.5 m
- 7 dm

Making Connections

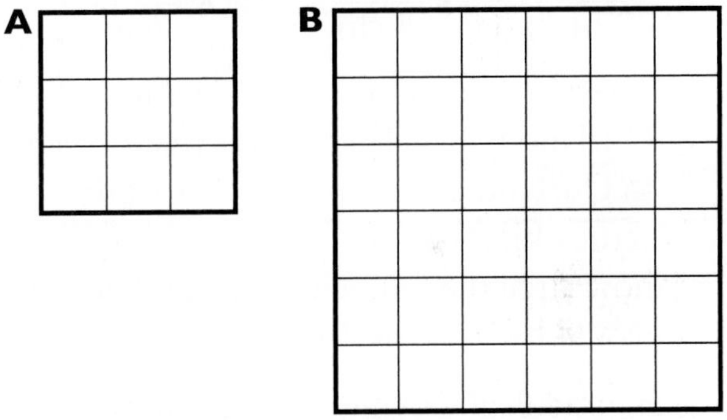

19. Count the smaller squares to figure out the sizes of squares A and B in square units.

a. What are the perimeter and the area of square A?

b. What are the perimeter and the area of square B?

20. Compare the two perimeters and the two areas. Describe each size relationship using a scale factor.

Homework 7

Housing and Feeding Gulliver

Applying Skills

Convert these fractions to like measurement units.

Example $\frac{4\text{ m}}{4\text{ cm}} = \frac{400\text{ cm}}{4\text{ cm}}$

1. $\frac{43\text{ cm}}{43\text{ mm}}$ **2.** $\frac{5\text{ m}}{5\text{ cm}}$ **3.** $\frac{6\text{ km}}{6\text{ m}}$

Use your answers from items 1–3 to show a scale factor that is less than one.
HINT: Reduce the larger number to one.

Example $\frac{400\text{ cm} \div 400}{4\text{ cm} \div 400} = \frac{1}{0.01} = 1{:}0.01$

4. $\frac{43\text{ cm}}{43\text{ mm}}$ **5.** $\frac{5\text{ m}}{5\text{ cm}}$ **6.** $\frac{6\text{ km}}{6\text{ m}}$

7. The scale factor of Teeny Town to Ourland is 1:6. That means that objects in Ourland are 6 times the size of the same objects in Teeny Town. Figure out how large the following Ourland objects would be in Teeny Town:

a. a book 30 cm high and 24 cm wide

b. a girl 156 cm tall

c. a table 1 m high, 150 cm wide, and 2 m long

Extending Concepts

8. Albert is using a scale factor of 3:1 for his school project. The height of the walls he measured are 3 m and the walls in the model he made are 1 m high. A 3-ft-high chair became a 1-ft-high chair in his project. Can he use both metric and U.S. customary measurement units in the same project? Why or why not?

9. The scale factor of Itty-Bittyville to Ourland is 1:4. That means that objects in Ourland are 4 times the size of the same objects in Itty-Bittyville. Estimate the sizes of each of the following Itty-Bittyville objects and find an object in Ourland that is about the same size:

a. an Itty-Bittyville textbook

b. an Itty-Bittyville double bed

c. an Itty-Bittyville two-story building

d. an Itty-Bittyville car

10. Estimate the scale factor of Peeweeopolis to Ourland if the area of an Ourland postage stamp is equal to the area of a Peeweeopolis sheet of paper.

Making Connections

For items 11–13 use the figure below.

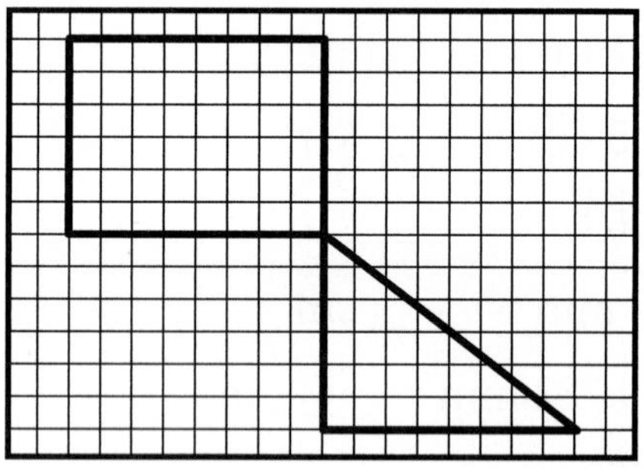

11. What is the area in square units of:

a. the rectangle? **b.** the triangle?

12. Enlarge each shape using a scale factor of 3:1. What is the area in square units of:

a. the rectangle? **b.** the triangle?

13. How did you figure out the area of each shape for items **11** and **12**?

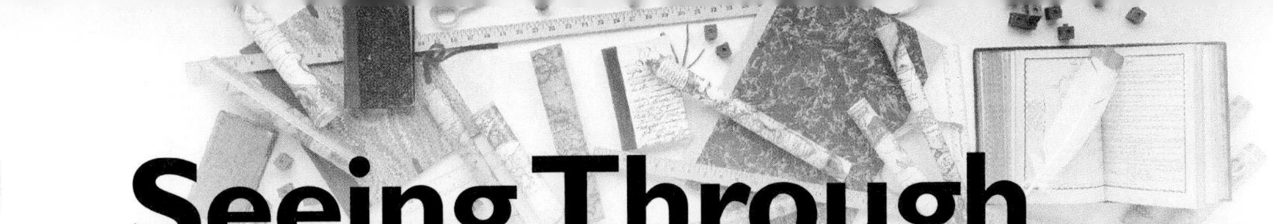

Seeing Through Lilliputian Eyes

Homework 8

Applying Skills

Reduce the following fractions to the lowest terms.

Example $\frac{36}{42} = \frac{6}{7}$

1. $\frac{81}{63}$ **2.** $\frac{4}{24}$ **3.** $\frac{16}{20}$

4. $\frac{5}{50}$ **5.** $\frac{27}{36}$ **6.** $\frac{36}{48}$

7. $\frac{90}{120}$ **8.** $\frac{12}{10}$ **9.** $\frac{75}{100}$

10. $\frac{11}{33}$ **11.** $\frac{14}{21}$ **12.** $\frac{80}{25}$

13. $\frac{9}{18}$ **14.** $\frac{4}{12}$

Making Connections

17. In science-fiction movies, miniatures and scale-factor models are used to create many of the special effects. In one case, the special effects team created several different scale models of the hero's spaceship. The life-size ship that was built to use for the filming was 60 ft long. One scale model was 122 cm long by 173 cm wide by 61 cm high.

a. What was the scale factor?

b. What were the width and height of the life-size ship?

Extending Concepts

15. Measure the length and width of each of the following shapes. Which measurement system, metric or U.S. customary, would be the easiest to use to enlarge each object using a scale factor of 2:1? Why?

a.

b.

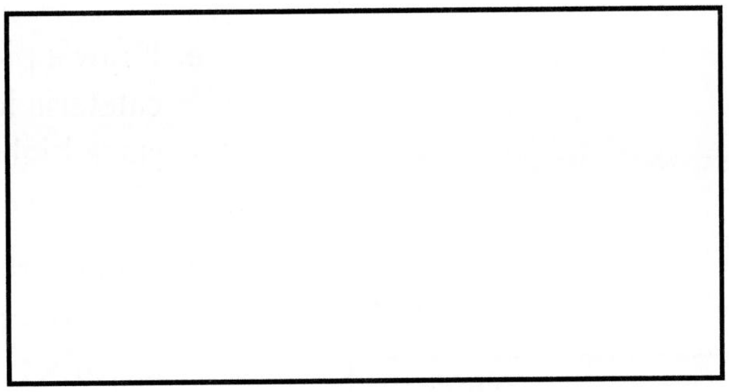

16. The scale factor of Miniopolis to Ourland is 1:7. That means that objects in Miniopolis are $\frac{1}{7}$ the size of the same objects in Ourland. Figure out how large the following Ourland objects would be in Miniopolis:

a. a building that is 147 ft high, 77 ft wide, and 84 ft long

b. a road that is 2 miles long

c. a piece of paper that is $8\frac{1}{2}$ in. by 11 in.

Homework 9

Lands of the Large

Applying Skills

In the following exercises provide equivalent decimals.

Example $\frac{1}{2} = 0.5$

1. $\frac{1}{20}$ **2.** $\frac{1}{3}$ **3.** $\frac{1}{4}$

4. $\frac{1}{5}$ **5.** $\frac{1}{10}$ **6.** $\frac{1}{8}$

7. $\frac{3}{4}$ **8.** $\frac{1}{7}$

9. The scale factor of Big City to Ourland is 6.5:1. That means that objects in Big City are 6.5 times the size of the same objects in Ourland. Figure out how large the following Ourland objects would be in Big City:

a. a tree that is 9 ft tall

b. a man that is 6 ft tall

c. a photo that is 7 in. wide and 5 in. long

Extending Concepts

10. Complete the following table by figuring out equivalent scale factors for each row.

	Decimals	Fractions	Whole Numbers
	1.5:1	$1\frac{1}{2}:1$	3:2
a.	6.5:1		
b.		$8\frac{1}{4}:1$	
c.			5:3

11. The scale factor of Big City to Hugeville is 3:2. That means that objects in Big City are 1.5, or $1\frac{1}{2}$, times the size of the same objects in Hugeville. How large would each of the Big City objects from item **9** be in Hugeville?

Making Connections

12. The scale factor is 5:1 for a giant ice cube in comparison to the school cafeteria's ice cubes.

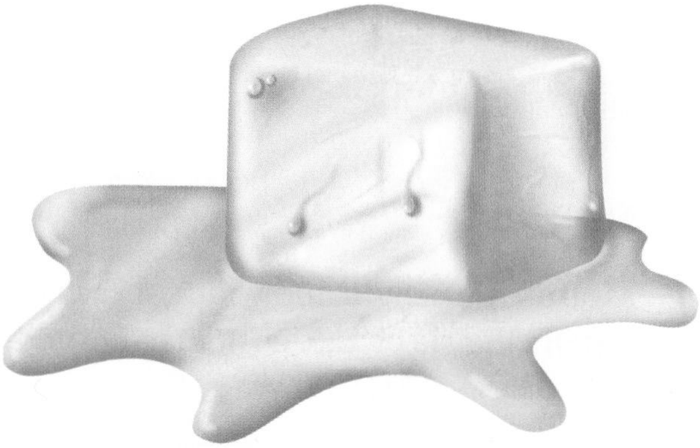

a. Draw a picture that shows how many cafeteria ice cubes you would have to stack high, wide, and deep to build a giant ice cube.

b. What is the total number of cafeteria ice cubes it would take to fill one giant ice cube?

10 Homework

Lands of the Little

Applying Skills

Complete the following chart by supplying the missing equivalents as decimals or fractions.

	Fractions	Decimals
1.	$\frac{1}{2}$	
2.		0.25
3.	$\frac{2}{3}$	
4.		0.7
5.	$\frac{3}{4}$	
6.		0.05
7.	$\frac{3}{8}$	
8.		0.125

Reduce the scale factor to a fraction. HINT: Divide each number by the largest number.

Example $10{:}7 = \frac{10}{10}{:}\frac{7}{10} = 1{:}\frac{7}{10}$

9. 3:2 **10.** 4:3 **11.** 5:3

Extending Concepts

12. The scale factor for Giantland to Ourland is 10:1. What is the scale factor from Ourland to Giantland? Write the scale factor using a decimal or fraction for Ourland to Giantland.

13. The scale factor of Wee World to Ourland is 0.5:1. That means that objects in Wee World are 0.5 the size of the same objects in Ourland. What is another way to write this scale factor without using a decimal?

14. Using the scale factor from item **13,** figure out how large the following Ourland objects would be in Wee World:

a. a house that is 15 ft high, 35 ft wide, and 60 ft long

b. a train that is 360 ft long and 20 ft high

c. a woman who is 5 ft 4 in. tall

Making Connections

15. Answer this Dr. Math letter:

> Dr. Math,
>
> When I was doing problems 9–11 in today's homework, my friend said there was a pattern between the whole-number scale factor and the fraction scale factor. I don't see it. Can you please explain it to me? Could I use this pattern to rescale objects more efficiently?
>
> D.S. Mall

Homework 11 Gulliver's Worlds Cubed

Applying Skills

Complete the following chart with equivalent expressions.

	Exponent	Arithmetic Expression	Value
	3^3	$3 \times 3 \times 3$	27
1.	2^2		
2.		$4 \times 4 \times 4$	
3.			25
4.	6^2		
5.			49
6.	8^3		
7.			81
8.	10^2		
9.	5^3		
10.		$6 \times 6 \times 6$	

Tell whether each unit of measurement would be used for area or volume.

11. yd^2 (square yard)

12. cm^3 (cubic centimeter)

13. m^2 (square meter)

14. in^2 (square inch)

15. ft^3 (cubic feet)

16. mm^3 (cubic millimeter)

Extending Concepts

17. Concrete To Go is going to pour a patio 4 yd long, 4 yd wide, and $\frac{1}{12}$ yd deep. Do they need to know the area or the volume to know how much concrete is needed? Come up with a strategy to figure out how much concrete they should pour.

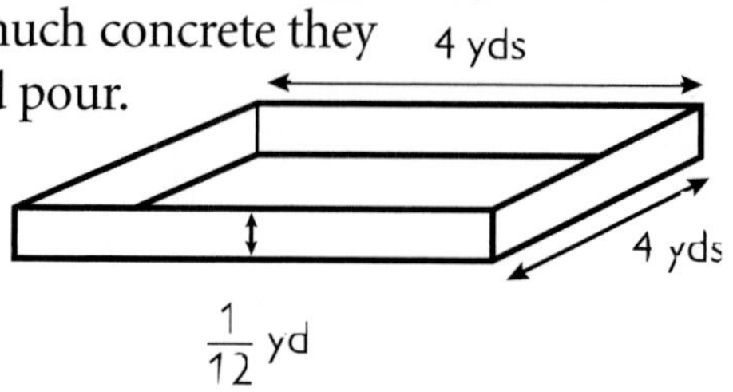

Making Connections

18. Imagine you have been hired by a famous clothing designer. It is your job to purchase the fabric for the upcoming designs. Your boss asks you to draw a smaller version of the designer's most successful scarf using a scale factor of 1:3. The original scarf is one yard long and one yard wide.

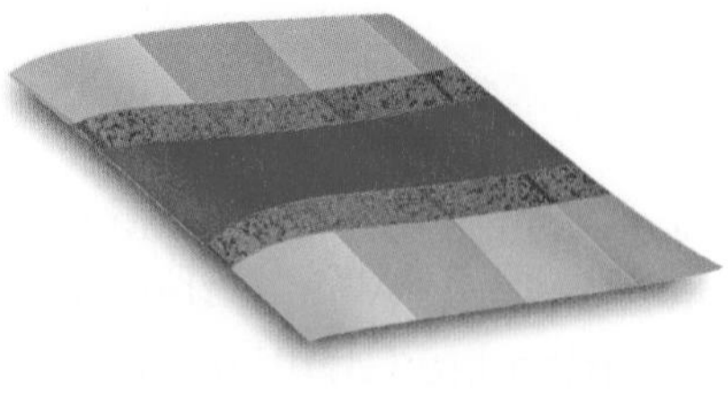

a. Draw a pattern with measurements for the new smaller scarf.

b. The company will make one hundred smaller scarves using your new pattern. How much fabric should you purchase?

Homework 12

Stepping into Gulliver's Worlds

Applying Skills

Complete the following chart by supplying the missing equivalents as decimals or fractions.

	Decimal	Fraction
1.	0.125	
2.		$\frac{3}{8}$
3.	0.75	
4.	0.67	
5.		$\frac{1}{2}$
6.	0.2	
7.		$\frac{1}{10}$
8.		$\frac{1}{20}$

9. Measure the sides of the square in inches. What is the:

a. perimeter? **b.** area?

10. Use the metric system to measure the height, width, and length of the cube below.

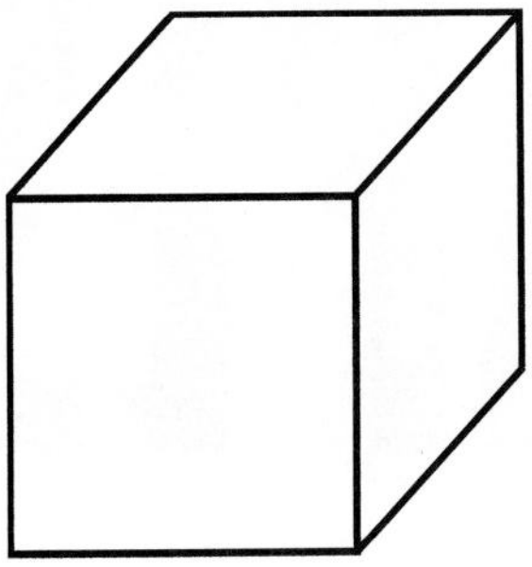

Extending Concepts

11. Shelley wants to cover a box with contact paper. The box is 1 ft high, 1 ft wide, and 1 ft deep.

a. Draw the box and show the measurements.

b. How many square feet of paper will she need to cover one face of the box?

c. How many square feet will she need to cover all sides of the box?

Making Connections

12. The creator of a famous theme park wanted children to feel bigger than life. The scale factor of objects in real life to objects in the park is 1:0.75.

a. Height of a street light?

Real life: 12 ft

Theme park:

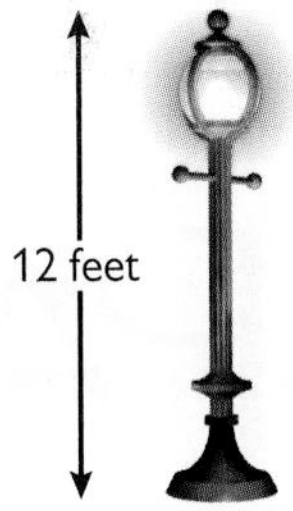

b. Length and width of a door?

Real life: 32″ × 80″

Theme park:

c. Height, width, and depth of a box?

Real life: 4 ft × 4 ft × 8 ft

Theme park:

PHASE**ONE**

Describing Patterns Using Tables

You will practice looking for patterns in different places—in drawings and numbers, sometimes in a story. You will even make up some patterns of your own. By making tables of the data you find, you will discover ways to extend the patterns to very large sizes.

How does math relate to patterns?

PATTERNS IN NUMBERS AND SHAPES

PHASE**TWO**

Describing Patterns Using Variables and Expressions

In this phase you look for patterns in letters that grow and chocolates in a box. You will begin to use the language of algebra by giving the rules for these patterns using variables and expressions. You will practice using expressions to compare different ways of describing a pattern to see if they give the same result.

PHASE**THREE**

Describing Patterns Using Graphs

In this phase you will plot points in all parts of the coordinate grid to make a mystery drawing for your partner. You will turn number rules into graphs and look at the patterns they make. By comparing graphs of some teenagers' wages for summer jobs, you will decide who has the better pay rate.

PHASE**FOUR**

Finding and Extending Patterns

After you have learned some ways to describe the rules for patterns, this phase gives you a chance to try out your skills in new situations. You look for patterns in a story about a sneaky sheep and in animal pictures that grow. You analyze three different patterns for inheriting some money in order to give some good advice.

PHASE ONE

In this phase you will look for patterns in three different situations. By making tables of the data, you can see how the patterns develop and discover how to extend them to greater size. Working with patterns in numbers, shapes, and a story helps you to develop math skills that carry over into algebra problems. You will begin to be able to work out a rule that will apply to all cases of a situation, and use your rule to solve problems.

Describing Patterns Using Tables

WHAT'S THE MATH?

Investigations in this section focus on:

PATTERN SEEKING

- Developing skills in looking for patterns in new situations
- Writing rules for patterns
- Extending pattern rules to apply to all cases

RECORDING DATA

- Making tables of data to describe patterns

NUMBER

- Exploring number relationships to look for patterns

MathScape Online
mathscape1.com/self_check_quiz

1 Calendar Tricks

FINDING AND DESCRIBING NUMBER PATTERNS

Finding a pattern can help you solve problems in surprising ways. In this activity, you will think about patterns as you test whether tricks using the numbers on a calendar will always be true. Then you will invent and test your own tricks.

Look for a Pattern

How can you know whether number patterns on a grid will always be true?

Paul invented three tricks for a block of four numbers. Find which of Paul's tricks are true for every possible two-by-two block of numbers on the calendar.

- Which tricks always worked? Which did not? How did you find out?
- For any tricks that did not always work, how can you revise them so that they do always work?

Paul's Box of Tricks

Sun	Mon	Tue	Wed	Thurs	Fri	Sat
1	2	3	4	5	6	7
8	9	10	11	12	13	14
15	16	17	18	19	20	21
22	23	24	25	26	27	28
29	30	31				

Trick One: The sums of opposite pairs of numbers will be equal.
For example: $2 + 10 = 3 + 9$.

Trick Two: If you add all four numbers, the sum will always be evenly divisible by 8.
For example:
$2 + 3 + 9 + 10 = 24; 24 \div 8 = 3$.

Trick Three: If you multiply opposite pairs of numbers, the two answers will always differ by 7.
For example:
$2 \times 10 = 20; 3 \times 9 = 27; 27 - 20 = 7$.

Invent Your Own Tricks

How can you use patterns to invent tricks that will always work?

The calendar on this page shows sequences of three numbers going diagonally to the right, shaded blue, and to the left, outlined in red. Can you make up tricks about diagonals of three numbers like these?

Make up at least two tricks for diagonals of three numbers like the numbers shaded blue in the example.

- Will your tricks be true for every diagonal of three numbers? How do you know?
- Which of your tricks will still be true if the diagonal goes in the opposite direction like the numbers in the red outline?

Make up at least two more tricks using your own shapes. Your shapes should be different from the ones shown so far.

- Are your tricks always true, no matter where on the calendar you put your shape?
- How do you know your tricks will always be true?

Write About Finding Patterns

Think about the patterns you explored and the tricks you invented in this lesson.

- How did you check whether your tricks will work everywhere on the calendar?
- What suggestions would you give to help another student find patterns?

Patterns on a Slant

Sun	Mon	Tue	Wed	Thurs	Fri	Sat
1	2	3	4	5	6	7
8	9	10	11	12	13	14
15	16	17	18	19	20	21
22	23	24	25	26	27	28
29	30	31				

hot words pattern

Homework

page 354

2 Painting Faces

EXPLORING PATTERNS BY ORGANIZING DATA

Here is a problem about painting all sides of a three-dimensional shape. Sometimes using objects is helpful in solving problems like this. You can record the data you get when you solve for shorter lengths and organize it into a table to help you find a rule for any length.

Make a Table of the Data

How can you set up a table to record data about a pattern?

A company that makes colored rods uses a paint stamping machine to color the rods. The stamp paints exactly one square of area at a time. Every outside face of each rod has to be painted, so this length 2 rod would need 10 stamps of paint.

How many stamps would you need to paint rods from lengths 1 to 10? Record your answers in a table and look for a pattern.

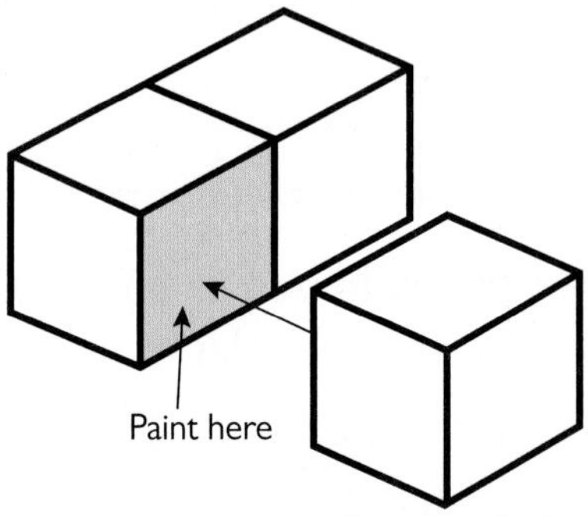

Length 2 rod; each end equals 1 square.

How to Organize Your Data in a Table

1. Use the first column to show the data you begin with. Write the numbers in order from least to greatest. In this example, the lengths of the rods go in the first column.
2. Use the second column to show numbers that give information about the first sequence. Here, the numbers of stamps needed to paint each length of rod go in the second column.

Length of Rod	Stamps Needed
1	
2	
3	

Extend the Pattern

Look at the table you made. Do you see a pattern in the data in the table?

1 Use what you learned from the table. Find the number of paint stamps you need to paint rods of lengths 25 and 66.

2 Write a rule you could use to extend the pattern to any length of rod.

Now look again at your table and your rule. Decide how you could use the number of stamps to find the size of the rod. Think about the operations you used in your rule.

3 What if it takes 86 stamps to paint the rod? How long is the rod?

4 If it takes 286 stamps to paint the rod, how long is the rod?

5 Write a rule you could use to find the length of any rod if you know the number of stamps needed to paint it.

Can you write a rule you could use to extend the pattern?

Write About Making a Table

Think about how you used the table you made to solve problems about painting the rods.

- How did making a table help you to find the pattern?
- Describe how far you think you would need to extend a table to be sure of a pattern.

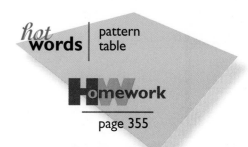

hot words | pattern table

Homework

page 355

3 Crossing the River

USING TABLES TO DESCRIBE AND PREDICT PATTERNS

Examining a pattern can help you develop a general rule that applies to any stage of the pattern. In this investigation you will look for a pattern to solve the problem of getting a group of hikers across a river using one small boat.

How can finding a pattern help solve for all cases?

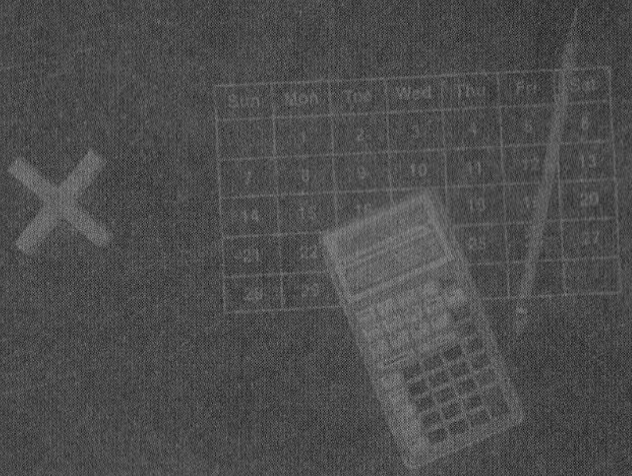

Find a Rule for Any Number

Think carefully about how the hikers could cross the river using just one boat. It may be helpful to act it out or use a diagram to solve the problem. As you work, make a table showing how many trips it takes for 1 to 5 adults and 2 children to cross. Look for the pattern, then use it to find how many trips are required for the other groups to cross to the other side.

1. How many one-way trips does it take for the entire group of 8 adults and 2 children to cross the river? Tell how you found your answer.
2. How many trips in all for 6 adults and 2 children?
3. 15 adults and 2 children?
4. 23 adults and 2 children?
5. 100 adults and 2 children?

Tell how you would find the number of one-way trips needed for any number of adults and two children to cross the river.

(Everyone can row the boat.)

Ten Hikers—One Boat

A group of 8 adults and 2 children needs to cross a river. They have a small boat that can hold either:

1 adult or 1 child or 2 children

Use Your Method in Another Way

Use the pattern to find the number of adults who need to cross the river for each case.

1 It takes 13 trips to get all of the adults and the 2 children across the river.

2 It takes 41 trips to get all of the adults and the 2 children across the river.

3 It takes 57 trips to get all of the adults and the 2 children across the river.

How can you work backward from what you know?

Tell How You Look for Patterns

Write a friend a letter telling how you look for patterns. Give examples from the patterns you have investigated so far. Answers to the following questions will help you write your letter.

- How can a table help you discover and describe a pattern?
- What other tools are helpful?
- How does finding a pattern help you solve problems?

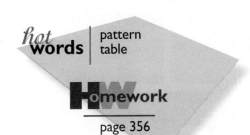

hot words | pattern table

Homework
page 356

PHASE TWO

As you look for patterns in this phase, you will think about how you can describe their rules in a way that applies to all situations. You will begin to use the language of algebra by writing rules using variables and expressions. You will compare your rules to those of other students to see if they give the same result. Some of the patterns change in more than one way and you will need to find a way to express that in your rule.

Describing Patterns Using Variables and Expressions

WHAT'S THE MATH?

Investigations in this section focus on:

PATTERN SEEKING

- Looking for patterns in numbers and shapes
- Examining patterns with two variables

ALGEBRA

- Using variables and expressions

EQUIVALENCE

- Exploring equivalence of expressions

MathScape Online
mathscape1.com/self_check_quiz

4 Letter Perfect

USING VARIABLES AND EXPRESSIONS TO DESCRIBE PATTERNS

Using variables and expressions gives you a shorthand way to describe a pattern. These tile letters grow according to different patterns. You will explore how to write a rule to predict the number of tiles needed to make letters of any size.

Find a Rule That Fits Every Case

How can a pattern help you predict the number of tiles used for any size?

Find a rule that will tell how many tiles it takes to build any size of the letter *I*.

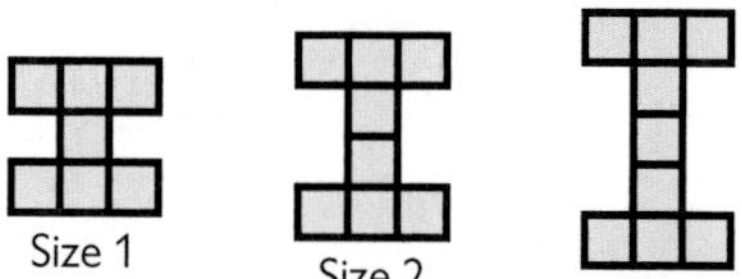

Size 1 Size 2 Size 3

1 Look for a pattern. Describe it clearly with words.

2 Describe the pattern using variables and expressions. This rule tells how the letter grows.

3 Use the rule to predict the number of tiles needed for each *I*:

a. size 12 **b.** size 15 **c.** size 22 **d.** size 100

Suppose you had 39 tiles. What is the largest size of *I* that you could make?

Using Variables and Expressions to Describe Patterns

These are the first three sizes of the letter *O*.

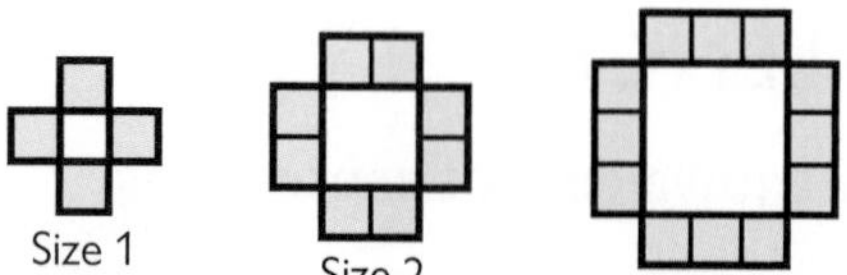

Size 1 Size 2 Size 3

How many tiles are needed to make each size? The pattern that tells how many can be described in words: The number of tiles needed is four times the size.

You can write this as 4 × *size*.

A shorter way to write the same thing is 4 × s or 4s.

In this example, the letter s is called a **variable** because it can take on many values.

4s is an **expression.** An expression is a combination of variables, numbers, and operations.

Relate the Rule to the Pattern

For one letter in the chart See How They Grow, the number of tiles is always $4s + 1$. The variable s stands for the size number. Decide how many tiles are added at each step. Look for the pattern.

1 Which letter do you think fits the pattern, *L*, *T*, or *X*? How many tiles are needed for size 16 of the letter?

2 For each of the other letters, give a rule that tells the number of tiles in any size.

How can variables and expressions describe a pattern?

Write About Your Own Letter Pattern

Make up your own letter shape with tiles. Figure out how you can make the letter grow into larger sizes.

- Draw your letter and tell how it grows.
- Give the rule for your letter using variables and expressions.
- Show how you can use the rule to predict the number of tiles that it would take to build size 100 of your letter.

See How They Grow

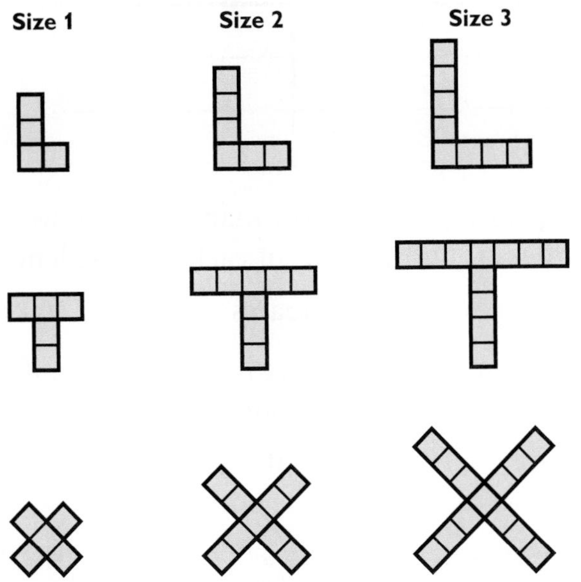

hot words | expression variable

Homework
page 357

5 Tiling Garden Beds

DESCRIBING PATTERNS WITH EQUIVALENT EXPRESSIONS

You have used tables and variables to describe different kinds of patterns. Here you will apply what you have learned to a new situation. You will show how each part of your solution relates to the situation. Then you can compare different ways of expressing the same idea.

What different rules or expressions can you write to describe a pattern?

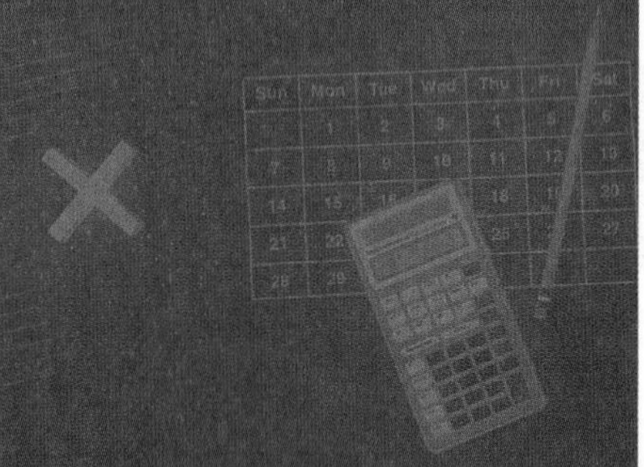

Find the Number of Tiles

Here are three sizes of gardens framed with a single row of tiles:

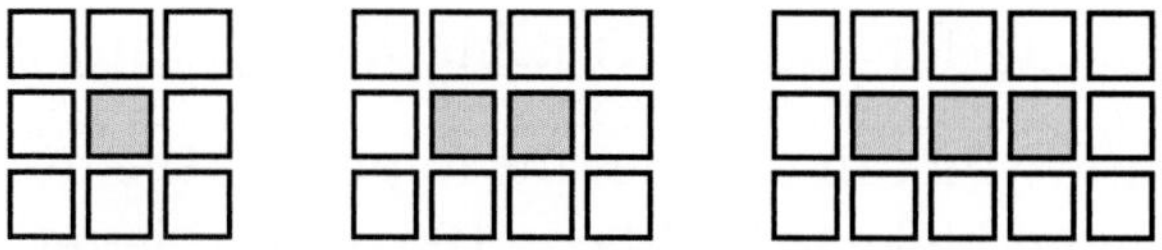

Length 1 Length 2 Length 3

1 Begin a table that shows the number of tiles for each length. Use the table to write an expression that describes the number of tiles needed for a garden of any length.

Length of Garden	Number of Tiles
1	8
2	
3	

2 Use your expression to find how many tiles you would need to make a border around gardens of each of these lengths.

a. 20 squares **b.** 30 squares **c.** 100 squares

3 Tell how you would find the length of the garden if you knew only the number of tiles in the border.

Test your method. How long is the garden if the following numbers of tiles are used for the border?

a. 68 tiles **b.** 152 tiles **c.** 512 tiles

4 Relate each part of your expression to the garden and the tiles.

Extend the Rule

Some gardens are two squares wide, and vary in length. For example:

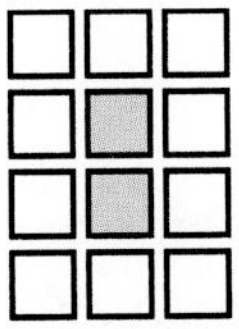

Length = 1
Width = 2

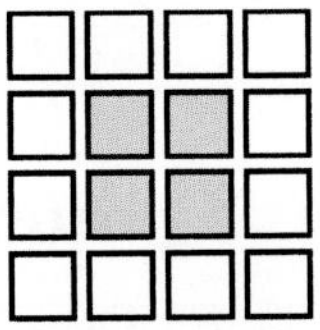

Length = 2
Width = 2

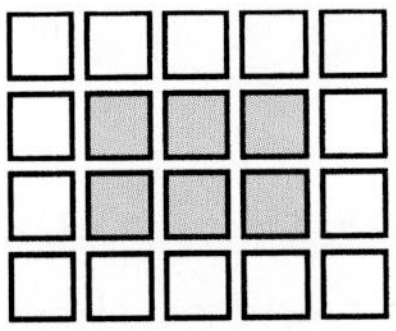

Length = 3
Width = 2

Can you figure out the number of tiles needed for gardens of any length and a width of 2? Use an expression that describes your method. Use the method to solve these problems.

How many tiles do you need to make a border around each of the following gardens?

1 $l = 5, w = 2$

2 $l = 10, w = 2$

3 $l = 20, w = 2$

4 $l = 100, w = 2$

Can you write a rule for tiling a garden of any length and any width?

Write About Equivalent Expressions

You and your classmates may use different expressions to describe these patterns. Compare your ideas with others.

- What equivalent expressions did you and your classmates write?
- How do you know they are equivalent?

Conventions for Algebraic Notation

Writing variables and expressions in standard ways avoids confusion. The numeral is placed before the letter representing a variable:

$2l$ not $l2$

The numeral 1 is not required before a variable:

Use l instead of $1l$

"Two times the length" can be stated several ways:

$2l \quad 2 \cdot l \quad 2 \times l$

Place parentheses carefully.

$l + 3 \times 2$ does not equal $(l + 3) \times 2$.

hot words | equivalent expression

Homework

page 358

6 Chocolates by the Box

EXPLORING PATTERNS WITH TWO VARIABLES

Some patterns can be described using more than one variable. The boxes of chocolates in this lesson are an example of this type of pattern. You will look for a way to write a rule that will apply to all sizes of chocolate boxes.

How can you use variables to describe the pattern?

Find the Contents for Each Size

Buy a Box of Chocolates—Get a Bonus

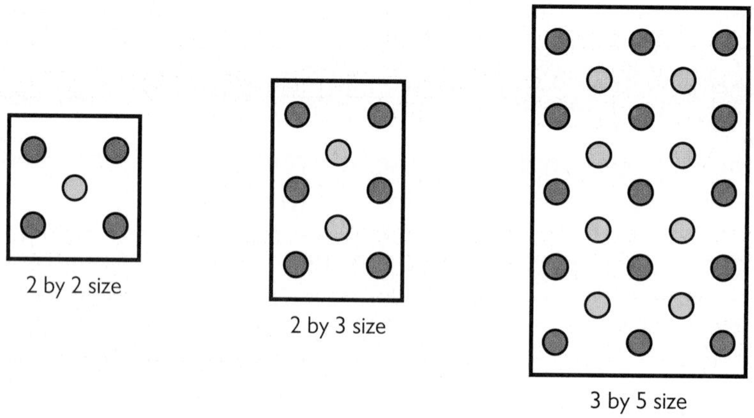

When you buy a box of Choco Chocolates, you get a bonus light chocolate between every group of four dark chocolates, as the diagrams show. The size of the box tells you how many columns and how many rows of dark chocolates come in the box.

1 How many *dark* chocolates will you get in each size of box?

a. 4 by 4 **b.** 4 by 8 **c.** 6 by 7

d. 12 by 25 **e.** 20 by 20 **f.** 100 by 100

2 How can you figure out the number of dark chocolates in any size box? Explain your method using words, diagrams, or expressions.

Rewrite Your Rule

How can you use your method to write another rule?

Now that you have developed a rule for finding the number of dark chocolates, how can you figure out the number of light chocolates in any size box?

1 Explain your method using words, diagrams, or expressions.

2 Test your method. How many *light* chocolates will you get in each box of these sizes?

a. 4 by 4 **b.** 4 by 8 **c.** 6 by 7

d. 12 by 25 **e.** 20 by 20 **f.** 100 by 100

3 Use variables and expressions to describe the total number of all chocolates in any box.

Write About Using Variables

Think about how you have used variables and expressions to describe patterns. Write a note to one of next year's students telling how you use these tools to solve problems.

- Give some examples of using expressions to describe patterns.
- Tell how you decide whether two expressions are equivalent.

A Pattern of *J*s with Two Variables

Choco Company uses chocolate squares to make letters you can eat. They make the letter *J* in different heights and widths.

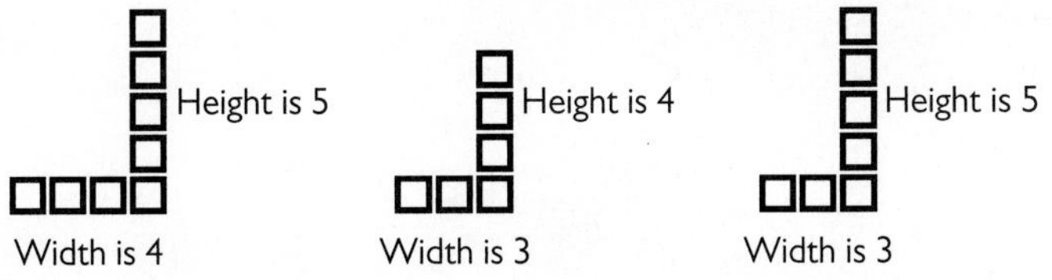

To find the total number of squares needed to make a letter *J*, you can add the height and width and subtract 1.

You can write this with words and symbols:
height + width − 1

Or you can use two variables (one for height and one for width) and say it this way: $h + w - 1$

hot words expression, variable

Homework page 359

PHASE THREE

In this phase, you will move around the coordinate plane to identify points and to graph data that you develop when you are looking for patterns. As you work with number rules for patterns, you begin to see the relationship of a list of ordered pairs to the line they form when they are graphed. You will use the information you find in the graphs to solve problems about some real-life situations involving summer jobs for two teenagers.

Describing Patterns Using Graphs

WHAT'S THE MATH?

Investigations in this section focus on:

PATTERN SEEKING

- Relating a pattern rule to a graphed line and the line to the rule

ALGEBRA

- Identifying points on the coordinate plane
- Making tables of ordered pairs
- Graphing ordered pairs on the coordinate plane

PROBLEM SOLVING

- Interpreting data from a graph
- Using patterns on a graph to solve problems

MathScape Online
mathscape1.com/self_check_quiz

7 Gridpoint Pictures

USING A COORDINATE PLANE

So far, you've described patterns using words, pictures, tables, variables, and expressions. Now you will look at another tool: the coordinate grid or plane. Before using this tool, review some of the basic definitions in "Finding Your Way Around the Coordinate Plane."

Describe a Picture on the Coordinate Plane

How can you make and describe a picture on a coordinate plane?

Draw a simple picture on a coordinate plane. You may want to write your initials in block letters or draw a house or other simple object. Be sure your picture has parts in all four quadrants.

Using coordinates, write a description of how to make your picture. You can also include other directions, but you may not use pictures.

Finding Your Way Around the Coordinate Plane

The coordinate plane is divided into four quadrants by the horizontal *x*-axis and the vertical *y*-axis. The axes intersect at the origin. You can locate any point on the plane if you know the coordinates for *x* and *y*. The *x*-coordinate is always stated first.

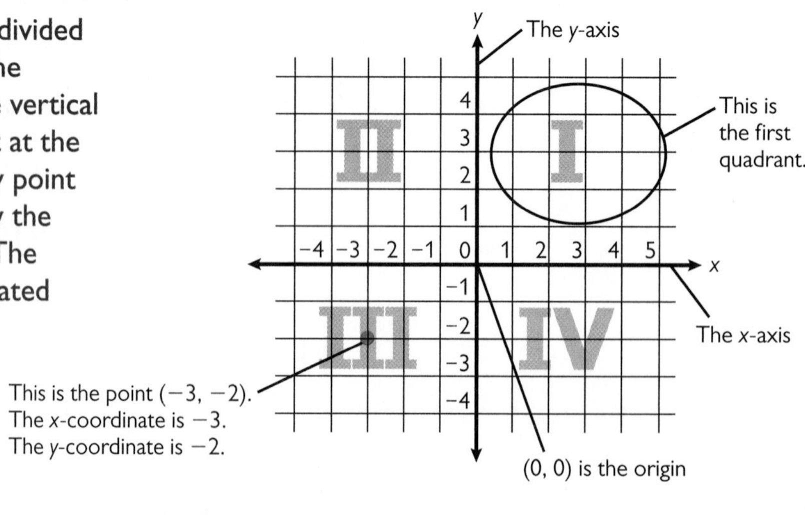

Decode a Gridpoint Picture

When your picture and list are finished, trade descriptions with a partner. Do not show your pictures until later.

1 Draw the picture your partner has described. Keep notes on anything that is not clear.

2 Return the description and your picture to your partner. Check the picture your partner drew from your description to see if it matches your original drawing. If there are differences, what caused them? If necessary, revise your description.

How can coordinates help you draw a gridpoint picture?

Write About What You See

Suppose you are given the coordinates for a set of points. How could you tell if the points are all on a straight vertical line? a straight horizontal line? Explain.

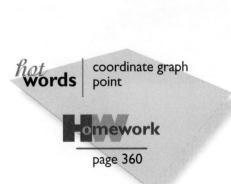

hot **words** | coordinate graph
point

Homework

page 360

8 Points, Plots, and Patterns

SHOWING A RULE ON THE COORDINATE PLANE

When you used points on a coordinate plane to describe a picture, you followed a visual pattern. Now you will see what happens when the points all spring from a number rule. Keep an eye out for patterns!

What patterns in the points on the coordinate grid fit a number rule?

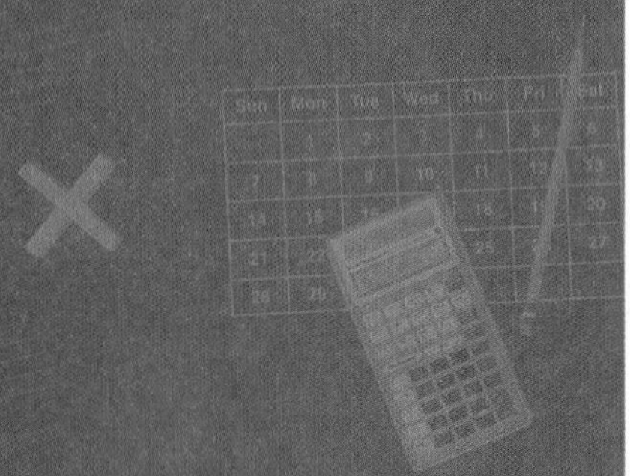

Find Some Patterns in the Plots

You may think of a number rule as being expressed only in words and numerals. Some surprising things develop when you plot a graph using ordered pairs that follow a rule.

1. Make up a number rule of your own.
2. Make a table of points that fit the rule.
3. Plot the points on the coordinate plane.
4. What patterns do you notice on the coordinate plane? Can you find points that fit the rule but do not fit the pattern?

Repeat the process with other number rules. Keep a record of your results.

Ordered Pairs and Number Rules

In the ordered pair (4, 8), 4 is the *x*-coordinate and 8 is the *y*-coordinate.

A number rule tells how the two numbers in an ordered pair are related. Here are some examples:

- The *x*-coordinate is half the *y*-coordinate.
- The *y*-coordinate is 6 more than the *x*-coordinate.
- The *x*-coordinate and the *y*-coordinate are the same.

Find the Rule for the Pattern

Think about the patterns that you have been plotting on the grid. Do you think you can find the rule for a line passing through two points?

Try this. Mark the two endpoints that are given. Carefully draw a line between the points. What is the pattern or rule for all the points falling exactly on the line?

1 (6, 4) and (−6, −8)

2 (3, 12) and (−1, −4)

How does a line show a relationship between coordinates on a grid?

Write About Graphing Number Rules

What do you notice about the graphs from your number rules and those of your classmates? Write a summary of as many generalizations as possible. Here are some possibilities that you might want to include:

- What can you say about rules that give a line that passes through (0, 0)?
- How can you make parallel lines move up or down the graph?
- How can you change your rule to make a line steeper?

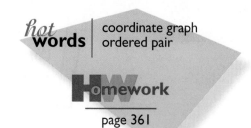

hot words | coordinate graph
ordered pair

Homework

page 361

9 Payday at Planet Adventure

COMPARING DIFFERENT PATTERNS ON A GRAPH

Rachel and Enrico have summer jobs at Planet Adventure, a local amusement park. Since Rachel gets bonus pay, it's not easy to make a quick comparison of their wages. You will use a graph to compare how much they earn for different lengths of time.

What can a graph tell about wages?

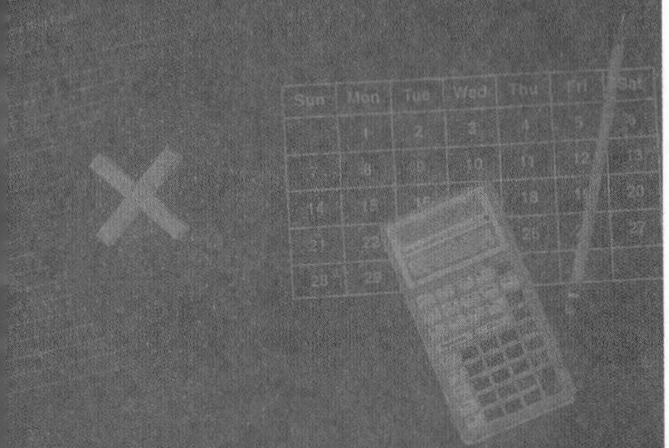

Make a Graph of Earnings

Rachel works in the Hall of Mirrors. Her rate of pay each day is $5 per hour. She also gets a daily $9 bonus for wearing a strange costume.

- Make a table to show what pay she should receive for different numbers of hours worked each day.
- Next, draw a graph of the data for Rachel's pay. Label the axes and choose an appropriate scale for the graph.

How would you find the amount of money Rachel earns for any number of hours worked? Use words, diagrams, or equations to explain your method.

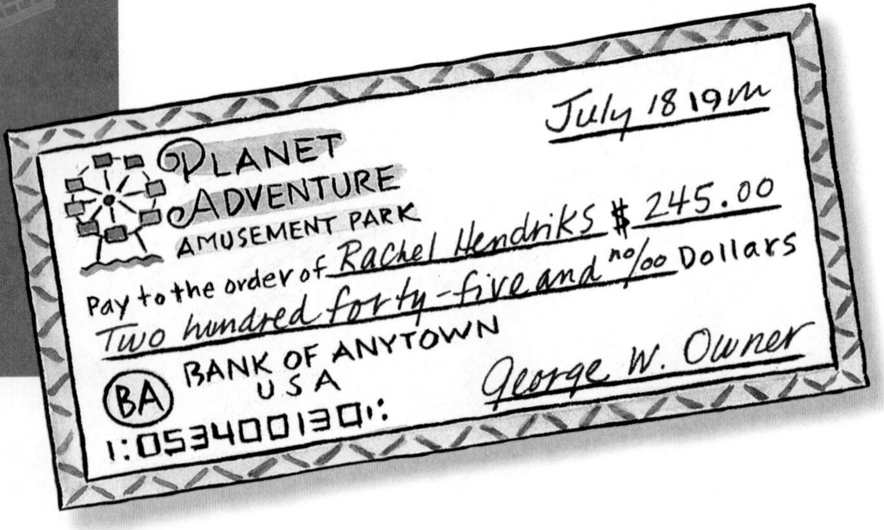

Compare Earning Rates

How can a graph help compare wages?

Enrico works at the Space Shot roller coaster. His rate of pay is $6.50 an hour.

- Make a table to show what pay he receives for different numbers of hours worked each day.
- On the same grid you used for Rachel's pay, draw a graph of the data for Enrico's pay.
- How would you find the amount of pay Enrico earns for any number of hours worked? Is the method different from the one you used for Rachel?

Write About the Graphs

Compare the graphs of Rachel's pay and Enrico's pay. Which job pays better? How did you decide?

- Tell how the graphs are the same and how they are different.
- For what numbers of hours worked does Rachel earn more than Enrico? Enrico more than Rachel?
- Do Rachel and Enrico ever earn the same amount for the same number of hours worked? How did you find out?

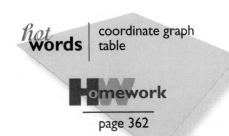

hot words | coordinate graph, table

Homework

page 362

PHASE FOUR

This phase gives you a chance to try out all the tools you have learned to use in describing patterns. You will examine the situations, decide how you will explain the pattern, and then write a rule to extend the pattern to any size. You will also use variables and expressions to describe patterns. You will soon see how much you have learned about looking for patterns.

Finding and Extending Patterns

WHAT'S THE MATH?

Investigations in this section focus on:

PATTERN SEEKING

- Identifying, describing, and generalizing patterns
- Choosing appropriate tools to describe patterns

NUMBER OPERATIONS

- Using inverse operations with pattern rules

ALGEBRA

- Using variables and expressions to describe patterns
- Making lists of ordered pairs to describe patterns
- Graphing ordered pairs

PROBLEM SOLVING

- Using patterns in problem situations

MathScape Online
mathscape1.com/self_check_quiz

10 Sneaking Up the Line

SOLVING A SIMPLER PROBLEM TO FIND A PATTERN

Finding a pattern in a simpler problem can help you understand a problem with greater values. Careful reasoning in a sample problem will help you make a rule about this sneaky situation.

Solve a Simpler Problem

How can you identify a pattern by solving a simpler problem first?

After you read "A Woolly Tale" and make your prediction, try some small problems to help look for a pattern.

1. Can you find how many sheep would be shorn before Eric if there are 6 sheep ahead of him? Use counters, diagrams, or any other method to solve the problem.

   What if there are 11 sheep ahead of him? What if the number in front of Eric is 4 to 10? 11 to 13? It may help to make a table, then graph the data.

Use what you learned in the simpler problems.

2. Find how many sheep would be shorn before Eric if there were 49 sheep in front of him. Does the answer match the prediction you made at first?

3. Describe a rule or expression you would use to find the number shorn before Eric for any number of sheep in front of him.

A Woolly Tale

Eric the Sheep is at the end of a line of sheep waiting to be shorn. But being an impatient sort of a sheep, every time the shearer takes a sheep from the front to be shorn, Eric sneaks up the line two places.

Think about how long it will take Eric to reach the head of the line. Before you begin to work, make a prediction. If there are 49 sheep ahead of him, how many of the sheep will be shorn before Eric?

Test Your Rule

In each case, find how many sheep are shorn before Eric.

1 There were 37 sheep in front of Eric.

2 There were 296 sheep in front of him.

3 There were 1,000 sheep in front of him.

4 There were 7,695 sheep in front of the sneaky sheep.

Now try using your rule to find how many sheep were lined up in front of Eric if:

5 13 sheep were shorn before him.

6 21 sheep were shorn before Eric.

Will your rule work for any number?

Write About Your Sneaky Rule

Eric's pattern of sneaking up the line follows a rule with some new things to think about.

- How do situations 5 and 6 above differ from 1–4?
- Describe how you used your rule to find how many sheep were ahead of Eric.

hot words | coordinate graph
table

Homework
page 363

11 Something Fishy

EXPLORING GEOMETRIC PATTERNS OF GROWTH

You have learned to identify the rule for many patterns from simple to complex. Here you find how to describe the pattern as an animal drawing changes in more than one way. Then you get to grow your own animals and describe their patterns.

How can you describe the growth of some geometric patterns?

Go Fishing for Some New Patterns of Growth

The fish in Patternville grow in a particular way. The diagram on dot paper shows the first four stages of this growth.

1. For fish in growth stages 1–6, figure out how many line segments and spots they would have in each stage. Show your answers using expressions, tables, and graphs.

2. Use what you observed about the growth in the number of *line segments* to answer these questions:
   a. How many line segments would a fish in stage 20 have?
   b. How many line segments would a fish in stage 101 have?
   c. In what stage of growth has the fish 98 line segments?
   d. In what stage of growth has the fish 399 line segments?

Something Fishy in Patternville

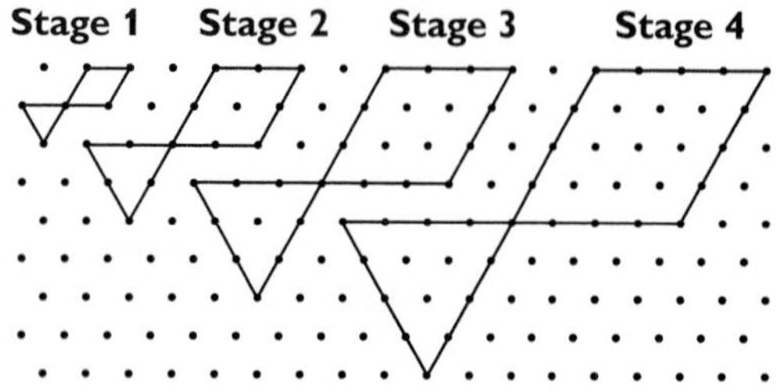

There are two ways to measure the growth of the fish: line segments and spots.

Line Segments
Count the line segments between dots needed to make the fish. It takes seven line segments to make the fish in stage 1.

Spots
Count the number of spots inside the fish's body (not including the tail). A fish in stage 1 has no spots and a fish in stage 2 has one spot.

3 Use what you observed about the growth in the number of *spots* to answer these questions:

a. How many spots would a fish have in stage 20?

b. How many spots would a fish have in stage 101?

How can you figure out how many line segments and spots a fish would have at any stage? Use words, diagrams, or equations to explain your method. Look for a short way to state it clearly.

Create Your Own Growing Patterns

Look at the examples from the Patternville Zoo. Then use graph paper, dot paper, or toothpicks to create your own "animal" and show how it grows in stages. Make sure you have a clear rule for the way it grows.

Draw your animal and write the rule for its growth at any stage on the back of the drawing. Create a table and a graph to show stages 1–5 of your animal's growth.

Write an explanation of why your rule will predict the number of dots or line segments for any size of your animal.

What are some different patterns that describe how a drawing can grow?

Examples from the Patternville Zoo

Stage	1	2	3
Dots	14	16	18
Area	6	7	8

Stage	1	2	3
Hexagons	1	4	7
Perimeter	6	18	30

Stage	1	2	3
Perimeter	11	22	33
Area in triangles	11	44	99

hot words expression variable

Homework page 364

12 The Will

USING PATTERNS TO MAKE A DECISION

The results of a growth pattern are not always obvious at first. In this situation you will project the patterns into the future in order to make a good recommendation for Harriet. You may choose any tools you wish to describe the patterns.

Make a Prediction

Which payment plan looks like the best choice?

Harriet's uncle has just died. He has left her some money in his will, but she must decide how it is paid. Read about the three plans in the will. You can help her choose which plan would be the best for her.

Before doing any calculations, predict which plan would give Harriet the greatest amount of money at the end of year 25. Give your reasons for your choice.

From Harriet's Uncle's Will

... and to my niece Harriet, I give a cash amount of money to spend as she pleases at the end of each year for 25 years. Knowing how much she likes a bit of mathematics, I give her a choice of three payment plans.

Plan A:

$100	at the end of year 1
$300	at the end of year 2
$500	at the end of year 3
$700	at the end of year 4
$900	at the end of year 5

(and so on)

Plan B:

$10	at the end of year 1
$40	at the end of year 2
$90	at the end of year 3
$160	at the end of year 4
$250	at the end of year 5

(and so on)

Plan C:

1 cent	at the end of year 1
2 cents	at the end of year 2
4 cents	at the end of year 3
8 cents	at the end of year 4
16 cents	at the end of year 5

(and so on)

Compare the Plans

How can you describe the patterns and extend them into the future?

Each plan pays a greater amount of money each year, but the amounts increase in different ways. In order to compare the plans, you will need to find the amount Harriet would receive for each plan in each year.

1 For each plan, what patterns did you see in the amount Harriet would receive each year? Use words, diagrams, or expressions to explain the patterns.

2 Make tables and graphs to show the amount she would receive for each plan in each year. What amount of money will Harriet receive at the end of the tenth year? the twentieth year? the twenty-fifth year?

3 Describe a method you could use to find the amount Harriet would get at the end of any year for each plan.

Give Some Good Advice

Write a letter to Harriet advising her as to which plan she should choose and why. Make sure to compare the three plans. You can choose words, diagrams, tables, graphs, and expressions to support your recommendation.

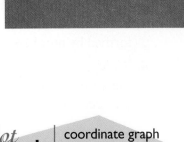

hot words | coordinate graph
expression

Homework

page 365

Homework 1

Calendar Tricks

Applying Skills

Read the four statements A–D. Then tell which are true for each two-by-two block of four numbers.

A. The sum of the four numbers is divisible by 4.

B. The sum of the four numbers is divisible by 8.

C. The sum of the bottom two numbers differs from the sum of the top two numbers by 14.

D. The number in the bottom right corner is three times as big as the number in the top left corner.

Sun	Mon	Tue	Wed	Thurs	Fri	Sat
1	2	3	4	5	6	7
8	9	10	11	12	13	14
15	16	17	18	19	20	21
22	23	24	25	26	27	28
29	30	31				

1. Which statements are true for the shaded block on the calendar?

2. Which statements are true for the block formed by numbers 6, 7, 13, and 14?

3. Which statements are true for the block formed by numbers 19, 20, 26, and 27?

4. Which statements do you think are true for every possible two-by-two block of four numbers on the calendar?

Extending Concepts

5. Choose two different blocks of nine numbers arranged 3 across and 3 down on a calendar, and check that this rule holds: For any block of nine numbers the average of the four corner numbers is equal to the middle number. Show your work and explain why the rule works.

6. Make up your own trick which works for any three-by-three block of nine numbers. Explain why your trick works.

Making Connections

7. It is said that the seven-day week was based originally on the idea of the influence of the planets. For a long time people believed that seven celestial bodies revolved around the earth. The early Romans observed an eight-day week based on the recurrence of market days.

a. What pattern or trick do you notice about numbers on a diagonal of a calendar such as 2, 10, 18,...? Why does this trick work? Revise the rule for the pattern so that it would work for a calendar with 8 days in each row.

b. Would the rule in item **5** work for a calendar with 8 days in each row? Why or why not?

Homework 2: Painting Faces

Applying Skills

A company that makes colored rods uses a paint stamping block to paint only the front and one end of each rod like this:

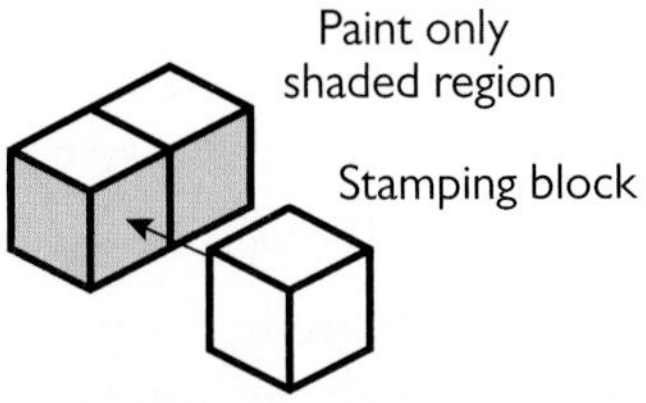

1. Copy and complete the table to show how many stamps would be needed to paint rods with lengths 1 to 10.

Length of Rod	Stamps Needed
1	2
2	3
3	
4	
5	
6	
7	
8	
9	
10	

2. What rule could you use to find the number of stamps needed for a rod of any length?

3. How many paint stamps are needed to paint a rod of length 23? 36? 64?

4. How long is the rod if the number of stamps needed is 23? 55? 217?

Extending Concepts

Suppose the company also makes cubes of different sizes and uses the stamping block to paint only the front face of each cube as shown.

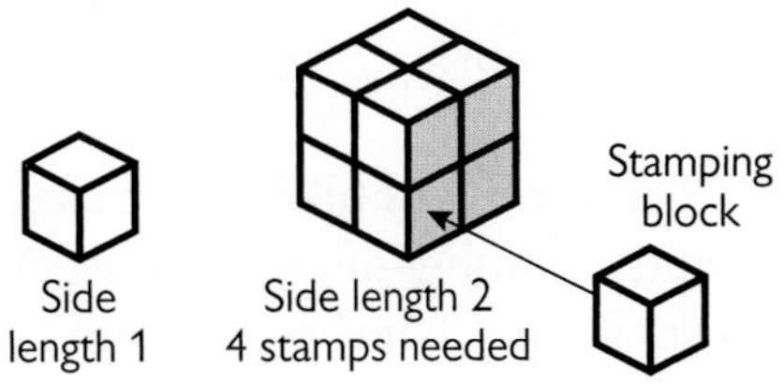

5. Make a table to show the number of paint stamps needed for cubes with side lengths 1 to 6. What pattern do you notice? Write a general rule for finding the number of paint stamps needed for any cube.

6. Use your rule from item **5** to find how many stamps you would need for a cube with side length 43.

Writing

7. Answer the letter to Dr. Math.

Dear Dr. Math,

I decided I didn't need to figure out the number of stamps that would completely cover every different size rod. Instead, I just tested some sample lengths and made a table like this:

Length of Rod	Stamps Needed
1	6
3	14
5	22
10	42
20	82
100	402

But I'm confused and can't see a pattern. Was this a good shortcut?

Pat Turn

Homework 3

Crossing the River

Applying Skills

Suppose that a group of hikers with exactly two children and a number of adults must cross a river in a small boat. The boat can hold either one adult, one child, or two children. Anyone can row the boat.

How many one-way trips are needed to get everyone to the other side if the number of adults is:

1. 3? **2.** 4? **3.** 5?

4. Make a table that records the number of one-way trips needed for all numbers of adults from 1 to 8.

How many trips are needed if the number of adults is:

5. 11? **6.** 37?

7. 93? **8.** 124?

How many adults are in the group if the number of trips needed is:

9. 25? **10.** 49?

11. 73? **12.** 561?

Extending Concepts

13. Find the number of one-way trips that would be needed for a group of 4 adults and 3 children to cross the river. Use a recording method that keeps a running total of the number of trips and the number of adults and children on each side of the river.

14. Make a table showing the number of trips needed for 1 to 5 adults and 3 children to cross the river. Describe in words a general rule that you could use to find the number of trips needed for any number of adults and 3 children. How did your table help you to find the rule?

15. How many trips would it take for 82 adults and 3 children to cross?

16. How many adults are in the group if 47 trips are needed to get all the adults and 3 children across the river?

Making Connections

17. For transportation, the Hupa Indians of Northwestern California used canoes hollowed out of half of a redwood log. These canoes could carry up to 5 adults. If one adult could row the canoe, how many one-way trips would be needed for 17 adults to cross a river using one of these canoes?

4 Homework

Letter Perfect

Applying Skills

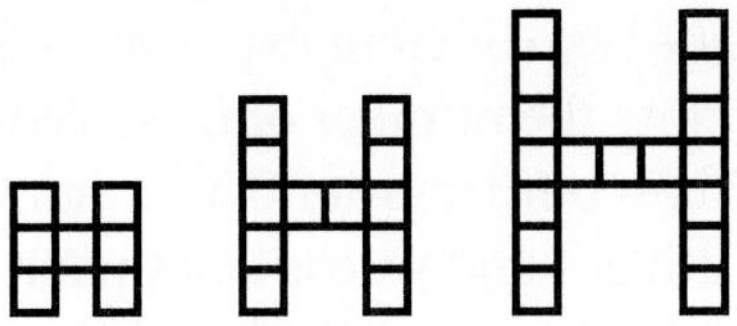

Size 1   Size 2   Size 3

1. Draw an *H* of size 4 and an *H* of size 5.

2. How many tiles are needed for each of the sizes 1 to 5 of the letter *H*?

3. How many tiles are added at each step to the letter *H*?

4. Which of these expressions tells how many tiles are needed for a letter *H*? The variable *s* stands for the size number.

$2s$ $5s$ $2s + 5$ $5s + 2$ $4s + 3$

5. Predict the number of tiles needed for a letter *H* of these sizes:

a. 11   **b.** 19   **c.** 28

d. 57   **e.** 129

6. What is the largest size of *H* you could make if you had:

a. 42 tiles?   **b.** 52 tiles?

c. 127 tiles?   **d.** 152 tiles?

Extending Concepts

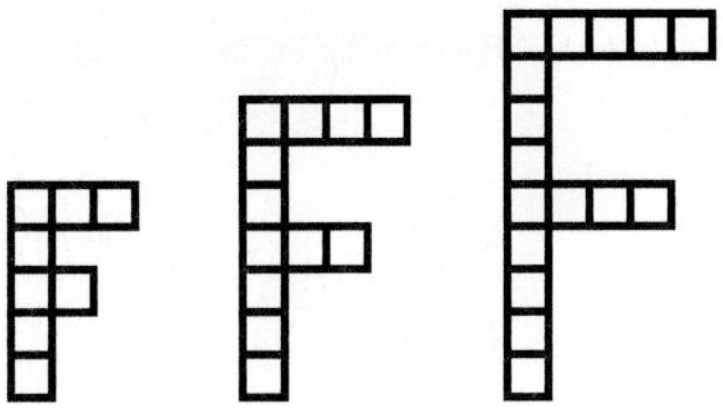

Size 1   Size 2   Size 3

7. a. Find how many tiles are needed for a letter *F* of sizes 1 to 5. Then describe in words a rule you could use to find the number of tiles needed for a letter *F* of any size.

b. Write an expression to describe your rule in item **7a.** Use the letter *s* to stand for the size number.

c. Explain why *s* is called a variable.

8. For a mystery letter, the number of tiles needed is $3s + 8$. The variable *s* stands for the size number. How many tiles are added at each step? How can you tell? Which letter grows faster, the mystery letter or the letter *F*?

Making Connections

9. The Celsius temperature scale uses the freezing point of water as 0 degrees Celsius and its boiling point as 100 degrees Celsius. If *c* stands for a known Celsius temperature, the expression $1.8c + 32$ can be used to find the Fahrenheit temperature. What Fahrenheit temperature corresponds to 16 degrees Celsius? to 25 degrees Celsius? How did you find your answers?

Homework 5: Tiling Garden Beds

Applying Skills

Find the value of these expressions if $l = 3$ and $w = 5$:

1. $6l$ **2.** $3l + w$ **3.** $4(l + w)$

4. $lw - 2$ **5.** $l(w - 2)$

For items **6–17**, assume that each garden is one square wide. Suppose you want to frame your garden with a single row of tiles like this:

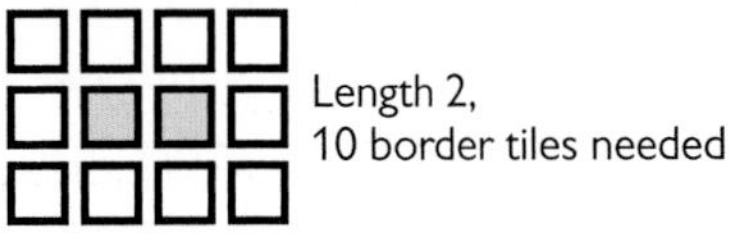

Find the number of border tiles needed for a garden of each length.

6. 4 **7.** 7 **8.** 13

9. 25 **10.** 57 **11.** 186

Find the length of the garden if the number of border tiles needed is:

12. 16 **13.** 38 **14.** 100

15. 370 **16.** 606

17. Which of these expressions could *not* be used to find the number of tiles needed to make a border around a garden with length l squares?

a. $(l + 3) \times 2$ **b.** $(l + 2) \times 2 + 2$

c. $2l + 6$ **d.** $(l + 2) \times 2$

e. $(l + 1) \times 2 + 4$

Extending Concepts

18. a. Write two different expressions for finding the number of tiles needed to make a border around a garden whose width is 4 and whose length may vary. Use l to represent the length.

b. Explain why each of the expressions makes sense by relating each part of the expression to the garden and the tiles.

c. How many tiles are needed to make a border around a garden with a width of 4 and a length of 35?

19. Suppose that one length of the garden is along a wall like this:

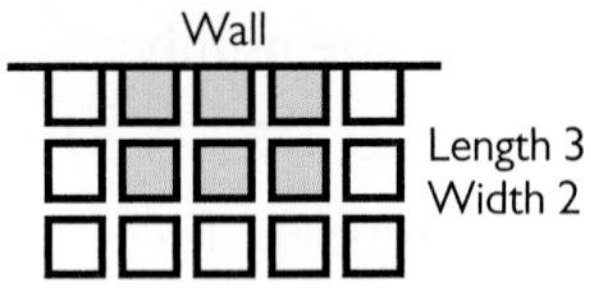

The length and the width may both vary. Write an expression for the number of tiles needed. Use l for the length and w for the width.

Making Connections

20. A Japanese garden is considered a place to contemplate nature. An enclosure such as a bamboo fence is often used to separate the garden from the everyday world outside and to create the feeling of a sanctuary. What length of fencing would be needed to enclose a rectangular garden 70 feet long and 30 feet wide?

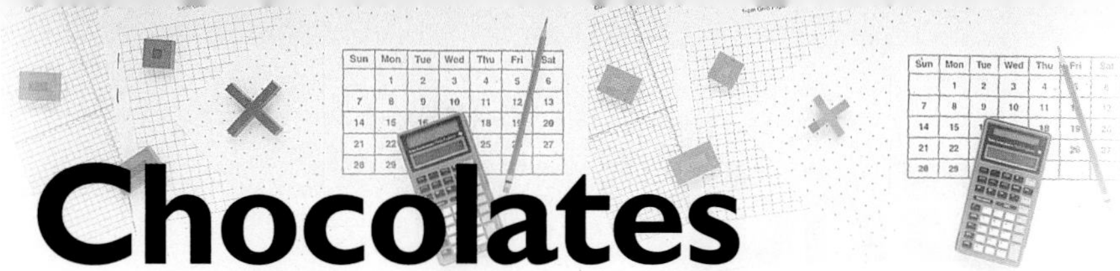

Chocolates by the Box

Applying Skills

In a box of Choco Chocolates there is a light chocolate between each group of 4 dark chocolates as shown. The size of the box tells the number of columns and rows of dark chocolates.

3 by 4 size

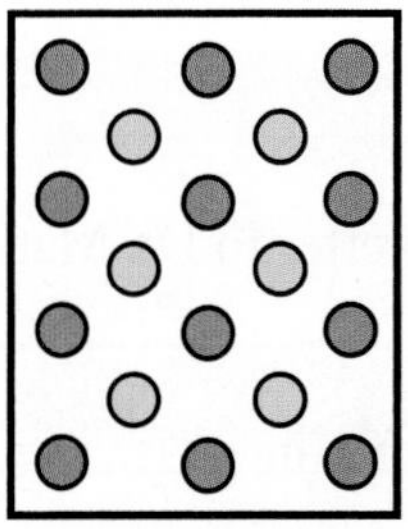

Find the number of dark chocolates and the number of light chocolates in boxes of these sizes:

1. 5 by 5 **2.** 8 by 10

3. 15 by 30 **4.** 40 by 75

Give the total number of chocolates in boxes of these sizes:

5. 3 by 5 **6.** 7 by 9

7. 18 by 20 **8.** 30 by 37

Extending Concepts

9. Either of the two equivalent expressions $lw + (l - 1) \times (w - 1)$ or $2lw - l - w + 1$ may be used to find the total number of chocolates in a box of Choco Chocolates.

a. Verify that both expressions give the same result for a 16 by 6 box.

b. Explain why the first expression makes sense.

10. Choco Chocolates wants to make new triangle-shaped boxes as shown. Write rules for finding the number of light chocolates, dark chocolates, and total chocolates in a triangular box of any size. How many dark and how many light chocolates are in a box of size 9? size 22?

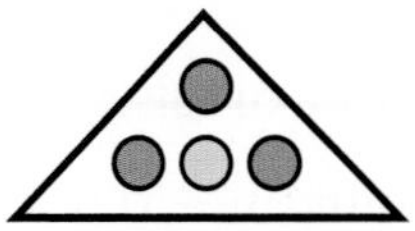

Size 2 triangle

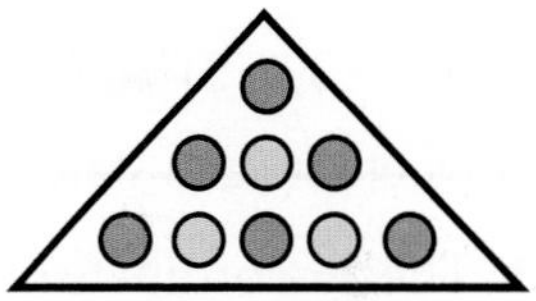

Size 3 triangle

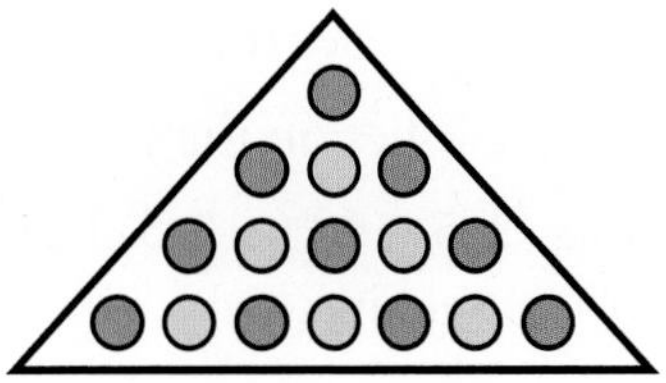

Size 4 triangle

Writing

11. Write a paragraph describing how you have used variables and expressions to describe patterns. Be sure to explain the meaning of the words *variable* and *expression.* Give some examples of using expressions to describe patterns.

Homework 7: Gridpoint Pictures

Applying Skills

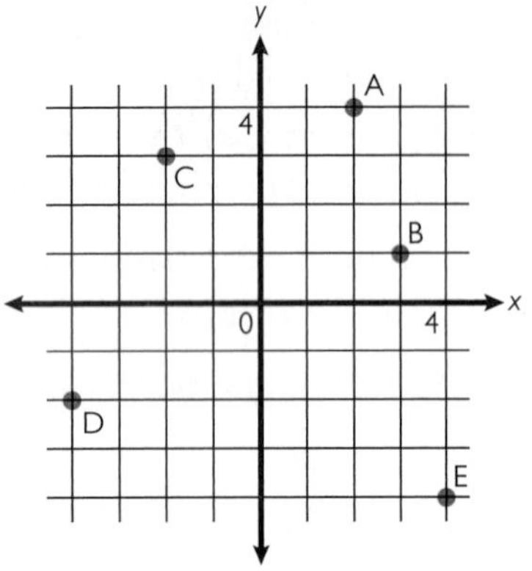

1. Give the coordinates of each of the five points shown on the coordinate plane.
2. Draw a grid and axes similar to those in item **1.** Place a dot at each point.

   a. $(1, 4)$ **b.** $(-3, 4)$ **c.** $(-1, 0)$
   d. $(-3, -2)$ **e.** $(1, -2)$

Tell whether each statement is true or false:

3. The y-coordinate of the point $(3, -5)$ is 3.
4. The x-coordinate of the point $(-9, 1)$ is -9.
5. The point $(0, -5)$ lies on the x-axis.
6. The point $(0, 4)$ lies on the y-axis.
7. The points $(2, 5)$ and $(-3, 5)$ lie on the same horizontal line.

Extending Concepts

8. Using coordinates, tell how to make this picture. You can include other directions but you may not use pictures.

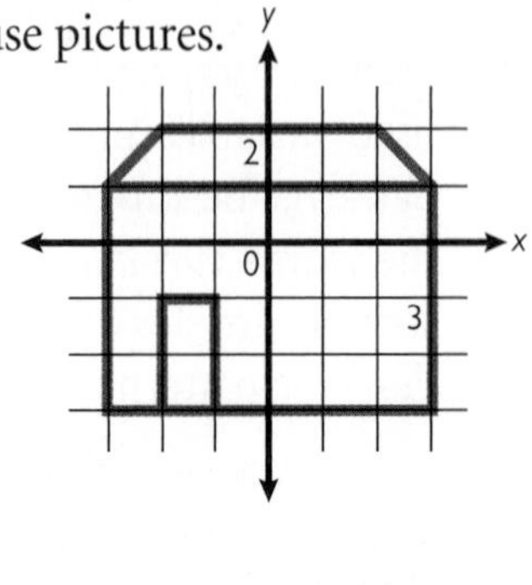

9. **a.** Give the coordinates of three points which lie on the x-axis. How can you tell if a point lies on the x-axis?

   b. Give the coordinates of two points which lie on the same vertical line. How can you tell if two points lie on the same vertical line?

Writing

10. Answer the letter to Dr. Math.

Dear Dr. Math,

Tom asked me to predict what I would get if I started at (3, −5) on a coordinate plane and drew a line to (3, 1), then another line to (3, 7). I noticed right away that all the points have the same x-coordinate. I know that the x-axis is horizontal, so I figured that I would get a horizontal line. Is this good reasoning? I have to know for sure because Tom will really gloat if I get this wrong.

K. O. R. Denate

Homework 8

Points, Plots, and Patterns

Applying Skills

Rule A: The y-coordinate is three times the x-coordinate.

Rule B: The y-coordinate is three more than the x-coordinate.

Read Rules A and B, then tell which of the ordered pairs below satisfy each rule.

1. $(2, 6)$ **2.** $(12, 4)$ **3.** $(3, 6)$

4. $(0, 3)$ **5.** $(-1, -3)$ **6.** $(5, 2)$

Read Rules C, D, and E, then answer items **7** and **8.**

Rule C: The y-coordinate is twice the x-coordinate.

Rule D: The y-coordinate is six more than the x-coordinate.

Rule E: The y-coordinate is five times the x-coordinate.

7. Of C, D, and E, which rule or rules produce a line passing through the origin?

8. Of C, D, and E, which rule produces the steepest line?

9. a. Copy and complete this table using **Rule F:** The x-coordinate is two less than the y-coordinate.

x	y
	4
	−2
3	
−1	

b. Plot the points from your table on a coordinate grid. Do the points lie on a straight line?

Extending Concepts

10. Make a table of points that fit this rule: The y-coordinate is twice the x-coordinate. Plot the points on a coordinate plane and draw a line through the points.

11. Pick a new point on the line you plotted in item **10**. What do you notice about its coordinates? Do you think that this would be true for any point on the line?

12. Make up two different number rules which would produce two parallel lines.

13. Make up a number rule which would produce a line that slopes downward from left to right.

Making Connections

In a *polygon,* the number of diagonals that can be drawn from one *vertex* is 3 fewer than the number of sides of the polygon.

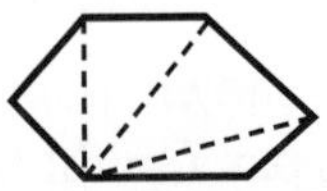

6 sides, 3 diagonals

In the table below, x represents the number of sides of a polygon and y represents the number of diagonals that can be drawn from one vertex.

14. Complete the table and plot the points on a coordinate plane. What do you notice?

x	y
5	2
6	
9	
	8

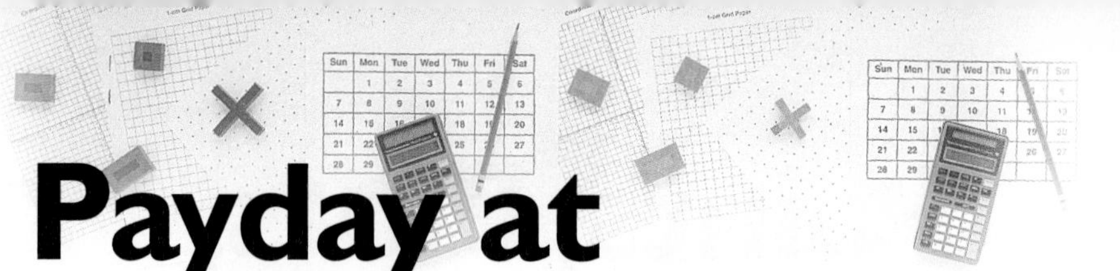

Homework 9

Payday at Planet Adventure

Applying Skills

At Planet Adventure, Lisa works at the waterslide. Her rate of pay is $9 per hour. Joel works at the Mystery Ride and makes $5 per hour plus a one-time $12 bonus for dressing up in a clown costume.

1. Make a table to show the pay Lisa would receive for 1–7 hours of work.

2. How many hours would Lisa have to work to earn:

a. $81? **b.** $99?

c. $31.50? **d.** $58.50?

3. Make a table to show the pay Joel would receive for 1–7 hours of work.

4. How much will Joel earn if he works 8 hours? 4.5 hours? 7.5 hours?

5. How many hours would Joel have to work to earn:

a. $57? **b.** $67?

c. $39.50? **d.** $54.50?

6. Copy the axes shown. Make a graph of the data for Lisa's pay. On the same grid, make a graph of the data for Joel's pay.

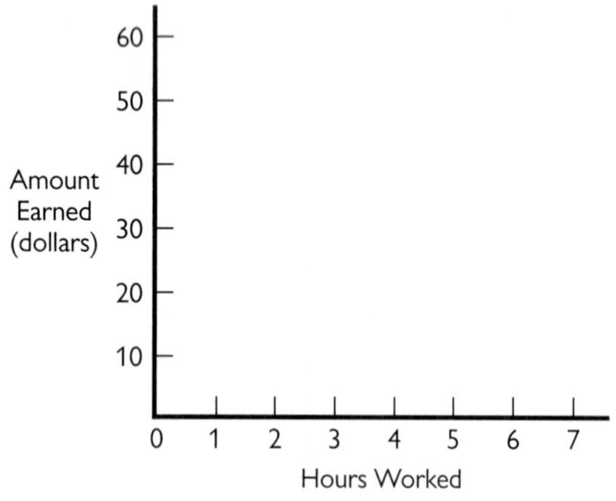

Extending Concepts

7. Write a rule that you could use to find the wage earned by Joel for any number of hours of work.

8. For what number of hours of work will Lisa and Joel earn the same amount? How can you figure this out using the graphs? without using the graphs?

9. For what number of hours worked will Joel earn more than Lisa?

10. Why is Lisa's graph steeper than Joel's?

11. Why does Joel's graph not pass through the origin?

12. If Joel works 4 hours, how much does he earn? What is his rate *per hour*? If he works more than 4 hours, is his average rate of income per hour higher or lower than this?

Writing

13. Answer the letter to Dr. Math.

Dear Dr. Math,

I've been offered two different jobs. The first one would pay $8 per hour. The second one would pay a fixed $24 each day plus $5 per hour. Which job should I take if I want to earn as much money as possible? How do I figure it out?

Broke in Brokleton

10 Homework

Sneaking Up the Line

Applying Skills

Eric the Sheep is waiting in line to be shorn. Each time a sheep at the front of the line gets shorn, Eric sneaks up the line two places.

How many sheep will be shorn before Eric if the number of sheep in front of him is:

1. 5?

2. 8?

3. 15?

4. 42?

5. 112?

6. 572?

How many sheep could have been in line in front of Eric if the number of sheep shorn before him is:

7. 7?

8. 12?

9. 25?

10. 39?

Extending Concepts

For the following questions, suppose that each time a sheep at the front of the line gets shorn, Eric sneaks up *three* places instead of two.

11. Complete the table for this situation. Then graph the data.

Number of Sheep In Front of Eric	Number of Sheep Shorn Before Eric
4	
5	
6	
7	
8	
9	
10	
11	

12. How many sheep will be shorn before Eric if there are 66 sheep in front of him? How did you figure this out?

13. How could you find the number of sheep shorn before Eric for *any* number of sheep in front of him? Describe two different methods you could use and explain why each method works.

14. If 28 sheep are shorn before Eric, how many sheep could have been in front of him? How did you figure it out? Why is the answer not unique?

Writing

15. Describe a method you could use to find the number of sheep shorn before Eric for any number of sheep in front of him and for any number of sheep that Eric sneaks past.

Homework 11: Something Fishy

Applying Skills

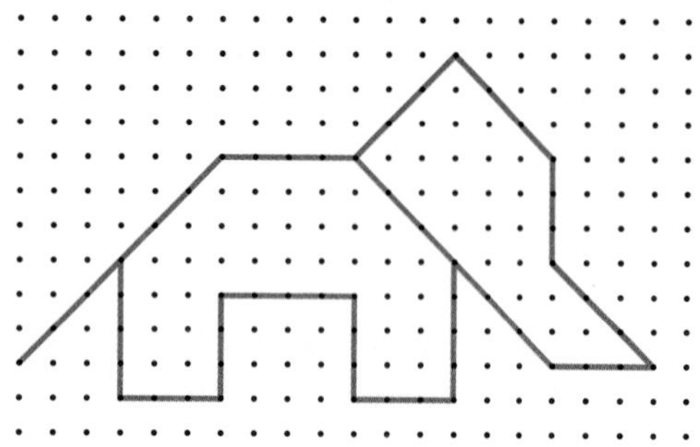

This elephant drawing in growth stage 3 has a height of 10 units, a length of 19 units, and a body area (excluding the head) of 49 square units.

1. Figure out what the height, length, and body area would be for elephants in growth stages 1–6. Show your answers in a table.

2. Make graphs to show the height, length, and body area for elephant drawings in growth stages 1–6.

Use the patterns that you observe in your table to find the height, length, and body area for elephant drawings in each of these growth stages:

3. 12

4. 20

5. 72

6. 103

7. In what stage of growth is the elephant if its height is 28? 49? 76? 211?

8. In what stage of growth is the elephant if its length is 67? 133? 325?

Extending Concepts

9. Use words and expressions to describe rules you could use to find the height, length, and body area for elephant drawings in any growth stage.

10. What growth stage is an elephant in if its body area is 529? Explain how you figured out your answer.

11. Create your own "animal" and draw pictures to show how it grows in stages. Make sure you have a clear rule for the way it grows. Make a table to show stages 1–5 of your animal's growth. Describe in words a rule for its growth. Describe your rule using variables and expressions.

Writing

12. Do you think that in reality, the height of an actual elephant is likely to increase according to the pattern you described in item **9**? Why or why not?

Homework 12 The Will

Applying Skills

Annie may receive money according to any one of three plans. The amounts of money that each plan would yield at the end of years 1, 2, 3, 4, and 5, respectively, are as follows:

Plan A: \$100, \$250, \$400, \$550, \$700, and so on.

Plan B: \$10, \$40, \$100, \$190, \$310, and so on.

Plan C: 1 cent, 3 cents, 9 cents, 27 cents, 81 cents, and so on.

1. Make a table showing the amount of money Annie would receive from each plan at the end of years 1–15.

2. For each plan, make a graph showing the amount of money Annie would receive at the end of years 1–10.

3. Which plan yields the most money at the end of year 8? 12? 17? 20?

4. Describe in words the growth pattern for each plan.

Extending Concepts

5. At the end of which year will Annie first receive more than \$5,000 if she uses Plan A? Plan B? Plan C?

6. Which plan would you recommend to Annie? Why?

7. Figure out a pattern which gives more money at the end of the fifteenth year than Plan A but not as much as Plan B.

Making Connections

8. The **half-life** of a radioactive substance is the time required for one-half of any given amount of the substance to decay. Half-lives can be used to date events from the Earth's past. Uranium has a half-life of 4.5 billion years! Suppose that the half-life of a particular substance is 6 days and that 400 grams are present initially. Then the amount remaining will be 200 grams after 6 days, 100 grams after 12 days, and so on.

a. What amount will remain after 18 days? after 24 days?

b. When will the amount remaining reach 3.125 grams?

c. Will the amount remaining ever reach zero? Why or why not?

Glossary/Glosario

Cómo usar el glosario en español:

1. Busca el término en inglés que desees encontrar.

2. El término en español, junto con la definición, se encuentra debajo del término en inglés.

A

actual size the true size of an object represented by a scale model or drawing
tamaño real el tamaño verdadero de un objeto representado por un modelo o dibujo a escala

additive system a mathematical system in which the values of individual symbols are added together to determine the value of a sequence of symbols
sistema aditivo sistema matemático en el cual los valores de los símbolos individuales se suman para determinar el valor de una secuencia de símbolos

Examples: The Roman numeral system, which uses symbols such as I, V, D, and M, is a well-known additive system.
Ejemplos: El sistema numérico romano, el cual usa símbolos como I, V, D y M es un sistema aditivo bien conocido.

This is another example of an additive system:
Este es otro ejemplo de un sistema aditivo:

If □ equals 1 and ▽ equals 7,
then ▽▽□ equals $7 + 7 + 1 = 15$.

Si □ es igual a 1 y ▽ es igual a 7,
entonces ▽▽□ es igual a
$7 + 7 + 1 = 15$.

algorithm a specific step-by-step procedure for any mathematical operation
algoritmo procedimiento específico, paso a paso, para cualquier operación matemática

area the size of a surface, usually expressed in square units
área el tamaño de una superficie, usualmente expresada en unidades cuadradas

Example/**Ejemplo:**

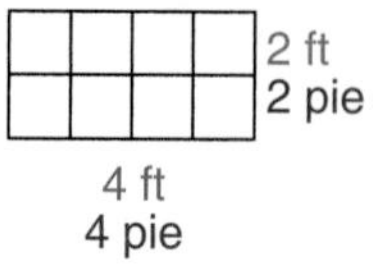

area = 8 ft^2
área = 8 pie^2

arithmetic expression a mathematical relationship expressed as a number, or two or more numbers with operation symbols
expresión aritmética una relación matemática expresada como un número, o dos o más números con signos de operación

average the sum of a set of values divided by the number of values
promedio la suma de un conjunto de valores dividido entre el número de valores

Example: The *average* of 3, 4, 7, and 10 is $(3 + 4 + 7 + 10) \div 4$ or 6.
Ejemplo: El promedio de 3, 4, 7 y 10 es $(3 + 4 + 7 + 10) \div 4$ ó 6.

B

bar graph a way of displaying data using horizontal or vertical bars
gráfica de barras una forma de mostrar información usando barras horizontales o verticales

base [1] the side or face on which a three-dimensional shape stands; [2] the number of characters a number system contains
base [1] el lado o cara donde descansa una figura tridimensional; [2] el número de caracteres que contiene un sistema numérico

base-ten system the number system containing ten single-digit symbols {0, 1, 2, 3, 4, 5, 6, 7, 8, and 9} in which the numeral 10 represents the quantity ten
sistema de base diez el sistema de números que contiene 10 símbolos de dígitos sencillos {0, 1, 2, 3, 4, 5, 6, 7, 8, 9} en el cual el numeral 10 representa la cantidad diez

base-two system the number system containing two single-digit symbols {0 and 1} in which 10 represents the quantity two
sistema de base dos el sistema de números que contiene dos símbolos de dígitos sencillos {0 y 1} en el cual el 10 representa la cantidad dos

binary system the base two number system, in which combinations of the digits 1 and 0 represent different numbers, or values
sistema binario sistema de números de base dos, en el cual las combinaciones de los dígitos 1 y 0 representa números o valores diferentes

broken-line graph a type of line graph used to show change over a period of time
gráfica de línea quebrada un tipo de gráfica lineal usada para mostrar el cambio en un periodo de tiempo

Example/Ejemplo:

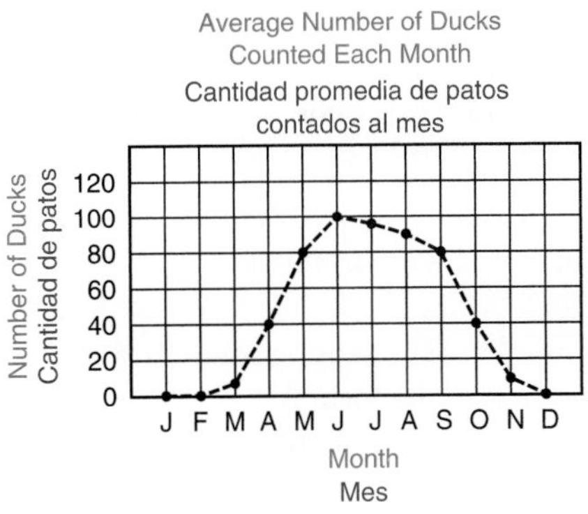

C

chance the probability or likelihood of an occurrence, often expressed as a fraction, decimal, percentage, or ratio
posibilidad la probabilidad de una ocurrencia, a menudo expresada como fracción, decimal, porcentaje o razón

common denominator a whole number that is the denominator for all members of a group of fractions
denominador común un número entero que es el denominador de todos los miembros de un grupo de fracciones

Example: The fractions $\frac{5}{8}$ and $\frac{7}{8}$ have a common denominator of 8.
Ejemplo: Las fracciones $\frac{5}{8}$ y $\frac{7}{8}$ tienen al 8 como denominador común.

common factor a whole number that is a factor of each number in a set of numbers
factor común un número entero que es el factor de cada número en un conjunto de números

Example: 5 is a *common factor* of 10, 15, 25, and 100.
Ejemplo: 5 es un *factor común* de 10, 15, 25 y 100.

coordinate graph the representation of points in space in relation to reference lines—usually, a horizontal x-axis and a vertical y-axis
gráficas de coordenadas la representación de puntos en el espacio en relación con las rectas de referencia—generalmente, un eje x horizontal y un eje y vertical

cubic centimeter the amount contained in a cube with edges that are 1 cm in length
centímetro cúbico la cantidad que contiene un cubo con aristas que tienen 1 cm de longitud

D

decimal system the most commonly used number system, in which whole numbers and fractions are represented using base ten
sistema decimal el sistema numérico más comúnmente usado, en el cual los números enteros y las fracciones se representan usando bases de diez

Example: Decimal numbers include 1,230, 1.23, 0.23, and −123.
Ejemplo: Los números decimales incluyen 1,230, 1.23, 0.23 y −123.

denominator the bottom number in a fraction
denominador el número que está debajo en una fracción

Example: for $\frac{a}{b}$, b is the denominator
Ejemplo: para $\frac{a}{b}$, b es el denominador

difference the result obtained when one number is subtracted from another
diferencia el resultado obtenido cuando un número es restado de otro

distribution the frequency pattern for a set of data
distribución el patrón de frecuencia para un conjunto de datos

double-bar graph a graphical display that uses paired horizontal or vertical bars to show a relationship between data
gráfica de doble barra un diseño gráfico que usa barras horizontales y verticales para indicar la relación entre los datos

Example:/Ejemplo:

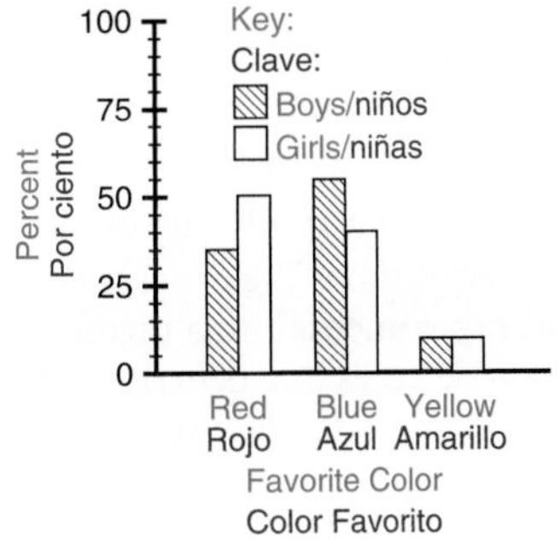

E

edge a line along which two planes of a solid figure meet
arista una línea en donde se unen dos planos de un cuerpo geométrico

equal angles angles that measure the same number of degrees
ángulos iguales ángulos que miden el mismo número de grados

equilateral a shape having more than one side, each of which is the same length
equilátero una figura que tiene más de un lado, los cuales son del mismo largo

equivalent equal in value
equivalente igual en valor

equivalent expressions expressions that always result in the same number, or have the same mathematical meaning for all replacement values of their variables
expresiones equivalentes expresiones que siempre tienen el mismo número como resultado o el mismo significado matemático para todos los valores sustitutos de sus variables

Examples/**Ejemplos:** $\frac{9}{3} + 2 = 10 - 5$

$2x + 3x = 5x$

equivalent fractions fractions that represent the same quotient but have different numerators and denominators
fracciones equivalentes fracciones que representan el mismo cociente, pero tienen numeradores y denominadores diferentes

Example/**Ejemplo:** $\frac{5}{6} = \frac{15}{18}$

estimate an approximation or rough calculation
estimado una aproximación o calculación aproximada

expanded notation a method of writing a number that highlights the value of each digit
anotación desarrollada un método para escribir un número que destaca el valor de cada dígito

Example/**Ejemplo:** $867 = 800 + 60 + 7$

experimental probability a ratio that shows the total number of times the favorable outcome happened to the total number of times the experiment was done
probabilidad experimental una razón que muestra el número total de veces que ocurrió un resultado favorable en el número total de veces que se realizó el experimento

exponent a numeral that indicates how many times a number or expression is to be multiplied by itself
exponente numeral que indica cuántas veces un número o expresión se debe multiplicar por sí mismo

Example: In the equation $2^3 = 8$, the *exponent* is 3.
Ejemplo: En la ecuación $2^3 = 8$, el *exponente* es 3.

expression a mathematical combination of numbers, variables, and operations
expresión una combinación matemática de números, variables y operaciones

Example/**Ejemplo:** $6x + y^2$

F

face a two-dimensional side of a three-dimensional figure
cara lado bidimensional de una figura tridimensional

factor a number or expression that is multiplied by another to yield a product
factor un número o expresión que se multiplica por otro para tener un producto

Example: 3 and 11 are *factors* of 33.
Ejemplo: 3 y 11 son *factores* de 33.

fraction a number representing some part of a whole; a quotient in the form $\frac{a}{b}$
fracción número que representa una parte de un entero; un cociente en la forma $\frac{a}{b}$

frequency graph a graph that shows similarities among the results so one can quickly tell what is typical and what is unusual
gráfica de frecuencia gráfica que muestra similitudes entre los resultados de forma que podamos notar rápidamente lo que es típico y lo que es inusual

G

greatest common factor (GCF) the greatest number that is a factor of two or more numbers
máximo factor común (MFC) el número más grande que sea un factor de dos o más números

Example: The *greatest common factor* of 30, 60, and 75 is 15.
Ejemplo: El *máximo factor común* de 30, 60 y 75 es 15.

I

improper fraction a fraction in which the numerator is greater than the denominator
fracción impropia una fracción en la cual el numerador es mayor que el denominador

Examples/**Ejemplos:** $\frac{21}{4}, \frac{4}{3}, \frac{2}{1}$

integers the set of all whole numbers and their additive inverses $\{\ldots -5, -4, -3, -2, -1, 0, 1, 2, 3, 4, 5 \ldots\}$
enteros el conjunto de todos los números enteros y sus aditivos inversos $\{\ldots -5, -4, -3, -2, -1, 0, 1, 2, 3, 4, 5 \ldots\}$

inverse operations operations that undo each other
operaciones inversas operaciones que se deshacen mutuamente

Examples: Addition and subtraction are inverse operations: $5 + 4 = 9$ and $9 - 4 = 5$. Adding 4 is the inverse of subtracting by 4.
Multiplication and division are inverse operations: $5 \times 4 = 20$ and $20 \div 4 = 5$. Multiplying by 4 is the inverse of dividing by 4.
Ejemplos: La adición y la resta son operaciones inversas: $5 + 4 = 9$ y $9 - 4 = 5$. Sumar 4 es el inverso de restar por 4.
La multiplicación y la división son operaciones inversas: $5 \times 4 = 20$ y $20 \div 4 = 5$. Multiplicar por 4 es el inverso de dividir por 4.

isometric drawing a two-dimensional representation of a three-dimensional object in which parallel edges are drawn as parallel lines
dibujo isométrico una representación bidimensional de un objeto tridimensional en el que las aristas paralelas están dibujadas como líneas paralelas

Example/Ejemplo:

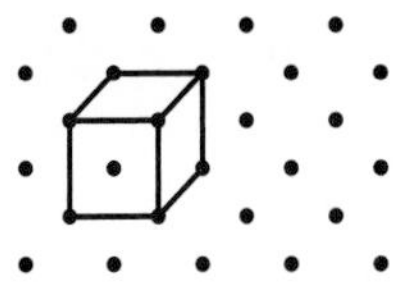

L

least common denominator (LCD) the least common multiple of the denominators of two or more fractions
mínimo común denominador (MCD) el mínimo común múltiplo de los denominadores de dos o más fracciones

Example: 12 is the *least common denominator* of $\frac{1}{3}$, $\frac{2}{4}$, and $\frac{3}{6}$.
Ejemplo: 12 es el *mínimo común denominador* de $\frac{1}{3}$, $\frac{2}{4}$, y $\frac{3}{6}$.

least common multiple (LCM) the smallest nonzero whole number that is a multiple of two or more whole numbers
mínimo común múltiplo (MCM) el número más pequeño que no sea cero que sea múltiplo de dos o más números enteros

Example: The *least common multiple* of 3, 9, and 12 is 36.
Ejemplo: El *mínimo común múltiplo* de 3, 9 y 12 es 36.

linear measure the measure of the distance between two points on a line
medida lineal la medida de la distancia entre dos puntos en una recta

M

mean the quotient obtained when the sum of the numbers in a set is divided by the number of addends
media el cociente que se obtiene cuando la suma de los números de un conjunto se divide entre el número de sumandos

Example: The *mean* of 3, 4, 7, and 10 is $(3 + 4 + 7 + 10) \div 4$ or 6.
Ejemplo: La *media* de 3, 4, 7 y 10 es $(3 + 4 + 7 + 10) \div 4$ o 6.

measurement units standard measures, such as the meter, the liter, and the gram, or the foot, the quart, and the pound
unidades de medidas las medidas estándares, así como el metro, el litro y el gramo, o el pie, el cuarto y la libra

median the middle number in an ordered set of numbers
mediana el número medio en un conjunto de números ordenado

Example: 1, 3, 9, 16, 22, 25, 27
16 is the *median.*
Ejemplo: 1, 3, 9, 16, 22, 25, 27
16 es la *mediana.*

metric system a decimal system of weights and measurements based on the meter as its unit of length, the kilogram as its unit of mass, and the liter as its unit of capacity
sistema métrico sistema decimal de pesos y medidas que tiene por base el metro como su unidad de longitud, el kilogramo como su unidad de masa y el litro como su unidad de capacidad

mixed number a number composed of a whole number and a fraction
número mixto un número compuesto de un número entero y una fracción

Example/Ejemplo: $5\frac{1}{4}$

mode the number or element that occurs most frequently in a set of data
moda el número o elemento que se presenta con más frecuencia en un conjunto de datos

Example: 1, 1, 1, 2, 2, 3, 5, 5, 6, 6, 6, 6, 8
6 is the *mode.*
Ejemplo: 1, 1, 1, 2, 2, 3, 5, 5, 6, 6, 6, 6, 8
6 es la *moda.*

multiple the product of a given number and an integer
múltiplo el producto de un número dado y un entero

> **Examples:** 8 is a *multiple* of 4. 3.6 is a *multiple* of 1.2.
> **Ejemplos:** 8 es un *múltiplo* de 4. 3.6 es un *múltiplo* de 1.2.

N

number sentence a combination of numbers and operations, stating that two expressions are equal
enunciado numérico combinación de números y operaciones, que expresa que dos expresiones son iguales

number symbols the symbols used in counting and measuring
símbolos numéricos símbolos usados para contar y medir

> **Examples/Ejemplos:** $1, -\frac{1}{4}, 5, \sqrt{}, -\pi$

number system a method of writing numbers. The Arabic *number system* is most commonly used today.
sistema numérico un método para escribir números. El *sistema numérico* arábico es el que usa más comúnmente en la actualidad.

numerator the top number in a fraction. In the fraction $\frac{a}{b}$, a is the numerator.
numerador el número de arriba en una fracción. En la fracción $\frac{a}{b}$, a es el numerador.

O

operations arithmetical actions performed on numbers, matrices, or vectors
operaciones acciones aritméticas hechas en números, matrices o vectores

opposite angle in a triangle, a side and an angle are said to be opposite if the side is not used to form the angle
ángulo opuesto en un triángulo, un lado y un ángulo están opuestos si el lado no se usa para formar el ángulo

> **Example/Ejemplo:**

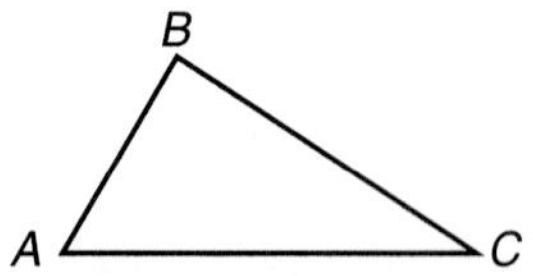

In ΔABC, $\angle A$ is opposite $\overline{BC}$.
En ΔABC, $\angle A$ está opuesto a $\overline{BC}$.

order of operations to find the answer to an equation, follow this four step process: 1) do all operations with parentheses first; 2) simplify all numbers with exponents; 3) multiply and divide in order from left to right; 4) add and subtract in order from left to right
orden de operaciones para encontrar el resultado de una ecuación, sigue este proceso de cuatro pasos: 1) haz primero todas las operaciones entre paréntesis; 2) simplifica todos los números con exponentes; 3) multiplica y divide en orden de izquierda a derecha; 4) suma y resta en orden de izquierda a derecha

ordered pair two numbers that tell the x-coordinate and y-coordinate of a point
par ordenado dos números que expresan la coordenada x y la coordenada y de un punto

> **Example:** The coordinates (3, 4) are an *ordered pair.* The x-coordinate is 3, and the y-coordinate is 4.
> **Ejemplo:** Las coordenadas (3, 4) son un *par ordenado.* La coordenada x es 3 y la coordenada y es 4.

orthogonal drawing always shows three views of an object—top, side, and front. The views are drawn straight-on.
dibujo ortogonal siempre muestra tres vistas de un objeto—pico, lado y frente

> **Example/Ejemplo:**

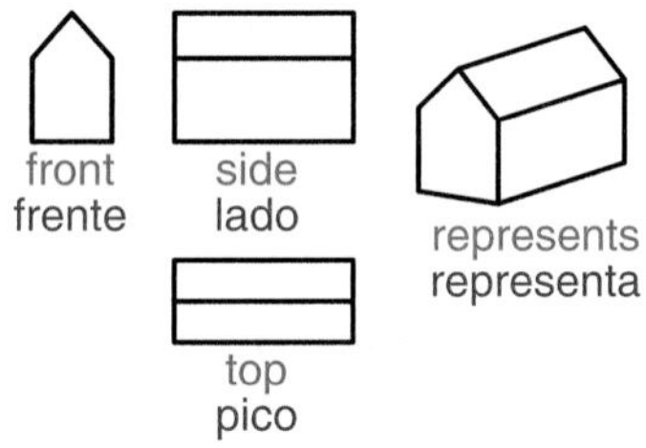

P

parallel straight lines or planes that remain a constant distance from each other and never intersect, represented by the symbol ||
paralela líneas rectas o planas que se mantienen a una distancia constante de cada una y nunca intersecan, representadas por el símbolo ||

> **Example/Ejemplo:**

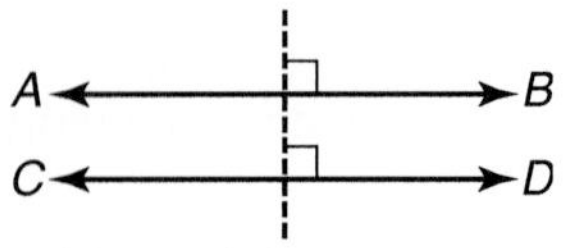

$\overleftrightarrow{AB}$ and $\overleftrightarrow{CD}$ are *parallel.*
$\overleftrightarrow{AB}$ y $\overleftrightarrow{CD}$ son *paralelas.*

pattern a regular, repeating design or sequence of shapes or numbers
patrón diseño regular y repetido o secuencia de formas o números

percent a number expressed in relation to 100, represented by the symbol %
por ciento un número expresado con relación a 100, representado por el signo %

Example: 76 out of 100 students use computers. 76 *percent* of students use computers.
Ejemplo: 76 de 100 estudiantes usan computadoras. El 76 *por ciento* de los estudiantes usan computadoras.

perimeter the distance around the outside of a closed figure
perímetro la distancia alrededor del contorno de una figura cerrada

Example/Ejemplo:

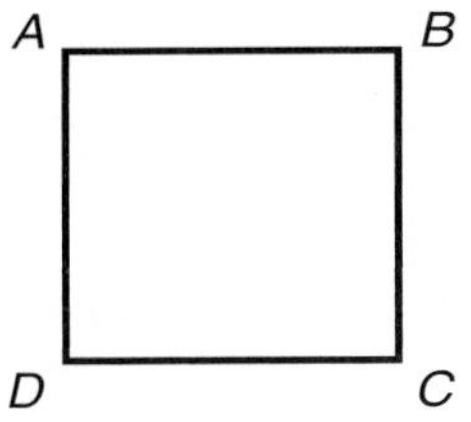

$AB + BC + CD + DA = \textit{perimeter}$
$AB + BC + CD + DA = \textit{perímetro}$

picture graph a graph that uses pictures or symbols to represent numbers
pictografía gráfica que usa dibujos o símbolos para representar números

place value the value given to a place a digit may occupy in a numeral
valor posicional el valor dado al lugar que un dígito pueda ocupar en un numeral

place-value system a number system in which values are given to the places digits occupy in the numeral. In the decimal system, the value of each place is 10 times the value of the place to its right.
sistema de valor posicional sistema numérico en el que se dan valores a los lugares que los dígitos ocupan en un numeral. En el sistema decimal, el valor de cada lugar es 10 veces el valor del lugar a su derecha.

point one of four undefined terms in geometry used to define all other terms. A *point* has no size.
punto uno de los cuatro términos sin definición en geometría que se usa para definir a todos los otros términos. Un *punto* no tiene tamaño.

polygon a simple, closed plane figure, having three or more line segments as sides
polígono una figura plana simple y cerrada que tiene tres o más líneas rectas como sus lados

Examples/Ejemplos:

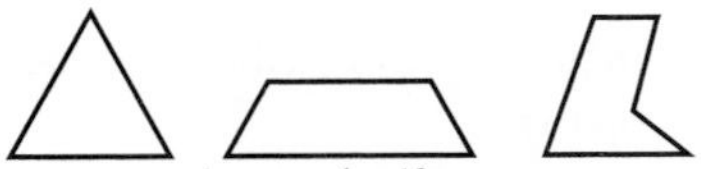

polygons/polígonos

polyhedron a solid geometrical figure that has four or more plane faces
poliedro cuerpo geométrico que tiene cuatro o más caras planas

Examples/Ejemplos:

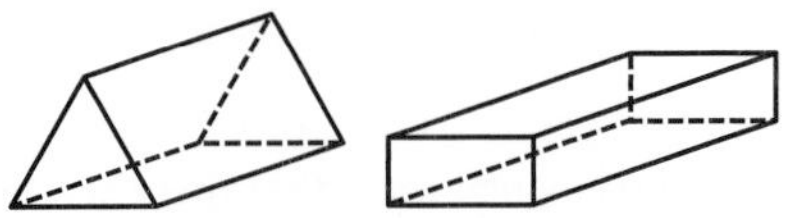

polyhedrons/poliedro

power represented by the exponent *n*, to which a number is raised by multiplying itself *n* times
potencia representada por el exponente *n*, por el cual un número aumenta al multiplicarse a sí mismo por *n* veces

Example: 7 raised to the fourth *power*
$7^4 = 7 \times 7 \times 7 \times 7 = 2{,}401$
Ejemplo: 7 elevado a la cuarta *potencia*
$7^4 = 7 \times 7 \times 7 \times 7 = 2{,}401$

predict to anticipate a trend by studying statistical data
predecir anticipar una tendencia al estudiar los datos estadísticos

prime factorization the expression of a composite number as a product of its prime factors
descomposición en factores primos la expresión de un número compuesto como un producto de sus factores primos

Examples/Ejemplos: $504 = 2^3 \times 3^2 \times 7$
$30 = 2 \times 3 \times 5$

prime number a whole number greater than 1 whose only factors are 1 and itself
número primo un número entero mayor que 1cuyos únicos factores son 1 y él mismo

Examples/Ejemplos: 2, 3, 5, 7, 11

prism a solid figure that has two parallel, congruent polygonal faces (called *bases*)
prisma cuerpo geométrico que tiene dos caras paralelas y poligonales congruentes (llamadas *bases)*

Examples/Ejemplos:

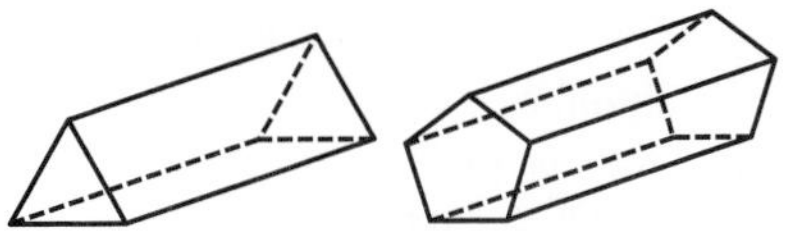

prisms/prismas

probability the study of likelihood or chance that describes the chances of an event occurring
probabilidad el estudio de las probabilidades que describen las posibilidades de que ocurra un suceso

product the result obtained by multiplying two numbers or variables
producto el resultado obtenido al multiplicar dos números o variables

proportion a statement that two ratios are equal
proporción igualdad de dos razones

pyramid a solid geometrical figure that has a polygonal base and triangular faces that meet at a common vertex
pirámide cuerpo geométrico que tiene una base poligonal y caras triangulares que se encuentran en un vértice común

Examples/Ejemplos:

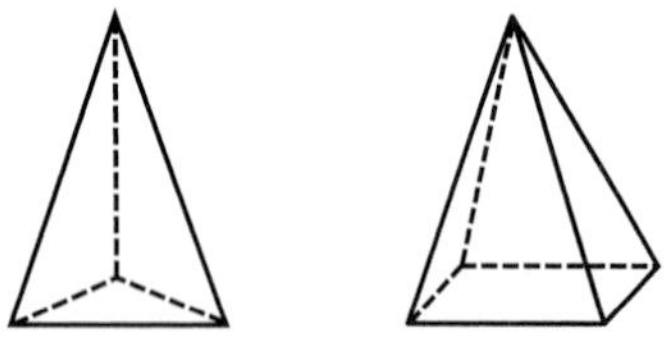

pyramids/pirámides

Q

quotient the result obtained from dividing one number or variable (the divisor) into another number or variable (the dividend)
cociente el resultado que se obtiene al dividir un número o variable (el divisor) por otro número o variable (el dividendo)

Example/Ejemplo:

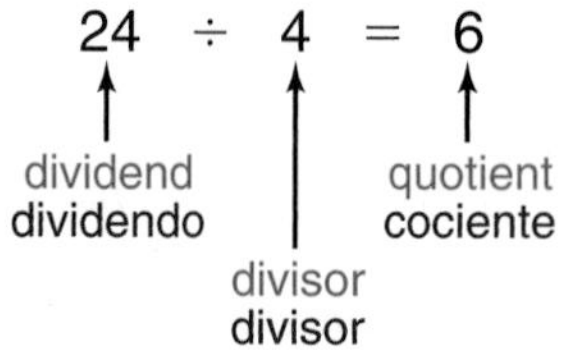

R

range in statistics, the difference between the largest and smallest values in a sample
rango en estadísticas, la diferencia entre los valores más grandes y menores en el ejemplo

ratio a comparison of two numbers
razón comparación de dos números

Example: The *ratio* of consonants to vowels in the alphabet is 21:5.
Ejemplo: La *razón* de las consonantes a las vocales en el abecedario es de 21:5.

reciprocal the result of dividing a given quantity into 1
recíproco el resultado de dividir una cantidad dada entre 1

Examples: The *reciprocal* of 2 is $\frac{1}{2}$; of $\frac{3}{4}$ is $\frac{4}{3}$; of x is $\frac{1}{x}$.
Ejemplos: El *recíproco* de 2 es $\frac{1}{2}$; de $\frac{3}{4}$ es $\frac{4}{3}$; de x es $\frac{1}{x}$.

regular shape a figure in which all sides are equal and all angles are equal
figura regular una figura en la que todos los lados son iguales y todos los ángulos son iguales

repeating decimal a decimal in which a digit or a set of digits repeat infinitely
decimal periódico un decimal en el que un dígito o un conjunto de dígitos se repiten infinitamente

Example/Ejemplo: 0.121212 . . .

right angle an angle that measures 90°
ángulo recto un ángulo que mide 90°

Example/Ejemplo:

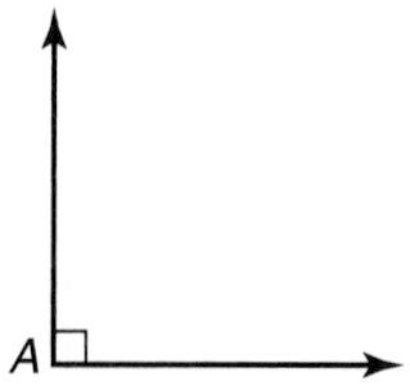

$\angle A$ is a *right angle.*
$\angle A$ es un *ángulo recto.*

Roman numerals the numeral system consisting of the symbols I (1), V (5), X (10), L (50), C (100), D (500), and M (1,000). When a Roman symbol is preceded by a symbol of equal or greater value, the values of a symbol are added (XVI = 16). When a symbol is preceded by a symbol of lesser value, the values are subtracted (IV = 4).
números romanos sistema numeral que consiste de los símbolos I (1), V (5), X (10), L (50), C (100), D (500), y M (1,000). Cuando un símbolo es precedido de un símbolo de menor valor, los valores se restan (IV = 4)

round to approximate the value of a number to a given decimal place
redondear aproximar el valor de un número a un lugar decimal

Examples: 2.56 rounded to the nearest tenth is 2.6
2.54 rounded to the nearest tenth is 2.5.
365 rounded to the nearest hundred is 400.
Ejemplos: 2.56 redondeado a la decena más cercana es 2.6.
2.54 redondeado a la decena más cercana es 2.5.
365 redondeado a la centena más cercana es 400.

rule a statement that describes a relationship between numbers or objects
regla un enunciado que describe a la relación entre números o objetos

S

sampling with replacement a sample chosen so that each element has the chance of being selected more than once
muestra por sustitución una muestra escogida para que el elemento tenga la oportunidad de ser seleccionado más de una vez

Example: A card is drawn from a deck, placed back into the deck, and a second card is drawn. Since the first card is replaced, the number of cards remains constant.
Ejemplo: Se saca una baraja, vuelve a colocarla en la pila y se saca una segunda baraja. Como la primera baraja volvió a la pila, el número de barajas permanece constante.

scale the ratio between the actual size of an object and a proportional representation
escala la razón entre el tamaño real de un objeto y una representación proporcional

scale drawing a proportionally correct drawing of an object or area at actual, enlarged, or reduced size
dibujo a escala un dibujo proporcionalmente correcto de un objeto o área en su tamaño real, ampliado o reducido

scale factor the factor by which all the components of an object are multiplied in order to create a proportional enlargement or reduction
factor de escala el factor por el cual todos los componentes de un objeto son multiplicados para crear una ampliación o reducción proporcional

scale size the proportional size of an enlarged or reduced representation of an object or area
tamaño de escala el tamaño proporcional de una representación aumentada o reducida de un objeto o área

signed number a number preceded by a positive or negative sign. Positive numbers are usually written without a sign.
número con signo un número precedido por un signo positivo o negativo. Los números positivos por lo general se escriben sin signo.

standard measurement commonly used measurements, such as the meter used to measure length, the kilogram used to measure mass, and the second used to measure time
medida estándar medidas usadas comúnmente, así como el metro para medir longitud, el kilogramo para medir masa y el segundo uso es para medir tiempo

sum the result of adding two numbers or quantities
suma el resultado de sumar dos números o cantidades

Example: $6 + 4 = 10$
10 is the *sum* of the two addends, 6 and 4.
Ejemplo: $6 + 4 = 10$
10 es la *suma* de dos sumandos, 6 y 4.

survey a method of collecting statistical data in which people are asked to answer questions
encuesta un método de recolectar datos estadísticos en el que se le pide a las personas que contesten preguntas

T

table a collection of data arranged so that information can be easily seen
tabla una colección de datos organizados de forma que la información pueda ser vista fácilmente

terminating decimal a decimal with a finite number of digits
decimal finito un decimal con un número limitado de dígitos

theoretical probability the ratio of the number of favorable outcomes to the total number of possible outcomes
probabilidad teórica la razón del número de resultados favorables en el número total de resultados posibles

three-dimensional having three measurable qualities: length, height, and width
tridimensional que tiene tres propiedades de medición; longitud, altura y ancho

two-dimensional having two measurable qualities: length and width
bidimensional que tiene dos propiedades de medición; longitud y ancho

V

variable a letter or other symbol that represents a number or set of numbers in an expression or an equation
variable una letra o otro símbolo que representa al número o al conjunto de números en una expresión o en una ecuación

Example: In the equations $x + 2 = 7$, the variable is x.
Ejemplo: En las ecuaciones $x + 2 = 7$, la variable es x.

vertex (pl. *vertices*) the common point of two rays of an angle, two sides of a polygon, or three or more faces of a polyhedron
vértice el punto común de las dos semirrectas de un ángulo, dos lados de un polígono o tres o más caras de un poliedro

Examples/Ejemplos:

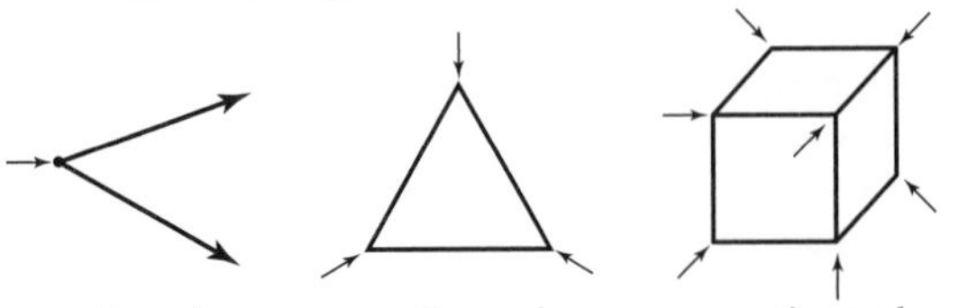

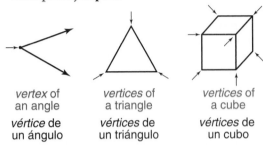

vertex of an angle
vértice de un ángulo

vertices of a triangle
vértices de un triángulo

vertices of a cube
vértices de un cubo

volume the space occupied by a solid, measured in cubic units
volumen el espacio que ocupa un cuerpo, medido en unidades cúbicas

Example/Ejemplo:

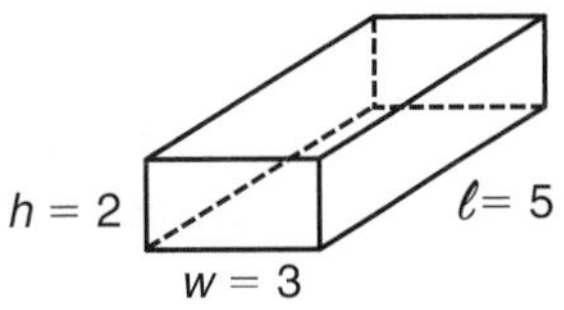

The *volume* of this rectangular prism is 30 cubic units.
$2 \times 3 \times 5 = 30$

El *volumen* de este prisma rectangular es 30 unidades cúbicas.
$2 \times 3 \times 5 = 30$

Z

zero-pair one positive cube and one negative cube used to model signed number arithmetic
par cero un cubo positivo y un cubo negativo usados para modelar el número con signo aritmético

INDEX

H

I

L

M

O

P

Q

R

S

T

PHOTO CREDITS

Unless otherwise indicated below, all photography by Chris Conroy and Donald B. Johnson.

3 (tr)courtesy KTVU-TV (Bill Martin); **16 17** Image Club Graphics; **17** (tr)courtesy KTVU-TV (Bill Martin); **49** (tc)Ken Whitmore/Getty Images, (tr)Will & Deni McIntyre/Getty Images; **60 61** Ken Whitmore/Getty Images; **70–71** Will & Deni McIntyre/Getty Images; **73 75 77 79** bar art courtesy Rosicrucian Egyptian Museum and Planetarium, owned and operated by the Rosicrucian Order, AMORC, San Jose, CA; **93** (b)Aaron Haupt; **94** Matt Meadows; **96** Photodisc/Getty Images; **105** ThinkStock LLC/Index Stock Imagery; **106** Aaron Haupt; **116** Images.com/CORBIS; **128** Matt Meadows; **162** Images; **163** (tl tr)Images; **164** (tc)Slyvian Grandadam/Photo Researchers, (b)Photodisc/Getty Images; **165** (t)Norris Taylor/Photo Researchers, (c)Andrea Moore; **167** Norris Taylor/Photo Researchers; **174** Photodisc/Getty Images; **179 through 193** bar art excerpted from Shelter © 1973 by Shelter Publications, Inc., P.O. Box 279, Bolinas, CA 94924/distributed in bookstores by Random House/reprinted by permission; **184** (background)Images, (foreground)Photodisc/Getty Images; **206** (tl)Glencoe file photo, (tr)Photodisc/Getty Images; **207** (tl tc tr)Photodisc/Getty Images, (b)Najlah Feanny/CORBIS; **208 218** Photodisc/Getty Images; **229** Matt Meadows; **230** Photodisc/Getty Images; **237** David Young-Wolff/PhotoEdit; **239** Chris Windsor/Getty Images; **241 242** Photodisc/Getty Images; **249** Aaron Haupt; **321** (tc)Bruce Stromberg/Graphistock, (tr)Jeremy Walker/Getty Images; **338–339** Bruce Stromberg/Graphistock; **345** (b)Chad Ehlers/Photo Network; **346–347** Jeremy Walker/Getty Images.